Interfacial Phenomena

SURFACTANT SCIENCE SERIES

CONSULTING EDITORS

Volume 1: NONIONIC SURFACTANTS, edited by Martin J. Schick

Volume 2: SOLVENT PROPERTIES OF SURFACTANT SOLUTIONS, edited by Kozo Shinoda (*out of print*)

Volume 3: SURFACTANT BIODEGRADATION, by Robert D. Swisher (*out of print*)

Volume 4: CATIONIC SURFACTANTS, edited by Eric Jungermann

Volume 5: DETERGENCY: THEORY AND TEST METHODS (*in three parts*), edited by W. G. Cutler and R. C. Davis

Volume 6: EMULSIONS AND EMULSION TECHNOLOGY (*in three parts*), edited by Kenneth J. Lissant

Volume 7: ANIONIC SURFACTANTS (*in two parts*), edited by Warner M. Linfield

Volume 8: ANIONIC SURFACTANTS–CHEMICAL ANALYSIS, edited by John Cross

Volume 9: STABILIZATION OF COLLOIDAL DISPERSIONS BY POLYMER ADSORPTION, by Tatsuo Sato and Richard Ruch

Volume 10: ANIONIC SURFACTANTS–BIOCHEMISTRY, TOXICOLOGY, DERMATOLOGY, edited by Christian Gloxhuber

Volume 11: ANIONIC SURFACTANTS–PHYSICAL CHEMISTRY OF SURFACTANT ACTION, edited by E. H. Lucassen-Reynders

Volume 12: AMPHOTERIC SURFACTANTS, edited by B. R. Bluestein and Clifford L. Hilton

Volume 13: DEMULSIFICATION: INDUSTRIAL APPLICATIONS, by Kenneth J. Lissant

Volume 14: SURFACTANTS IN TEXTILE PROCESSING, by Arved Datyner

Volume 15: ELECTRICAL PHENOMENA AT INTERFACES: FUNDAMENTALS, MEASUREMENTS, AND APPLICATIONS, edited by Ayao Kitahara and Akira Watanabe

Volume 16: SURFACTANTS IN COSMETICS, edited by Martin M. Rieger

Volume 17: INTERFACIAL PHENOMENA: EQUILIBRIUM AND DYNAMIC EFFECTS, by Clarence A. Miller and P. Neogi

OTHER VOLUMES IN PREPARATION

Interfacial Phenomena

Equilibrium and Dynamic Effects

CLARENCE A. MILLER

Department of Chemical Engineering
Rice University
Houston, Texas

P. NEOGI

Department of Chemical Engineering
University of Missouri
Rolla, Missouri

MARCEL DEKKER, INC. New York and Basel

Library of Congress Cataloging-in-Publication Data

Miller, Clarence A., [date]
Interfacial phenomena.

(Surfactant science series ; v. 17)
Includes bibliographies and index.
1. Surface chemistry. I. Neogi, Partho, [date]
II. Title. III. Series.
QD506.M55 1985 541.3'453 85-20579
ISBN 0-8247-7490-6

MARCEL DEKKER, INC.
270 Madison Avenue, New York, New York 10016

Current printing (last digit):
10 9 8 7 6 5 4

PRINTED IN THE UNITED STATES OF AMERICA

Preface

Flow and transport near interfaces strongly influence the performance of many operations in the processing industries. For instance, transfer rates during distillation, liquid extraction, and other mass transfer processes typically depend on the behavior of drops or bubbles which continually move and deform. Yet nonequilibrium effects are usually considered only briefly in textbooks on interfacial phenomena. A notable exception is *Interfacial Phenomena* by J. T. Davies and E. K. Rideal (Academic Press, 1963) which, however, appeared over twenty years ago and still relatively soon after the great increase in interest in dynamic interfacial phenomena began with the entry of greater numbers of chemical engineers into the field in the 1950s. And, of course, V. G. Levich's pioneering *Physiochemical Hydrodynamics*, published in English by Prentice-Hall in 1962, analyzes various situations involving dynamic behavior of interfaces although it is not primarily a book on interfacial phenomena.

Our purpose here has been to combine in one text an account of nonequilibrium interfacial phenomena such as wave motion and Marangoni flow with enough background on the fundamentals of interfaces to enable the dynamic analyses to be understood and to provide an initial overview of the field. To keep the book of a reasonable length, we have chosen to emphasize interfacial tension, contact angles, the forces between colloidal particles, and surfactants in our four background chapters; other topics covered in existing textbooks on interfacial phenomena have been given less emphasis, e.g., statistical mechanics of interfaces, adsorption, and experimental techniques for studying colloidal dispersions. We believe that our approach provides a useful means for introducing the study of dynamic phenomena relatively early in a course on interfaces. Of course, those planning to specialize in the field will also wish to take a thorough course on conventional surface and colloid chemistry or study carefully some of the existing textbooks on the subject.

The last three chapters, which deal with interfacial dynamics, assume an aquaintance with fluid mechanics and transport phenomena at the level of an introductory course. The techniques of linear stability analysis are introduced in Chapter V and used to consider break-up of interfaces, thin films, and jets under various conditions as well as interfacial wave motion with and without surfactants present. Marangoni flow and other phenomena involving interfacial transport are analyzed in Chapter VI; and Chapter VII presents a selection of situations involving flow near interfaces, including an introduction to the use of matched asymptotic expansions as a technique for analysis of such problems. Selected problems have been included at the end of each chapter for students to use in sharpening their understanding of the material presented.

We wish to thank Professor John C. Berg for reviewing the entire manuscript, Professor Dennis C. Prieve for reviewing the chapter on colloidal dispersions, and Dr. Surajit Mukherjee for reviewing the discussion of micelles; all provided very useful comments and suggestions. Ideas for presentation of various topics were gleaned from many sources over a period of years, but Professor L. E. Scriven merits special mention for his stimulating approaches which initiated one of us (CAM) into the realm of interfacial dynamics and stability. We acknowledge the patience and persistence of Minerva McCauley, who typed and prepared the manuscript in camera-ready form, and we thank the various publishers and authors who have allowed us to reproduce diagrams and photographs from previous studies.

Clarence A. Miller

P. Neogi

Contents

Interfacial Phenomena

I

Fundamentals of Interfacial Tension

1. INTRODUCTION TO INTERFACIAL PHENOMENA

An "interface" is, as the name suggests, a boundary between phases. Because interfaces are very thin -- in most cases only a few molecular diameters thick -- we sometimes tend to think of them as two-dimensional surfaces and neglect their thickness. But the third dimension is of great significance as well. Indeed, the rapid changes in density and/or composition across interfaces give them their most important property, an excess free energy or lateral stress which is usually called interfacial tension.

When three phases are present, three different interfaces are possible, one for each pair of fluids. Sometimes all three interfaces meet, the junction forming a curve known as a contact line. If one phase is a solid, the contact line lies along its surface. In this case the angle that the fluid interface makes with the solid surface is called the contact angle. Since it determines the wetting properties of liquids on solids, the contact angle is a second fundamental property important in interfacial phenomena.

Chapters I and II are devoted to these two fundamental properties which have roles in many diverse situations involving interfaces. Interfacial tension is a key factor influencing the shape of fluid interfaces, and it controls their deformability. Contact angles and wetting properties strongly affect the arrangement of the various phases in multiphase systems. When two fluid phases are present in a porous medium, for instance, as in soils or underground oil reservoirs, contact angle effects determine the positions of the two fluids, i.e., which pores are occupied by which fluids. Interfacial tension, on the other hand, determines whether individual globules of one of the fluids, if present, can deform sufficiently to pass through the tortuous pore structure when a pressure gradient is applied.

Interfacial effects are especially important in systems where interfacial area is large. This condition is met when one phase is dispersed in another as small drops or particles. With spherical particles, for example, the area to volume ratio of the dispersed phase is (3/R), where R is the particle radius. Clearly as R decreases with a given volume of the dispersed material present, interfacial area increases. When at least one dimension of each drop or particle falls to a value in the range of 1 μm or less, we say that a "colloidal dispersion" exists. Foams, aerosols, and emulsions are colloidal dispersions involving fluid interfaces which are familiar from everyday life and which are important in applications ranging from food products to cosmetics to drug delivery to detergency. Chapter III deals with the behavior of colloidal dispersions.

In multicomponent systems composition in the interfacial region can differ dramatically from that of either bulk phase. This difference is especially striking when surface-active materials or "surfactants" are present. As their name implies, these substances find it energetically favorable to be located at the interface rather than in the bulk phases. As discussed in Chapter IV, surfactants produce significant decreases in interfacial tensions and alter wetting properties as well, the latter property being of particular importance in detergency. Moreover, surfactant molecules tend to aggregate in solution, forming phases such as micellar solutions, microemulsions, and lyotropic liquid crystals with interesting and unusual properties. Surfactants and the phases they form are a major factor influencing the stability and behavior of some colloidal dispersions such as emulsions and foams.

The first four chapters thus provide a general background on interfacial phenomena, colloidal dispersions, and surfactants with emphasis on their equilibrium properties. The remaining chapters deal with dynamic behavior of interfaces, emphasizing this subject to a much greater degree than do most previous books on interfacial phenomena.

Chapter V considers the stability of fluid interfaces, a subject pertinent both to the formation of emulsions and aerosols and to their destruction by coalescence of drops. The closely related topic of wave motion is also discussed along with its implications for mass transfer. In both cases boundary conditions applicable at an interface are derived, a significant matter because it is through boundary conditions that interfacial phenomena influence solutions to the governing equations of flow and transport in fluid systems.

Interfacial phenomena involving heat and mass transport are described and analyzed in Chapter VI. Much of the chapter again deals with stability, in this case the "Marangoni" instability produced by interfacial tension gradients associated with temperature and concentration gradients along the interface. Time dependent variation of interfacial tension resulting from diffusion, adsorption, and desorption of various species is also discussed.

Finally, some interesting problems in fluid flow where there are significant interfacial effects are analyzed in Chapter VII. Topics considered are the effect of surface viscosity on flow, motion of drops and bubbles, simple coating flows, and spreading of liquid drops on solids. The method of matched asymptotic expansions is used in some of the analyses, and the likely utility of this method in dealing with other flow problems involving interfaces is emphasized.

2. INTERFACIAL TENSION - QUALITATIVE CONSIDERATIONS

We begin with the interface between a pure liquid and its vapor. As indicated in Figure I-1, the boundary between phases is not a surface of discontinuity where density changes abruptly but a region of finite thickness where density changes continuously. To be sure, this thickness is, except near the critical point, only a few molecular diameters. Nevertheless, it is an essential feature of interfacial structure.

Surface or interfacial tension of a fluid interface can be viewed in two quite different ways. From a thermodynamic point of view it is an additional free energy per unit area caused by the presence of the interface. As Figure I-1 indicates, the density is lower in the interfacial region than in the bulk liquid phase. As a result, the average distance between molecules is greater. Because the molecules attract one another, energy must be supplied to move them apart. System energy per molecule is thus greater at larger average separation distances, and we conclude that the energy per molecule is greater in the interfacial region than in the bulk liquid.

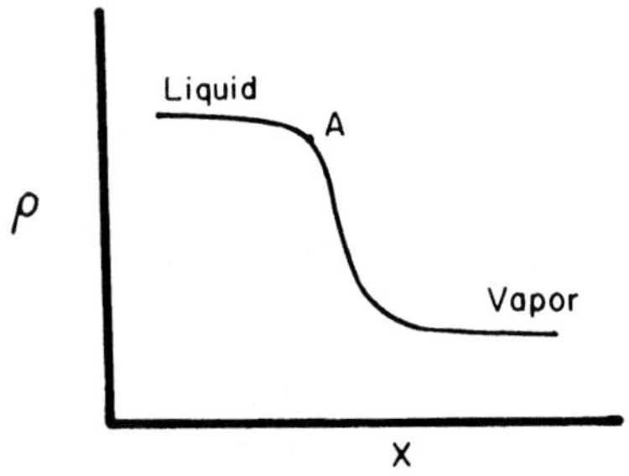

Figure I-1. Density variation in interfacial region.

At a liquid-liquid interface, such as exists between a pure hydrocarbon and water, the situation is slightly more complex because the separation distance between adjacent molecules does not vary much in the interfacial region and thus is not the main source of the excess energy of the interface. For simplicity, let us consider a binary system where the bulk phases are nearly pure component A and nearly pure component B. It is clear that in the interfacial region a molecule of A will have more B molecules and fewer A molecules as nearest neighbors than in bulk liquid A. A similar statement can be made about a molecule of B. Now thermodynamics teaches that for phase separation to occur in the first place, the attraction between an A and a B molecule must be less than the average of that between two A molecules and two B molecules. Hence the total attractive interaction per molecule is less in the interfacial region than in the bulk phases, the same result as found for the liquid-vapor interface. We conclude again that the energy per molecule must be greater in the interfacial region than in the bulk fluids.

A brief remark on terminology is in order at this point. In this book the term "interfacial tension" is used as an all-inclusive term applicable to liquid-gas, liquid-liquid, and solid-fluid interfaces. This usage differs from that of some authors who restrict interfacial tension to situations where neither phase is a gas or vapor. On the other hand, the term "surface tension" is used here only when one phase is a gas or vapor, in agreement with the usage of most authors.

Interfacial tension of a fluid interface also has a mechanical interpretation. From this point of view the tension is a force per unit length parallel to the interface, i.e., perpendicular to the local density or concentration gradient. At a given point, this force is the same in all lateral directions along the interface. Accordingly, interfacial tension is the two-dimensional counterpart of pressure in a bulk fluid, which has the same magnitude in all directions in three-dimensional space although, of course, it is compressive instead of tensile. We note that the concepts of interfacial tension obtained from the thermodynamic and mechanical approaches -- an energy per unit area and a force per unit length -- are dimensionally equivalent. In a similar way pressure may be thought of as an energy per unit volume or a force per unit area.

Why does a lateral force arise along an interface and why is it a tensile instead of a compressive force? The answer is not a simple one, and we provide a qualitative explanation here only for the basic liquid-vapor interface of Figure I-1. At such an interface two factors influence the local stress. One is the familar kinetic effect due to thermal motion of molecules which is responsible for pressure in a dilute gas. Its contribution to the stress is an isotropic pressure which is proportional to the local molecular density and to the absolute temperature.

The second factor is attractive interaction among molecules. Each pair of interacting molecules makes a tensile contribution to the stress which is directed along the line joining their centers. The overall tensile stress due to the sum of all such contributions is isotropic in a bulk phase but anisotropic in an interfacial region, i.e., different in directions parallel to and perpendicular to the density gradient. The reason for this anisotropy is simply that the number and distribution of molecules with which a given molecule can interact is different in the two directions. At point A in Figure I-1, for example, the molecular density falls more rapidly toward the vapor phase than it increases toward the liquid phase. As a result, we anticipate that there will be fewer pairs of molecules with their lines of centers nearly parallel to the density gradient than with their lines of centers nearly aligned with some lateral direction (perpendicular to the density gradient). In other words the tensile stress at A due to molecular interaction will be greater in the lateral direction than in the normal direction.

Now the net pressure in each bulk fluid is the difference between the kinetic and interaction contributions. Both contributions are much larger in the liquid than in the vapor owing to the higher molecular density of the former, but their difference is the same in both phases for a plane interface. Similarly, the net stress at any point in the interfacial region is the difference between the local kinetic and interaction contributions. It is anisotropic with its normal component always equal to the bulk fluid pressures for a plane interface as required for static equilibrium. In view of the argument of the preceding paragraph, we anticipate that the average lateral stress in the interfacial region is less compressive and more tensile than in the bulk fluids. The overall result of this effect is what we know as interfacial tension. We note that the anisotropy of stress in the interfacial region implies that, unlike a bulk fluid, an interface can support certain types of shear stress at static equilibrium.

It is noteworthy that, as shown in Section 4, the lateral stress normally becomes tensile in nature at some points in the interfacial region and is not simply a pressure with a smaller magnitude than in the bulk fluids. Careful experiments have shown that even bulk liquids can sometimes be subjected to tensile stresses (1). It has been proposed, moreover, that negative pressures (i.e., tensile stresses) are responsible for the ability of sap to rise to great heights in trees and for the ability of mangrove roots, whose sap contains little salt, to absorb water from the salty waters in which the mangrove tree thrives. This latter situation may be an example of nature's use of the reverse osmosis principle. Of course, negative pressures in bulk liquids are metastable conditions while interfacial tension is a true equilibrium property.

Whether interfacial tension is developed from thermodynamic (energy) or mechanical (force) considerations, its main effect is that a system acts to minimize its interfacial area. This tendency for interfacial contraction is the reason that a small drop of one fluid in another will, provided gravitational effects are small, be spherical, the shape which minimizes drop area for a given drop volume. But it is essential to recognize that the energy and force arguments lead not simply to qualitatively similar concepts but to the same quantitative value of interfacial tension, a point which is demonstrated below.

Both approaches are useful. The energy approach relates interfacial tension to thermodynamics and thus allows useful results to be derived, e.g., the Kelvin equation of Example I-1, which gives the effect of drop size on vapor pressure. The force approach is needed to justify using interfacial tension in boundary conditions involving forces and stresses at interfaces. Such boundary conditions are employed in solving the governing equations of fluid mechanics when fluid interfaces are present.

Some values of interfacial tension in common systems are listed in Table I. The SI system units of milli-Newtons per meter (mN/m) are used, 1 mN/m being equal to 1 dyne/cm, the units of interfacial tension found in all but recent papers. Note that liquids with higher cohesive energies

Table I. Representative Values of Surface and Interfacial Tensions

Fluid	Temperature(°C)	Surface Tension (mN/m)
Silver	1100	878
Mercury	20	484
Sodium nitrate	308	117
Water	20	72.8
Propylene carbonate	20	41.1
Benzene	20	29.0
n-Octanol	20	27.5
Propionic acid	20	26.7
n-Octane	20	21.8
Ethyl ether	20	17.0
Argon	-183	11.9
Perfluoropentane	20	9.9
Fluid Pair	**Temperature (°C)**	**Interfacial Tension (mN/m)**
Mercury-water	20	415
Mercury-benzene	20	357
n-Octane-water	20	50.8
Benzene-water	20	35.0
Ethyl ether-water	20	10.7
n-Octanol-water	20	8.5
n-Butanol-water	20	1.8

have larger surface tensions. Thus, surface tensions of liquid metals are higher than those of hydrogen bonded liquids such as water, which in turn are higher than those of nonpolar liquids such as pure hydrocarbons. Phenomenological methods of estimating surface tension are discussed in Section 5 below and in Section 2 of Chapter II. Interfacial tensions between immiscible liquids decrease as the liquids become more similar in chemical nature.

3. INTERFACIAL TENSION - THERMODYNAMIC APPROACH

As we have seen, the interfacial region is, in fact, three dimensional, i.e., it has a finite thickness. It is very convenient, however, to represent an interface as a mathematical surface of zero thickness because such properties as area and curvature are well defined and because the differential geometry of surfaces is well understood. How can a thermodynamic analysis be developed which reconciles the use of mathematical surfaces with the actual three-dimensional character of the interface?

Over a century ago Gibbs (2) introduced surface excess quantities as a first step toward resolving this problem. The basic idea is to choose a reference surface S somewhere in the interfacial region. This surface is everywhere perpendicular to the local density or concentration gradient. Consider a property such as internal energy in the region between surfaces S_A and S_B of Figure I-2 which are parallel to S but located in the respective bulk phases. Because the transition between bulk compositions occurs over a finite thickness, the actual internal energy U of this region differs from the value $(U_A + U_B)$ which would be calculated by assuming that bulk phases A and B extend unchanged all the way to S. The difference is called the surface excess internal energy U^s and is assigned to S:

$$U^s = U - U_A - U_B \qquad [I\text{-}1]$$

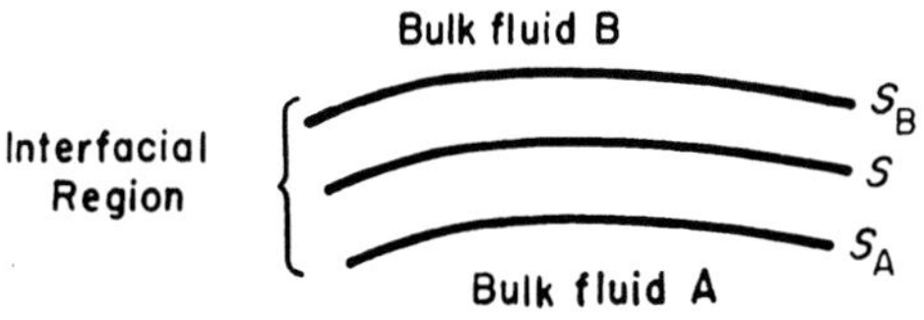

Figure I-2. Interfacial region bounded by parallel surfaces. S is the reference surface.

Similarly, surface excess values of other thermodynamic properties can be defined, as can the surface excess number of moles n_i^s of species i:

$$n_i^s = n_i - n_{iA} - n_{iB} \qquad [I\text{-}2]$$

Here n_i represents the actual number of moles of i in the region between S_A and S_B, n_{iA} represents the moles of i that would be present in the region between S_A and S if it were occupied by bulk fluid A, and n_{iB} is defined similarly for the region between S_B and S. While independent of the exact positions of S_A and S_B as long as these surfaces are in the bulk fluids, the values of surface excess quantities do depend on the position of S. They may also be either positive or negative.

It is clear from Eqs. [I-1] and [I-2] that surface excess quantities do take account of the variation of composition and properties across an interfacial region of finite thickness. As we shall see shortly, they can be used to define interfacial tension. Moreover, since all surface excess properties are assigned to the reference surface S, the area and curvature of S can be identified as the corresponding properties of the interface and used, for example, to describe interfacial deformation.

Let us consider further the interfacial region between S_A and S_B under equilibrium conditions. If its shape remains fixed, we suppose that its internal energy U is a function only of its entropy S and the number of moles n_i of species i in the region. Then we can write

$$dU = TdS + \sum_i \mu_i dn_i \qquad [I\text{-}3]$$

where $T = (\partial U/\partial S)_{n_i,\text{shape}}$ and $\mu_i = (\partial U/\partial n_i)_{S,n_j,\text{shape}}$

Now consider a process where bulk fluid A remains unchanged but some energy and/or mass is transferred between bulk fluid B and the interfacial region. The entire system consisting of the interfacial region and bulk fluids A and B is presumed to be completely isolated from its surroundings. In this case, an energy balance (First Law of Thermodynamics) shows that the total energy of the system remains unchanged. Recalling that there is no change in the energy of bulk fluid A, we have

$$dU_{tot} = 0 = TdS + \sum_i \mu_i dn_i + T_B dS_B + \sum_i \mu_{iB} dn_{iB} \qquad [I\text{-}4]$$

Mass balances for the various species require that $dn_i = -dn_{iB}$. In addition, the entropy of such an isolated system must be a maximum if it is in equilibrium, according to the Second Law of Thermodynamics. As a result, we have

$$dS_{tot} = 0 = dS + dS_B \qquad [I\text{-}5]$$

From [I-4],[I-5], and the mass balances, we find that

$$0 = (T-T_B)dS + \sum_i (\mu_i - \mu_{iB}) dn_i \qquad [I\text{-}6]$$

If this equation is to be satisfied for all possible changes, it is clear that we must have

$$T = T_B \quad , \quad \mu_i = \mu_{iB} \qquad [I\text{-}7]$$

A similar argument can be made for energy or mass transfer between the interfacial region and bulk fluid A. The overall conclusion is that the temperature T and the chemical potential μ_i of each component must be uniform throughout the system.

With the interfacial region still maintained at constant shape, we can write Eq. [I-3] and subtract from it the analogous equations which would apply for its two parts if they were occupied by bulk fluids A and B, respectively. The result is

$$dU^S \text{ (fixed shape)} = TdS^S + \sum_i \mu_i dn_i^S \qquad [I\text{-}8]$$

We now consider how dU^S might change for a fluid interface if the reference surface *S* is deformed. Both the area and curvature of *S* can change; but if the radii of curvature are much greater than interfacial thickness, we might expect curvature effects to be small, so that

$$dU^s = TdS^s + \sum_i \mu_i dn_i^s + \gamma dA \qquad [I-9]$$

Here γ is the interfacial tension defined as $(\partial U^s/\partial A)_{S^s, n_i^s}$

The Helmholtz free energy F for the interfacial region is defined in the usual way:

$$F = U - TS \qquad [I-10]$$

Subtracting from [I-10] the analogous equations which would apply if the two parts of the interfacial region were occupied by bulk fluids A and B, we obtain

$$F^s = U^s - TS^s \qquad [I-11]$$

Differentiating Eq. [I-11] and invoking Eq. [I-9], we find

$$dF^s = -S^s dT + \sum_i \mu_i dn_i^s + \gamma dA \qquad [I-12]$$

It is clear from this equation that the interfacial tension can be written in terms of the surface excess free energy as

$$\gamma = \left(\frac{\partial F^s}{\partial A}\right)_{T, n_i^s} \qquad [I-13]$$

That is, γ is the change in surface excess free energy produced by a unit increase in area.

Let us consider a given interfacial region and investigate the effect of shifting the reference surface *S* uniformly toward S_B by some small amount λ. Naturally, the free energy F of the overall region is unchanged since no change occurs in its physical state. We have

$$\begin{aligned} 0 = dF &= dF^s + dF_A + dF_B \\ &= -S^s dT + \sum_i \mu_i dn_i^s + \gamma dA \\ &\quad -S_A dT + \sum_i \mu_i dn_{iA} - p_A\, dV_A \\ &\quad -S_B dT + \sum_i \mu_i dn_{iB} - p_B dV_B \end{aligned} \qquad [I-14]$$

Now both the total volume V of the interfacial region and the total number of moles n_i of each species are constant. Hence,

$$dV = dV_A + dV_B = 0 \qquad [I\text{-}15]$$

$$dn_i = dn_i^s + dn_{iA} + dn_{iB} = 0 \qquad [I\text{-}16]$$

Invoking Eqs. [I-15] and [I-16] and recognizing that there is no temperature change, we find that [I-14] simplifies to

$$0 = \gamma dA - (p_A - p_B)dV_A \qquad [I\text{-}17]$$

Let A be the initial area of the reference surface S. Then it is clear that a simple shift in the reference surface gives

$$dV_A = A\lambda \qquad [I\text{-}18]$$

Moreover, it can be shown from geometrical considerations that

$$dA = -2H\ A\lambda \qquad [I\text{-}19]$$

where (-2H) is the sum of the reciprocal radii of curvature of S as measured in any two perpendicular planes containing the local normal to S, i.e.,

$$2H = -\left(\frac{1}{r_1} + \frac{1}{r_2}\right) \qquad [I\text{-}20]$$

The quantity H is often called the mean curvature. The minus sign in Eq. [I-20] is a convention; a radius of curvature is deemed positive if its center of curvature is on the A side of S and negative if its center of curvature is on the B side. Evaluation of r_1 and r_2 is considered in Section 6.

While Eq. [I-19] will not be proved here, it may be helpful to demonstrate its validity for the special case of a sphere which increases in radius by an amount dr. In this case we have

$$dA = d\,(4\pi r^2) = 8\pi r\ dr \qquad [I\text{-}21]$$

As 2H is (-2/r), A is $4\pi r^2$ and λ is dr, it is clear that Eq. [I-19] is satisfied.

When Eqs. [I-18] and [I-19] are substituted into Eq. [I-17], the result is

$$p_A - p_B = -2H\gamma \qquad [I\text{-}22]$$

This relationship among the two bulk phase pressures, the mean curvature, and the interfacial tension is fundamental to the study of fluid interfaces. It is often referred to as the Laplace or the Young-Laplace equation and is the basis of several methods of measuring interfacial tension. Note that with the interface curved in the manner shown in Figure I-2, H is negative and Eq. [I-22] implies that $p_A > p_B$. Thus, the pressure inside a drop or bubble always exceeds the pressure without.

A second fundamental equation of interfacial thermodynamics can be derived starting with Eq. [I-9]. Now U^s, S^s, n_i^s and A are all extensive variables, i.e., they are proportional to the area of $\mathcal{S}$ if the intensive variables of the system, such as temperature and pressure, remain constant. Accordingly, Euler's theorem of homogeneous functions can be applied just as in ordinary bulk phase thermodynamics to obtain:

$$U^s = TS^s + \sum_i \mu_i n_i^s + \gamma A \qquad [I\text{-}23]$$

Differentiating [I-23] and applying Eq. [I-9], we find

$$0 = S^s dT + \sum_i n_i^s d\mu_i + A d\gamma \qquad [I\text{-}24]$$

This equation is known as the Gibbs adsorption equation and is the analog of the Gibbs-Duhem equation for bulk fluids.

For a two-component system at constant temperature with $\mathcal{S}$ chosen in such a way that n_1^s vanishes, Eq. [I-24] simplifies to

$$\left(\frac{\partial \gamma}{d\mu_2} \right)_T = -\Gamma_2 = -\frac{n_2^s}{A} \qquad [I\text{-}25]$$

Further simplifying to the case where species 2 is a solute which exhibits ideal behavior in a solvent (species 1), we find

$$\Gamma_2 = -\frac{x_2}{RT} \left(\frac{\partial \gamma}{\partial x_2} \right)_T \qquad [I\text{-}26]$$

where x_2 is the bulk phase solute mole fraction. We see that the surface excess concentration Γ_2 can be calculated from measurements of interfacial tension as a function of composition. We see, moreover, that if interfacial tension decreases with increasing solute concentration, Γ_2 is positive, i.e., the solute tends to concentrate near the interface. Such solutes are called surface active materials.

The derivation leading to Eq. [I-26] involves a particular choice of the reference surface S. We may naturally ask about the sensitivity of quantities such as interfacial tension γ and the surface excess concentrations Γ_i to the location of the reference surface. For a plane interface the value of γ is independent of the position of S. This result follows from Eq. [I-39] (to be derived shortly) and from the equality of bulk phase pressures p_A and p_B in this case. We expect, therefore, that even for an interface which is slightly curved, γ should not be very sensitive to small shifts in the position of S.

The surface excess concentrations Γ_i are another matter, however. Let us consider the simple liquid-vapor interface of Figure I-1 and suppose that the reference surface S is at position A. If the bulk vapor density is small in comparison with the bulk liquid density, we see that shifting S by even one molecular diameter changes the value of Γ by an amount comparable to the surface concentration along an imaginary plane through the bulk liquid. Clearly, this is a very significant change, and it raises questions about results involving the Γ_i based on a particular (even though reasonable) choice of S.

Fortunately, such problems can usually be avoided -- at least for plane interfaces. The Gibbs-Duhem equations for the two bulk phases in a binary system at constant temperature are given by

$$\begin{aligned} dp_A &= c_{1A}\, d\mu_1 + c_{2A} d\mu_2 \\ dp_B &= c_{1B} d\mu_1 + c_{2B} d\mu_2 \end{aligned} \qquad \text{[I-27]}$$

For a plane interface $dp_A = dp_B$ and these equations can be used to eliminate $d\mu_1$ from Eq. [I-24]. The result is

$$\left(\frac{\partial \gamma}{\partial \mu_2} \right)_T = -\Gamma_2 + \left(\frac{c_{2A} - c_{2B}}{c_{1A} - c_{1B}} \right) \Gamma_1 = -\Gamma_{2,1} \qquad \text{[I-28]}$$

No assumption about the position of the reference surface has been made in deriving this equation. Moreover, since $(\partial\gamma/\partial\mu_2)_T$ is independent of the location of S for a plane interface, the "relative adsorption" $\Gamma_{2,1}$ must be as well. It is, of course, numerically equal to Γ_2 for the choice of S which makes Γ_1 vanish. Similar definitions of the various relative adsorptions $\Gamma_{i,1}$ can be made for multicomponent systems with i replacing 2 as a subscript in the definition of $\Gamma_{2,1}$(3). It can be shown that the $\Gamma_{i,1}$ are all independent of the reference surface position in this case.

Another useful relationship is obtained by dividing Eq. [I-23] by A and rearranging:

$$\gamma = \frac{F^s}{A} - \sum_i \mu_i \Gamma_i \qquad [I-29]$$

This equation is used to *define* the interfacial tension of a solid-fluid interface because it avoids complications involving strain energy effects associated with extending the previous definition (given following Eq. [I-9]) to the case of solids. We note from Eq. [I-29] that the interfacial tension is not, in general, equal to the Helmholtz free energy per unit area. The two are the same, however, for the special case of a single-component system with the reference surface chosen to make Γ_1 vanish.

Example I.1. Vapor Pressure of a Drop

Calculate the vapor pressure of a liquid drop as a function of its radius. Find the increase in vapor pressure for water at 100°C for drop radii of 100 μm, 1 μm, and 0.01 μm. Take the surface tension to be 60 mN/m and the density to be 0.994 gm/cm^3 for water at this temperature.

Solution The Young-Laplace equation [I-22] for the drop takes the form

$$p_L - p_V = \frac{2\gamma}{r} \qquad [I-E1-1]$$

where r is the drop radius. When r varies at constant temperature, this equation can be differentiated to obtain

$$dp_L - dp_V = d\left(\frac{2\gamma}{r}\right) \qquad [I-E1-2]$$

Since equilibrium is maintained during this process, the changes in the liquid and vapor chemical potentials must be equal

$$d\mu_L = v_L dp_L = d\mu_V = v_V dp_V \qquad [I-E1-3]$$

From this equation it is clear that

$$dp_L = \frac{v_V}{v_L} dp_V \qquad [I-E1-4]$$

Substituting this expression into Eq. [I-E1-2], invoking the ideal gas law since vapor pressure is low, and further noting that $v_L << v_V$ under these conditions, we find

$$\frac{RT}{v_L} \frac{dp_V}{p_V} = d\left(\frac{2\gamma}{r}\right) \qquad \text{[I-E1-5]}$$

When this equation is intergrated using the condition that p_V is the ordinary vapor pressure p_V^0 for a large pool of liquid in the limit $r \rightarrow \infty$, the result is the Kelvin equation

$$p_V = p_V^0 \exp\left(\frac{2\gamma v_L}{r\,RT}\right) \qquad \text{[I-E1-6]}$$

The vapor pressure is thus enhanced with decreasing drop radii though the effect is appreciable only for extremely small drops (see calculation below). This effect is of importance in homogeneous nucleation of a new phase.

For water at 100°C, we find the following values using the properties given above

r(μm)	p_V/p_V^0
100	1.000007
1	1.0007
0.01	1.073

4. INTERFACIAL TENSION - MECHANICAL APPROACH

Our next step is to analyze the interfacial region at static equilibrium from a mechanical point of view. It is necessary to specify the lateral boundary of the region, which will be taken to be a surface S_0 everywhere normal to S, S_A, and S_B of Figure I-2, and to all other surfaces parallel to these three which could be constructed within the region. The result is a closed interfacial region bounded by S_A, S_B, and S_0 (Figure I-3).

Within the interfacial region tangential and normal pressures p_T and p_N may differ, as discussed previously. In the bulk fluids, of course, the pressure is isotropic so that p_N and p_T are equal. Using this information, we apply Newton's Second Law to the interfacial region to obtain

$$\int_V \rho \hat{F}\, dV + \int_{S_A} p_A \mathbf{n}\, dS - \int_{S_B} p_B\, \mathbf{n}\, dS - \int_{S_0} p_T\, \mathbf{M}\, dS = \int_V \rho \mathbf{a}\, dV \qquad \text{[I-30]}$$

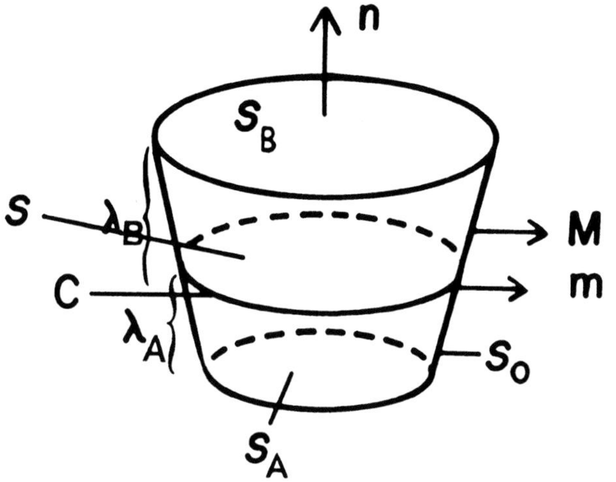

Figure I-3. "Pillbox" control volume. S_0 is perpendicular to S, S_A, and S_B.

Note that since no shear stresses act on the outer boundary of the interfacial region, the local force on this boundary is everywhere parallel to the local normal, viz., **n** for S_A and S_B, **M** for S_0 (see Figure I-3). Accordingly, the net force on the outer boundary is given by the three surface integrals of Eq. [I-30]. Also $\hat{\mathbf{F}}$ is the local body force per unit mass, most commonly gravity, and **a** is the local acceleration, which must be either a uniform translational acceleration or a uniform rotation to be consistent with static equilibrium.

Using the basic surface excess concept described previously, we wish to subtract from Eq. [I-30] the corresponding equations which would apply if the region between S and S_A were occupied by bulk fluid A and that between S and S_B by bulk fluid B. These equations are

$$\int_{V_A} \rho_A \hat{\mathbf{F}}\, dV + \int_{S_A} p_A \mathbf{n}\, dS - \int_{S} p_A \mathbf{n}\, dS - \int_{S_{oA}} p_A \mathbf{M}\, dS = \int_{V_A} \rho_A \mathbf{a}\, dV \qquad \text{[I-31]}$$

$$\int_{V_B} \rho_B \hat{\mathbf{F}}\, dV + \int_{S} p_B \mathbf{n}\, dS - \int_{S_B} p_B \mathbf{n}\, dS - \int_{S_{oB}} p_B \mathbf{M}\, dS = \int_{V_B} \rho_B \mathbf{a}\, dV \qquad \text{[I-32]}$$

When they are subtracted from Eq. [I-30], the result is

$$\int_{V} \Delta\rho\, \hat{\mathbf{F}}\, dV + \int_{S} (p_A - p_B)\, \mathbf{n}\, dS - \int_{S_0} \Delta p_T\, \mathbf{M}\, dS = \int_{V} \Delta\rho\, \mathbf{a}\, dV \qquad \text{[I-33]}$$

where $\Delta\rho = \begin{cases} \rho-\rho_A & \text{in } V_A \\ \rho-\rho_B & \text{in } V_B \end{cases}$

$\Delta p_T = \begin{cases} p_T-p_A & \text{in } V_A \\ p_T-p_B & \text{in } V_B \end{cases}$

If interfacial thickness is small in comparison with the radii of curvature of S, and if the body force at any lateral position is approximately uniform between S_A and S_B, we can write

$$\int_V \Delta\rho\, \hat{\mathbf{F}}\, dV \simeq \int_S \left[\int_{\lambda_A}^{\lambda_B} \Delta\rho\, d\lambda \right] \hat{\mathbf{F}}\, dS \qquad [I-34]$$

Similarly, if C is the closed curve which forms the outer boundary of S and if m is its outward pointing normal, we have

$$\int_{S_0} \Delta p_T\, \mathbf{M}\, dS \simeq \int_C \left[\int_{\lambda_A}^{\lambda_B} \Delta p_T\, d\lambda \right] \mathbf{m}\, ds \qquad [I-35]$$

Finally, a theorem from the differential geometry of surfaces (4) can be used to transform the integral along C to an integral over S:

$$\int_C f\, \mathbf{m}\, ds = \int_S [\nabla_s f + 2H f \mathbf{n}]\, dS \qquad [I-36]$$

where $(\nabla_s f)$ is the gradient of f within the surface, e.g., $[\mathbf{e}_x(\partial f/\partial x) + \mathbf{e}_y(\partial f/\partial y)]$ for a plane surface. This "surface divergence" theorem will not be proved here although demonstration of its validity in one simple situation is the objective of Problem I-6.

Substitution of Eqs. [I-34] - [I-36] into Eq. [I-33] yields

$$\int_S [\Gamma \hat{\mathbf{F}} - \Gamma \mathbf{a} + (p_A - p_B)\mathbf{n} + 2H\gamma\mathbf{n} + \nabla_s \gamma]\, dS = 0 \qquad [I-37]$$

$$\Gamma = \text{surface excess mass} = \int_{\lambda_A}^{\lambda_B} \Delta\rho\, d\lambda \qquad [I-38]$$

$$\gamma = \text{interfacial tension} = -\int_{\lambda_A}^{\lambda_B} \Delta p_T\, d\lambda \qquad [I-39]$$

As the extent of S is arbitrary, the integrand of Eq. [I-37] must itself vanish, which leads to the differential equation of fluid statics:

$$\Gamma \hat{\mathbf{F}} - \Gamma \mathbf{a} + (p_A - p_B)\,\mathbf{n} + 2H\gamma\mathbf{n} + \nabla_s \gamma = 0 \qquad \text{[I-40]}$$

In most cases of interest the surface excess mass Γ is small, so that the acceleration and body force terms may be neglected. Then Eq. [I-40] simplifies to two conditions. One of them, $\nabla_s \gamma = 0$, requires that interfacial tension be uniform. The other is the Young-Laplace equation [I-22], which was obtained previously from thermodynamics for situations where body force and acceleration terms were unimportant. That the same equation [I-22] results from independent thermodynamic and mechanical derivations implies that interfacial tension must have the same value whether it is defined as at Eq. [I-9] from energy considerations or as at Eq. [I-39] from force considerations. Simply put, the force and energy definitions of interfacial tension are equivalent, a conclusion emphasized in the work of Buff (5).

Equation [I-39] also allows us to demonstrate that tangential pressure p_T ordinarily must have negative values at some locations within the interfacial region. Consider a case where interfacial tension γ is about 10 mN/m and interfacial thickness is about 1 nm. It is clear from Eqs. [I-39] and [I-33] that the average value of $(p_N - p_T)$ in the interfacial region must be of the order of 10^7 N/m^2 or about 100 atm. As p_N must always lie between the bulk fluid pressures and hence is typically not more than a few atmospheres, it is evident that p_T must have appreciable negative values at some locations within the interfacial region, i.e., the lateral stress must be tensile there.

5. DENSITY AND CONCENTRATION PROFILES

Let us return to the density profile shown in Figure I-1 for a simple liquid-vapor interface. We may ask what determines the shape of the profile, i.e., whether it is broad or sharp. Clearly the answer is that the system chooses the shape which minimizes its free energy. But what factors influence the free energy?

One factor, which favors a sharp profile, is the intrinsically high free energy of material having densities intermediate between those of bulk liquid and bulk vapor. As Figure I-4 shows, these intermediate free

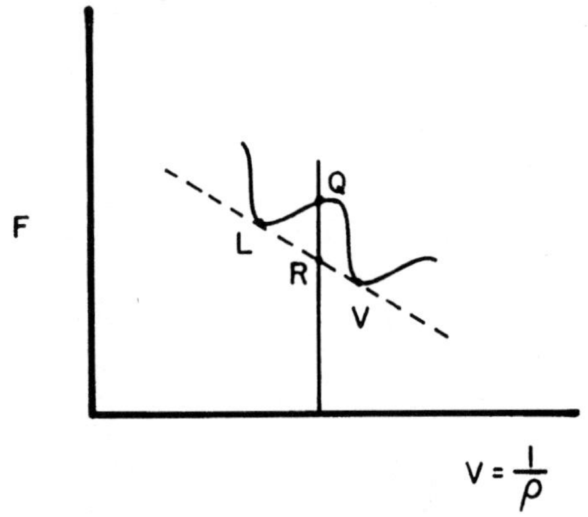

Figure I-4. Dependence of free energy on molar volume v for a fluid below its critical temperature. L and V are the states of the liquid and vapor phases in equilibrium.

energies, e.g., point Q on the curve between liquid and vapor densities L and V, are higher than those of corresponding mixtures such as R of liquid and vapor having the same overall specific volume v or density ρ. Indeed, it is precisely this situation which is responsible for the phase separation. As a result of the high intermediate free energies, the system has a tendency to minimize the amount of material with intermediate densities and hence to form a thin interfacial region.

The curve of Figure I-4 represents the free energy of a large volume of material with a uniform specific volume v. Each element of material interacts only with other material having the same specific volume v. In the interfacial region, however, density varies rapidly so that each element of material has elements with other values of v within the effective range of intermolecular forces. A correction to the free energy curve of Figure I-4 is required to account for this effect.

Near a point such as A of Figure I-1 the average density decrease over the effective range of intermolecular forces is greater in the direction of the vapor phase than is the average density increase in the opposite direction. The result is an increase in free energy of the material at A over that predicted by Figure I-4. The magnitude of the increase is greater for a thinner interface with a sharper density profile. Hence this effect opposes sharp profiles and favors thick interfaces with broad profiles.

These two factors, the intrinsically high free energy of intermediate densities and the correction due to rapid density changes within the interfacial region, thus have opposite effects on the interfacial profile. The actual profile represents a balance between them and corresponds to an interface having a thickness that is finite but typically quite small. The exception is near a critical point where the interface is thick, gradient energy effects are small, and interfacial tension approaches zero.

Many years ago van der Waals developed a phenomenological approach for determining the interfacial density profile and interfacial tension, taking into account these two factors. A recent treatment along the same lines is that of Cahn and Hilliard (6). The local Helmholtz free energy per unit volume f in the interfacial region is written as follows for a flat interface:

$$f = f_o(n) + k\left(\frac{dn}{dz}\right)^2 \qquad [I-41]$$

Here $f_o(n)$ is applicable to a large region having uniform molecular density n, while the second term is the correction necessary to account for the density gradient. The constant k is presumed to have a positive value.

As indicated in the discussion following Eq. [I-29], the interfacial tension γ is equal to the surface excess Helmholtz free energy per unit area (F^s/A) when the reference surface is chosen to make the surface excess mass Γ vanish. But for this reference surface the total amounts of liquid and vapor present in the extrapolated bulk phases must equal the amounts of the two phases that would form upon complete separation of the material in the interfacial region at constant total volume. Hence (F^s/A) can be evaluated by integrating the energy change required for complete separation at each local position:

$$\gamma = \frac{F^s}{A} = \int_{-\infty}^{\infty} (f - \Phi_L f_L - \Phi_V f_V)\, dz \qquad [I-42]$$

where Φ_L and Φ_V are the volume fractions of liquid and vapor that would be formed by material at position x and f_L and f_V are the Helmholtz free energy densities of the bulk phases. Invoking Eq. [I-41], we can write

$$\gamma = \int_{-\infty}^{\infty} \left[\Delta f + k\left(\frac{dn}{dx}\right)^2\right] dz \qquad [I-43]$$

$$\Delta f(n) = f_o(n) - \Phi_L f_L - \Phi_V f_V \qquad [I-44]$$

It is readily seen that Δf is simply the distance between the free energy curve and the line LV in Figure I-4, e.g., the vertical distance QR.

The density profile realized in actuality is that which minimizes the the integral in Eq. [I-43]. Since Δf and n are functions of z, the calculus of variations must be invoked (7). The resulting condition which must be satisfied throughout the interfacial region is given by

$$2k \frac{d^2n}{dz^2} - \frac{d\Delta f}{dn} = 0 \qquad \text{[I-45]}$$

Multiplying this expression by (dn/dz), integrating the resulting expression, and recalling that Δf and (dn/dz) vanish in the bulk fluids, we obtain

$$\Delta f(n) = k \left(\frac{dn}{dz}\right)^2 \qquad \text{[I-46]}$$

This equation provides information about the steepness of the density gradient at a position where the density is n. Substituting Eq. [I-46] into Eq. [I-43] and changing the variable of integration from x to n, we finally obtain the following expression for the interfacial tension.

$$\gamma = 2 \int_{\rho_V}^{\rho_L} [k \; \Delta f(n)]^{1/2} \, dn \qquad \text{[I-47]}$$

Evaluation of γ from Eq. [I-47] requires information on both $\Delta f(n)$ and k. For the former it is necessary to have an equation of state, for the latter some knowledge of how forces between molecules depend on the distance of separation. Attempts to include intermolecular forces and thus predict density profiles and interfacial tensions go back at least as far as van der Waals, who used his equation of state in a calculation of density profiles. Davis and Scriven (8) have recently summarized work in this area very clearly, including some of their own significant contributions of recent years. Among the topics they discuss is the use of density and composition profiles to calculate the stress distribution within the interfacial region. Rowlinson and Widom (9) also deal with various aspects of the molecular theory of interfacial tension in their recent book.

We note that the approach described above can be used for thin films and other "nonuniform" systems where density or composition varies rapidly with position at equilibrium (see Chapter II). It contrasts sharply with the classical treatment of interfacial thermodynamics given in Section 3 above which deals with overall properties of the interfacial region without considering density or concentration profiles or even the existence of molecules and intermolecular forces. An advantage of the earlier approach is that its results are independent of molecular properties. The principal disadvantage is that values of interfacial tension cannot be predicted. Hence the role of molecular theory is, as in other areas of thermodynamics, to provide information beyond that obtainable with the classical approach.

6. EQUILIBRIUM SHAPES OF FLUID INTERFACES

As indicated above, the Young-Laplace equation [I-22] is one fundamental result obtained from the theory of interfaces. Because this equation relates interfacial tension to the pressure difference between fluids at each point along an interface, it can be used with the equations of hydrostatics to calculate the shape of a static interface. Or, if interfacial shape can be determined experimentally, Eq. [I-22] can be used to obtain the interfacial tension. This latter procedure is the basis of several methods for measuring interfacial tension.

Because the mean curvature H appears in Eq. [I-22], its application to an interface of arbitrary shape requires considerable knowledge of differential geometry. A few general formulas from differential geometry are given in Chapter VII. Here we restrict attention to relatively simple shapes with considerable symmetry. In particular we consider the sessile drop method for measuring surface or interfacial tensions. A drop of fluid A is placed on a solid S which is otherwise in contact with fluid B (see Figure I-5). The materials must be chosen so that A is denser than B and does not spread spontaneously on S, indeed preferably so that the contact angle λ shown in Figure I-5 is relatively large at the final equilibrium position. For a given λ (and λ is basically a property of the materials present as discussed in Chapter II), two factors determine the shape of the fluid interface. One is gravity, which favors a large diameter drop of small height, so that the denser fluid A is at its lowest possible position. The other is interfacial tension, which favors a smaller drop of greater height where the surface area is minimized. Indeed, in the absence of gravity the interface would be a portion of a spherical surface. The actual drop shape reflects a balance between these two effects.

According to the usual equations of hydrostatics, we have in bulk fluids A and B

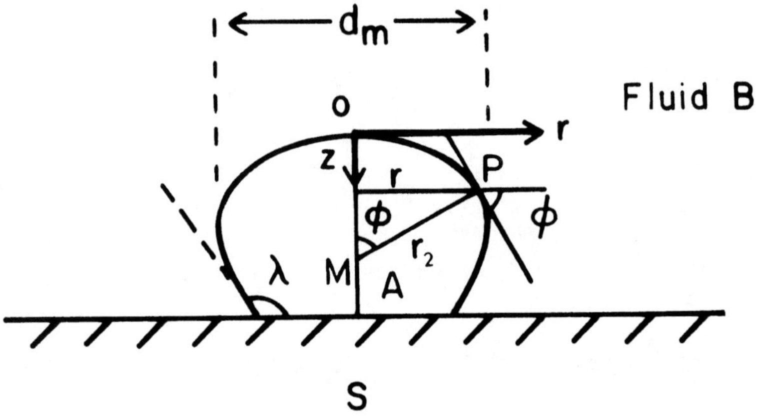

Figure I-5. Sessile drop.

$$p_A = p_{Ao} + \rho_A g z \qquad [I\text{-}48]$$

$$p_B = p_{Bo} + \rho_B g z \qquad [I\text{-}49]$$

Here the reference pressures p_{Ao} and p_{Bo} have been chosen for convenience as those at the drop apex 0 of Figure I-5. Also, a cylindrical coordinate system with its origin at 0 has been chosen as shown. If the drop is axisymmetric, the radius of curvature b at 0 is the same for all orientations, and Eq. [I-22] requires that

$$p_{Ao} - p_{Bo} = \frac{2\gamma}{b} \qquad [I\text{-}50]$$

We now want to apply Eq. [I-22] to a general point P on the drop where the radii of curvature are r_1 and r_2. Using Eqs. [I-48]-[I-50] with Eq. [I-22], we have

$$\gamma \left(\frac{1}{r_1} + \frac{1}{r_2} \right) = \frac{2\gamma}{b} + (\rho_A - \rho_B)\, gz \qquad [I\text{-}51]$$

The principal radii of curvature for an axisymmetric drop are given by (10)

$$\frac{1}{r_1} = \frac{d^2z/dr^2}{[1 + (dz/dr)^2]^{3/2}} \qquad [I\text{-}52]$$

$$\frac{1}{r_2} = \frac{dz/dr}{r[1 + (dz/dr)^2]^{1/2}} \qquad [I\text{-}53]$$

Here r_1, the radius of curvature in the plane of Figure I-5, is given by a well known expression from analytic geometry. Also r_2 is the radius of curvature around the drop as measured in a plane perpendicular to that of Figure I-5 but containing the local normal at P. The center of curvature for r_2 is at point M of Figure I-5 where the normal meets the drop axis. It is clear from the figure that

$$\sin\phi = \frac{r}{r_2} \qquad \text{[I-54]}$$

Invoking an identity from trigonometry, we can rewrite this equation in the following form

$$\frac{1}{r_2} = \frac{1}{r}\frac{\tan\phi}{(1+\tan^2\phi)^{1/2}} \qquad \text{[I-55]}$$

Since the tangent to the drop profile at P makes an angle ϕ with the horizontal, we can replace $\tan\phi$ with (dz/dr) in Eq. [I-55], thus obtaining Eq. [I-53].

Substituting Eqs. [I-52] and [I-53] into Eq. [I-51] and using the dimensionless coordinates $Z = (z/b)$ and $R = (r/b)$, we find

$$\frac{Z''}{[1+(Z')^2]^{3/2}} + \frac{Z'}{R[1+(Z')^2]^{1/2}} = 2 + \beta Z \qquad \text{[I-56]}$$

where

$$\beta = (\rho_A - \rho_B)\, gb^2/\gamma \qquad \text{[I-57]}$$

For this second order ordinary differential equation two boundary conditions are required. They are that both Z and its derivative Z' with respect to R must vanish at the origin $R = 0$.

Equation [I-56] cannot be solved analytically except for certain limiting cases. Numerical solution is straightforward, however, and leads to knowledge of $Z(R)$ for any specified value of β. Bashforth and Adams (11) first carried out a numerical solution more than a century ago, although complete and accurate tables are now available from more recent calculations using a high-speed computer (see 12). In practice one can find interfacial tension by measuring the maximum diameter d_m of the drop (see Figure I-5) and the vertical distance between the plane of the maximum diameter and the apex and then referring to the tables. For present purposes, however, the main point of the discussion is to illustrate that interfacial shape can be calculated from Eq. [I-22] if appropriate properties such as interfacial tension and densities are known.

Example I-2 Dimensions of a Sessile Drop

(a) A sessile drop of water is to be formed on a solid surface where the contact angle λ is 90°. Find the maximum diameter, height, and volume of the drop if the parameter β of Eq. [I-57] is to have a value of 20. Bashforth and Adams' numerical results for β = 20 are summarized in Table II. Take γ = 72 mN/m and ρ_A = 1 gm/cm^3 for water. Compare the drop dimensions you obtain with those which would exist if drop volume and contact angle were the same but gravitational effects were negligible.

(b) Repeat the calculation of drop dimensions for β = 20 and $\lambda = 90^0$ for a drop of an aqueous surfactant solution in oil. Take ρ_A = 1.0 gm/cm^3, ρ_B = 0.8 gm/cm^3, and γ = 0.01 mN/m.

Solution. (a) From Eq. [I-57] with ρ_B = 0 and the other values as given we find b = 1.212 cm for the radius of curvature at the drop apex. From the tables with $\phi = 90^0$, we obtain for the coordinates at the base of the drop, R = 0.51153, Z = 0.29623. Hence, the maximum drop radius is (0.51153 x 1.212) = 0.620 cm and the height is (0.29623 x 1.212) = 0.359 cm. The ratio of height to radius is about 0.58, which gives some feeling for the degree of flattening caused by gravity. Dimensionless drop volume (V/b^3) is 0.16502, so that the actual volume is (0.16502 x 1.212^3) = 0.294 cm^3.

In the absence of gravity the drop would be a hemisphere of radius r. Since the volume is ($2\pi r^3/3$) in this case, we can easily find r = 0.520 cm. Of course, the ratio of height to radius is now unity.

(b) Repeating the calculations of (a), one finds b = 3.19 x 10^{-2}cm, V = 5.38 x 10^{-6} cm^3, a maximum radius of 1.63 x 10^{-2} cm, and a drop height of 9.46 x 10^{-3} cm. Clearly, much smaller drops must be used to measure very low tensions. Systems having such low tensions are of interest in certain processes being developed for increasing oil recovery from underground reservoirs (see Chapter IV).

Example I-3 Shape of a Soap Film Between Parallel Rings

A soap film is formed between parallel rings of radius R as shown in Figure I-6. The line connecting the centers of the rings is of length 2L and is perpendicular to the planes of the rings. The rings are open so that the space within the film is in communication with the atmosphere.

(a) Will the profile of the soap film most nearly resemble curve 1, 2 or 3 in the diagram? Explain.

(b) Derive an equation for the film profile if effects of gravity can be neglected.

Table II. Dimensionless Coordinates of Sessile Drop for $\beta = 20$ from Bashforth and Adams Tables (11).

ϕ	$R = \frac{r}{b}$	$Z = \frac{z}{b}$	$\frac{V}{b^3}$
5°	0.08558	0.00370	0.00004
10	0.16261	0.01377	0.00058
15	0.22768	0.02811	0.00235
20	0.28143	0.04498	0.00584
25	0.32571	0.06327	0.01117
30	0.36236	0.08230	0.01828
35	0.39283	0.10167	0.02699
40	0.41824	0.12113	0.03706
45	0.43941	0.14050	0.04827
50	0.45698	0.15965	0.06037
55	0.47145	0.17849	0.07314
60	0.48322	0.19694	0.08635
65	0.49259	0.21492	0.09981
70	0.49984	0.23240	0.11334
75	0.50518	0.24931	0.12676
80	0.50880	0.26561	0.13993
85	0.51087	0.28126	0.15272
90	0.51153	0.29623	0.16502
95	0.51091	0.31049	0.7672
100	0.50914	0.32400	0.19809
105	0.50632	0.33674	0.19809
110	0.50255	0.34869	0.20764
115	0.49795	0.35982	0.21640
120	0.49258	0.37012	0.22435
125	0.48656	0.37959	0.23147
130	0.47995	0.38820	0.23780
135	0.47285	0.39596	0.24333
140	0.46533	0.40285	0.24810
145	0.45746	0.40889	0.25214
150	0.44933	0.41407	0.25549
155	0.44100	0.41841	0.25819
160	0.43254	0.42191	0.26029
165	0.42402	0.42460	0.26184
170	0.41550	0.42649	0.26289
175	0.40704	0.42761	0.26348
180	0.39870	0.42797	0.26367

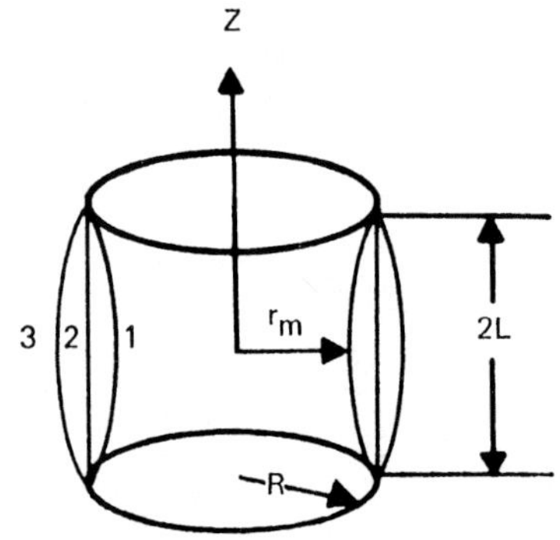

Figure I-6. Hypothetical profiles for a soap film between parallel rings.

Solution

(a) Since the rings are open, the pressure drop across the film is zero. From the Young-Laplace equation [I-22] we conclude that

$$\frac{1}{r_1} + \frac{1}{r_2} = 0 \qquad \text{[I-E3-1]}$$

That is, the centers of curvature for the two radii r_1 and r_2 at a point on the film must lie on opposite sides of the film. Since the film is axisymmetric, we define r_1 and r_2 as at Eqs. [I-52] and [I-53]. Now the center of curvature for r_2 must, by symmetry, lie on the axis joining the centers of the rings. Hence the center of curvature for r_1 must lie outside the film and profile 1 is the appropriate choice.

(b) This problem is most easily solved by using expressions other than those given by Eqs. [I-52] and [I-53] for r_1 and r_2. Owing to symmetry, we need consider only the upper half of the film, i.e., the region $z > 0$ in Figure I-6. From Eqs. [I-53] and [I-54] we have

$$\sin \phi = \frac{dz/dr}{[1 + (dz/dr)^2]^{1/2}} \qquad \text{[I-E3-2]}$$

When this expression is differentiated, the result is

$$\frac{d(\sin \phi)}{dr} = \frac{d^2z/dr^2}{[1 + (dz/dr)^2]^{3/2}} = \frac{1}{r_1} \qquad \text{[I-E3-3]}$$

Hence, we can write

$$\frac{1}{r_1} + \frac{1}{r_2} = \frac{1}{r}\frac{d}{dr}(r \sin \phi) = 0 \qquad \text{[I-E3-4]}$$

Integration of this equation yields

$$r \sin \phi = r_m \qquad [I-E3-5]$$

where r_m is the minimum radial coordinate of the film. It is reached at the position $z = 0$ exactly halfway between the two rings.

Now, as shown previously in deriving Eq. [I-53]

$$\frac{dz}{dr} = \tan \phi = \frac{\sin \phi}{(1-\sin^2\phi)^{1/2}} \qquad [I-E3-6]$$

Replacing $\sin \phi$ by (r_m/r) in accordance with Eq. [I-E3-5], we find

$$\frac{dz}{dr} = \frac{r_m/r}{[(1 - (r_m/r)^2]^{1/2}} \qquad [I-E3-7]$$

If we let $u = (r_m/r)$, this equation becomes

$$\frac{dz}{du} = - \frac{r_m}{u(1 - u^2)^{1/2}} \qquad [I-E3-8]$$

Integrating with the boundary condition $r = r_m$ $(u = 1)$ for $z = 0$, we find

$$z = r_m \ \ln \left\{ \frac{r}{r_m} + [(\frac{r}{r_m})^2 - 1]^{1/2} \right\} \qquad [I-E3-9]$$

The unknown quantity r_m can be found by imposing a second boundary condition, viz., $r = R$ for $z = L$. The resulting equation is

$$L = r_m \ \ln \left\{ \frac{R}{r_m} + [(\frac{R}{r_m})^2 - 1]^{1/2} \right\} \qquad [I-E3-10]$$

The profile shape in this case is called a catenary, the shape of the soap film itself a catenoid.

7. METHODS OF MEASURING INTERFACIAL TENSION

Besides the sessile drop method just discussed, various techniques have been used for measuring interfacial tensions. They are discussed extensively by Padday (12). Closely related to the sessile drop method is the pendant drop technique (see Figure I-7). Here a drop of the denser fluid is suspended from a capillary tube. Gravity acts to elongate the drop while interfacial tension opposes elongation because of the associated increase in interfacial area. Eq. [I-56] again governs drop shape except

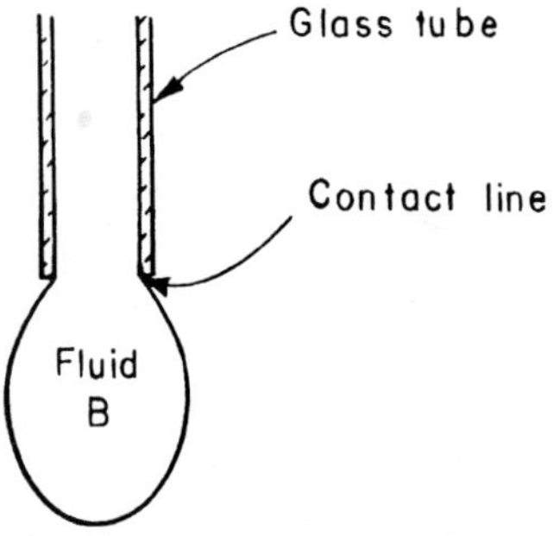

Figure I-7. Pendant drop.

that β now assumes negative values (since $\rho_A < \rho_B$). Measurements of parameters indicative of drop shape such as the maximum diameter and the height of the position of maximum diameter above the base of the drop can be used with tables based on numerical solution of Eq. [I-56] to obtain the interfacial tension (13).

Both sessile and pendant drops are suitable for use in liquid-gas and liquid-liquid systems, provided, of course, that the liquid surrounding the drop is transparent. Changes in drop shape can be followed to determine time-dependent effects on interfacial tension. Both methods have been used to measure low interfacial tensions in liquid-liquid systems, but only the sessile drop works well for tensions below about 0.01 mN/m. Finally, it should be noted that sessile and pendant "bubbles" of the less dense phase can be employed where the whole apparatus is, in effect, turned upside down (Figure I-8).

A related technique employs a "spinning drop" (Figure I-9). In this case a drop of the less dense fluid is injected into a container of the denser fluid, and the whole system is rotated as shown. In the resulting centrifugal field, the drop elongates along the axis of rotation. Interfacial tension opposes the elongation because of the increase in area and a configuration which minimizes system free energy is reached. The analysis is, in principle, similar to that for the sessile drop with the gravitational accelaration g replaced by the appropriate acceleration term for a centrifugal field (see Problem I-10). If the fluid densities ρ_A and ρ_B and the angular velocity ω of rotation are known, interfacial tension can be calculated from the measured drop profile. When

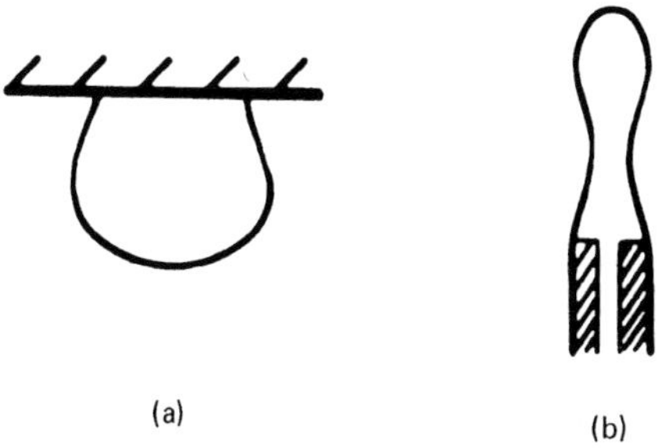

Figure I-8. (a) Sessile bubble. (b) Pendant bubble.

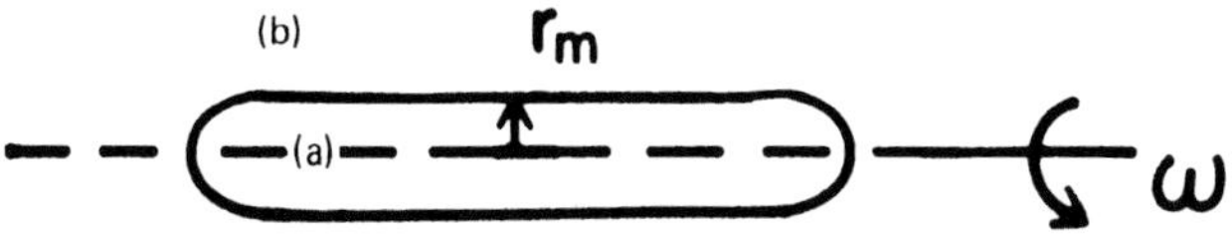

Figure I-9. Schematic diagram of a spinning drop.

drop length is much greater than the radius r_m, the following approximate expression holds:

$$\gamma = \frac{(\rho_B - \rho_A)\ \omega^2 r_m^3}{4} \tag{I-58}$$

The spinning drop device has been widely used in recent years to measure very low interfacial tensions in liquid-liquid systems of interest in processes for increasing recovery of petroleum from underground formations. Unlike the sessile and pendant drop schemes, no contact between the fluid interface and a solid surface is required. Both the drop and the surrounding fluid layer can also be made rather thin, so that results can be obtained even when the surrounding fluid is somewhat turbid, a frequent occurrence in practical systems. Finally, interfacial tension can, as with the sessile and pendant drops, be followed as a function of time.

A classical technique for measuring interfacial tension is the capillary rise method. As shown in Figure I-10, a capillary tube of small diameter is partially inserted into a liquid A. Fluid B may be either a gas or an immiscible liquid. If A is the wetting fluid, i.e., if its attractive interaction with the solid is stronger than that of B (see Chapter II), A begins to rise in the tube. Gravity, of course, opposes

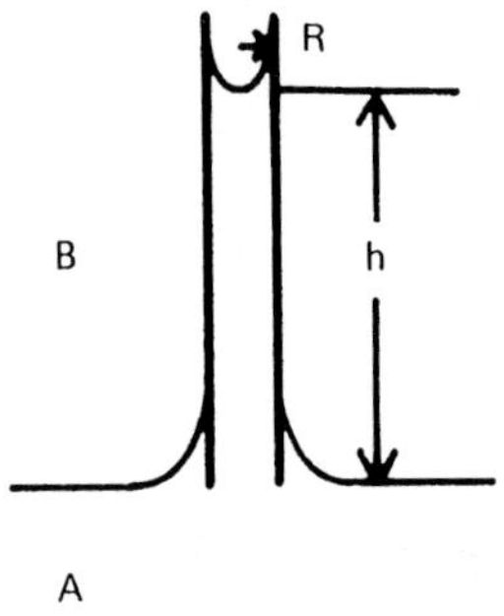

Figure I-10. Capillary rise in a tube of radius R.

this rise, and an equilibrium height h is ultimately reached where the overall free energy of the system is minimized.

Provided that the contact angle λ measured through A is zero and that the tube radius R is sufficiently small, the meniscus has the shape of a hemisphere of radius R. Hence from the Young-Laplace equation [I-22] we have

$$p_{Bh} - p_{Ah} = \frac{2\gamma}{R} \qquad \text{[I-59]}$$

If the interface between A and B in the large region outside the tube is flat, we can again invoke the Young-Laplace equation at the height of this interface

$$p_{Bo} - p_{Ao} = 0 \qquad \text{[I-60]}$$

The result of subtracting these equations is

$$(p_{Bh} - p_{Bo}) - (p_{Ah} - p_{Ao}) = \frac{2\gamma}{R} \qquad \text{[I-61]}$$

As in the sessile drop analysis discussed previously, the equations of hydrostatics also apply, so that

$$p_{Bh} - p_{Bo} = -\rho_B gh \qquad \text{[I-62]}$$

$$p_{Ah} - p_{Ao} = -\rho_A gh \qquad \text{[I-63]}$$

Substituting these equations into Eq. [I-61] and solving for γ, we find

$$\gamma = \frac{Rgh}{2}(\rho_A - \rho_B) \qquad \text{[I-64]}$$

The capillary rise method provides very accurate values of interfacial tension if used carefully. The capillary tubes must not deviate significantly from circular shape, must be of known and uniform radius, and must be carefully aligned in the vertical position. They must be cleaned or otherwise prepared to assure that the contact angle is zero. Finally, Eq. [I-64] must be corrected in experiments of high precision to account for meniscus deviation from the hemispherical shape due to gravity effects. Eq. [I-56] describes meniscus shape in this case since the situation is basically the same as that for a sessile bubble. Hence, the numerical solution as obtained for the sessile drop and bubble can be used to make the correction (see Problem I-13).

Some variants of the capillary rise method have been employed. In the differential capillary rise technique two tubes of different diameters are used, and the difference in height between the two menisci is measured. In the inclined capillary rise scheme, the tube is inclined at a known angle so that the length ℓ of tube occupied by the wetting fluid is greater than would be the height h of rise in a vertical tube. This method is useful when the interfacial tension is low because under these conditions the vertical rise h is small and difficult to measure accurately.

A relatively simple way to measure interfacial tensions is the drop weight or drop volume technique. The size of a pendant drop (Figure I-7) is slowly increased until the drop can no longer be prevented from falling and breaks off. The total weight or volume of a known number of drops is measured.

There is no exact theoretical method for calculating the interfacial tension from the known values of drop volume V and fluid densities ρ_A and ρ_B. Dimensional analysis can be used, however, to find suitable dimensionless variables for an empirical correlation. If viscous effects on drop size are assumed negligible, we may suppose that interfacial tension γ is a function F of the following variables

$$\gamma = F\,(V, (\rho_B - \rho_A), g, r) \qquad [I\text{-}65]$$

where r is the radius of the capillary tube from which the drop is formed. The outside radius should be used when the drop liquid wets the tube material, the inner radius when it does not. Application of dimensional analysis (see Problem I-11) yields two dimensionless variables, $(\gamma/(\rho_B-\rho_A)gr^2)$ and $(r/V^{1/3})$. If the relationship between these two variables is determined by experiment for one pair of fluids A and B, e.g., air and water, the resulting correlation may be used for other pairs.

Use of a micrometer syringe to feed the drop liquid to the tube makes determination of drop volume straightforward. The tip of the tube should be carefully ground so that is free from nicks and other irregularities. As with all other methods, evaporation must be prevented in liquid-gas-systems to prevent cooling of the interface and a resulting change in interfacial tension. With these provisions use of the drop volume method is convenient, and it is widely used. It should be noted that, by its very nature, this method is unsuitable when diffusion or adsorption effects dictate that considerable time is needed for the equilibrium value of interfacial tension to be attained.

Somewhat different are methods of measuring interfacial tension in which the vertical force acting on a solid body is measured during its withdrawal from a liquid. Objects of various shapes have been used, but the most common techniques are those employing a vertical plate (Wilhelmy plate method) and a horizontal circular ring (DuNouy ring method). In the former case, illustrated in Figure I-11, the force is measured when the lower edge of the plate is at the same elevation H as the large flat interface far from the plate. Under these conditions there is no buoyant force in a liquid-gas system, i.e., the liquid pressure acting on the bottom of the plate is equal to the (uniform) pressure in the gas phase. Provided that the contact angle is zero, the total force f on the plate is given by

$$f = mg + \gamma P \qquad [I\text{-}66]$$

where m is the mass of the plate and P its perimeter. Since all quantities except γ in Eq. [I-66] are known or can be measured, the tension is readily calculated.

The situation is more complex for the ring (Figure I-12) since the force on the ring due to the meniscus is not vertical at most positions of the ring above the surface. What is done is to measure the maximum force f_m during withdrawal and to employ dimensional analysis neglecting viscous effects in the same manner as was done above for the drop volume method. In this case we have

$$\gamma = F\,(f_m, (\rho_A - \rho_B)\,, g, r, R) \qquad [I\text{-}67]$$

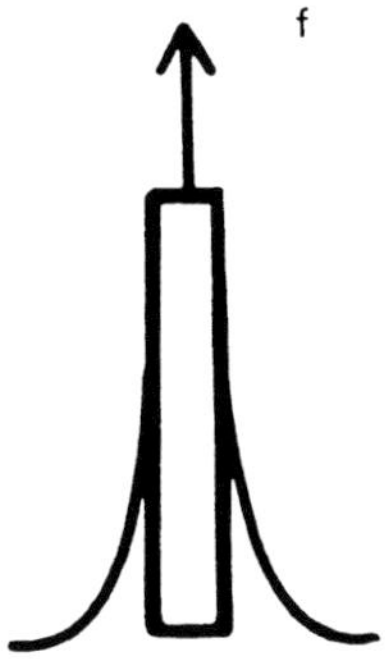

Figure I-11. Schematic diagram of the Wilhelmy plate method for measuring interfacial tension.

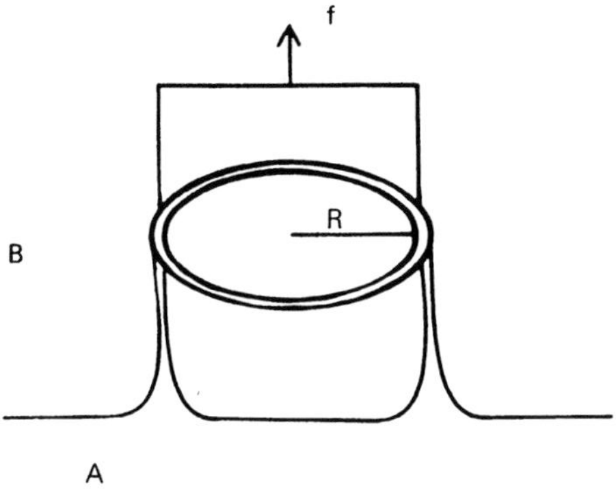

Figure I-12. Schematic digram of the Du Nouy ring method for measuring interfacial tension.

Here R is the radius of the ring and r the radius of the wire used to make the ring. Application of dimensional analysis yields three dimensionless groups:
$(\gamma/(\rho_A - \rho_B)\ g\ R^2)$, $(f_m/(\rho_A - \rho_B)\ gR^3)$, and (r/R). As before, experiments in one (or a few) systems can be undertaken to determine how the first of these groups, a dimensionless interfacial tension, depends on the other two. The results can then be applied in any system of interest.

In using the Wilhelmy plate and DuNouy ring methods it is important to ensure that the plate is indeed vertical and the ring horizontal during the measurement. One advantage of the Wilhelmy plate method is that, unlike the other techniques discussed in this section, it does not require knowledge of fluid densities. Since the plate is never removed from the interface and indeed a measurement requires very little interfacial distortion, it can also be used to follow time-dependent changes in interfacial tension. The DuNouy ring method cannot be used for this purpose, however, since relatively large distortions of the interface are involved in detaching the ring.

The actual calculation of interfacial tension from experimental data using some of these methods is considered in several of the problems at the end of this chapter. For further discussion of the various techniques, including some not described here, and for extensive tables of the numerical solution of Eq. [I-56] and of the empirical correlations for the drop volume and ring methods using dimensionless variables, the reader is referred to Padday's excellent review (12).

Example I-4. Capillary Rise in Air-Water and Oil-Water Systems

Calculate the height h of capillary rise in the air-water system ($\gamma = 72$ mN/m) for tube radii of 0.1, 0.01, and 0.001 mm. Repeat for an oil-water system with $\gamma = 30$ mN/m and $\Delta\rho = 0.15$ gm/cm^3. Assume that water completely wets the solid surface in both cases.

Solution Eq. [I-64] applies. Results are tabulated below. Note that the smaller tube radii are comparable to pore sizes in many soils and underground oil reservoirs. Since pores of various sizes may be anticipated in a given soil or reservoir, the results suggest that an underground boundary between air and water (water table) or between oil and water is probably not very sharp but has a substantial thickness.

R(mm)	h(cm)(Air-Water)	h(cm)(Oil-Water)
0.1	14.7	40.8
0.01	147	408
0.001	1470	4080

8. SURFACE TENSION OF BINARY MIXTURES

In the previous discussion of interfacial thermodynamics (Section 3) the number of moles n_i^s of each species at the interface was distinguished from the number of moles n_{iA} and n_{iB} in the adjacent bulk phases. We anticipate, therefore, that the corresponding interfacial mole fractions x_i^s differ from the bulk phase values x_{iA} and x_{iB}. To confirm this expectation and develop a quantitative expression for the difference, we consider an ideal liquid mixture whose surface is in contact with air or some other dilute gas.

Let us define a surface Gibbs free energy G^s as follows:

$$G^s = F^s - \gamma A \qquad [I-68]$$

If we recall that interfacial tension may be regarded as a negative two-dimensional pressure, the analogy between Eq. [I-68] and the usual definition of the bulk phase Gibbs free energy as $(F + pV)$ is evident. Differentiating Eq. [I-68] and invoking Eq. [I-12], we obtain

$$dG^s = -S^s dT - A d\gamma + \sum_i \mu_i^s \, dn_i^s \qquad [I-69]$$

For convenience here we have used μ_i^s to designate the surface chemical potential of species i although, as shown previously, μ_i^s is equal to the chemical potential μ_i in the bulk liquid.

From Eqs. [I-68] and [I-69] we conclude that

$$\mu_i^s = \left(\frac{\partial G^s}{\partial n_i^s} \right)_{T,\gamma,n_j^s} = \left(\frac{\partial F^s}{\partial n_i^s} \right)_{T,\gamma,n_j^s} - \gamma \left(\frac{\partial A}{\partial n_i^s} \right)_{T,\gamma,n_j^s} \qquad [I-70]$$

The last derivative in Eq. [I-70] may be termed the partial molar area $\bar{a}_i$ of species i because it is the two-dimensional counterpart of the usual definition of the partial molar volume in a bulk phase. If the surface layer of molecules is incompressible, we may assume that $\bar{a}_i$ has a constant value a_i independent of the value of the surface tension γ. The analogous assumption in a bulk phase would be that the partial molar volume $\bar{v}_i$ did not depend on pressure.

From the ordinary thermodynamics of bulk phases it can be readily shown that for an ideal solution where the partial molar volumes $\bar{v}_i$ are constant

$$\left(\frac{\partial F}{\partial n_i} \right)_{T,p,n_j} = \mu_i^o (T) + RT \ln x_i \qquad [I-71]$$

where the standard chemical potential $\mu_i^o (T)$ is a function of temperature but not of pressure (see Problem I-14). If the corresponding expression in two dimensions is substituted into Eq. [I-70], the result is

$$\mu_i^S = \mu_i^{OS}(T) + RT \ln x_i^S - a_i \gamma \qquad [I-72]$$

We naturally can use the well known expression for the chemical potentials in a bulk ideal solution

$$\mu_i = \mu_i^0 (T,p) + RT \ln x_i \qquad [I-73]$$

Let us consider the special case where the liquid is pure component i. In this case $x_i = x_i^S = 1$ and the surface tension has the pure component value γ_i. Equating the right hand sides of Eqs. [I-72] and [I-73] for this case, we find

$$\mu_i^{OS} = \mu_i^0 + \gamma_i a_i \qquad [I-74]$$

If this expression is used to replace μ_i^{OS} in Eq. [I-72] and if μ_i^S is set equal to μ_i for a mixture with some arbitrary value of x_i, the result is

$$\ln \frac{x_i^S}{x_i} = \frac{(\gamma - \gamma_i) a_i}{RT} \qquad [I-75]$$

As the surface tension of a mixture normally differs from that of any of its pure components, we conclude that the surface composition does differ from the bulk liquid composition.

For simplicity let us now restrict our attention to a binary mixture where $a_1 = a_2 = a$. Writing Eq. [I-75] for both components, we have

$$\frac{\gamma a}{RT} = \frac{\gamma_1 a}{RT} + \ln \frac{x_1^S}{x_1} = \frac{\gamma_2 a}{RT} + \ln \frac{x_2^S}{x_2} \qquad [I-76]$$

This equation may be rearranged to obtain

$$\frac{x_1^S}{1-x_1^S} = \frac{x_1}{1-x_1} \exp \left[\frac{(\gamma_2 - \gamma_1)a}{RT} \right] \qquad [I-77]$$

Thus the surface composition x_1^S can be calculated for any bulk phase composition x_1 provided that the molecular area a and the pure component surface tensions γ_1 and γ_2 are known. We note that if $\gamma_1 < \gamma_2$, Eq. [I-77] indicates that $x_1^S > x_1$. That is, the surface is enriched in the species having the lowest surface tension or surface free energy, the expected result.

Once x_1^S has been found, the surface tension γ of the mixture is readily obtained from Eq. [I-76]. Indeed, use of Eq. [I-77] to eliminate x_1^s yields the following explicit expression:

$$\exp\left(\frac{-\gamma a}{RT}\right) = x_1 \exp\left(\frac{-\gamma_1 a}{RT}\right) + x_2 \exp\left(\frac{-\gamma_2 a}{RT}\right) \quad [I-78]$$

When $[|\gamma_1-\gamma_2|a/RT] \ll 1$, this equation may be divided by exp $(-\gamma_1 a/RT)$ and the resulting expression expanded in a series. The result in this special case is

$$\gamma \sim x_1\gamma_1 + x_2\gamma_2 \quad [I-79]$$

This formula is often useful because two species which form an ideal mixture are normally very similar chemically and hence have nearly equal surface tensions.

The basic approach described here for ideal solutions can be extended to regular solutions, i.e., those where mixing is random but where interaction energies differ significantly for the various species present. For a binary mixture of two species with equal molar volumes it is well known that the regular solution approximation leads to the following expression for the bulk phase chemical potentials:

$$\mu_i = \mu_i^o\,(T,P) + RT\,\ell n\, x_i + \alpha\,(1-x_i)^2 \quad [I-80]$$

where α is a constant proportional to the change in interaction energy accompanying exchange of a molecule in pure liquid 1 with a molecule in pure liquid 2. It seems plausible and can be shown using a lattice model (3) that the corresponding expression for the surface chemical potential is given by

$$\mu_i^s = \mu_i^{os}\,(T) + RT\,\ell n\, x_i^s + \alpha\ell\,(1-x_i^s)^2 + \alpha m\,(1-x_i)^2 - \gamma a \quad [I-81]$$

Here ℓ is the fraction of the nearest neighbor sites of a surface molecule which are also at the surface and m is the fraction which are in the underlying liquid. A similar fraction m is in the overlying gas layer and presumed to be unoccupied. The quantities ℓ and m must therefore satisfy the relationship

$$\ell + 2m = 1 \quad [I-82]$$

Note that it has been assumed that $a_1 = a_2 = a$ in writing Eq. [I-81], a reasonable step in view of the previous assumption of equal molar volumes.

As before the quantities μ_i^{os} can be eliminated using the condition that $\gamma = \gamma_i$ for each pure liquid. Then equating the right hand sides of Eqs. [I-80] and [I-81] for a general mixture, we find after some manipulation that

$$\gamma = \gamma_1 + \frac{RT}{a} \ln \frac{x_1^s}{x_1} + \frac{\alpha \ell}{a} [(1-x_1^s)^2 - (1-x_1)^2] - \frac{\alpha m}{a} (1-x_1)^2$$

$$= \gamma_2 + \frac{RT}{a} \ln \frac{1-x_1^s}{1-x_1} + \frac{\alpha \ell}{a} [(x_1^s)^2 - x_1^2] - \frac{\alpha m}{a} x_1^2 \qquad [I-83]$$

The surface composition x_1^s can be found by equating the two expressions in Eq. [I-83] and the resulting value then used to calculate the surface tension γ. We note that the basic method can be extended to predict the interfacial tension between liquid phases in a binary system described by regular solution theory (3).

While the above discussion has dealt with liquid systems, it is noteworthy that surface solution theory is also of interest for solids. For example, the surface composition of metal alloy catalyst particles can differ significantly from the bulk composition. The proper design of the catalyst thus requires knowledge of how bulk and surface compositions are related.

Example I-5. Surface Tension of Ideal Binary Solution

Chlorobenzene (1) and bromobenzene (2) form ideal solutions. At 293°K the surface tensions are γ_1 = 33.11 and γ_2 = 36.60 mN/m. The area per molecule may be taken as 0.37 nm^2. Calculate the surface composition x_1^s and the surface tension γ when x_1 has values of 0.25 and 0.50. In the latter case compare the result with the experimental surface tension of 34.65 mN/m.

Solution. Equations [I-77] and [I-78] apply. The partial molar area a, calculated by multiplying the given area per molecule by Avogadro's number, is 2.228 x 10^{15} m^2/g-mol. Since $[(\gamma_2 - \gamma_1)a/RT]$ is about 0.32, Eq. [I-79] should give fairly good values for γ, but they should differ slightly from those of the general equation [I-78]. Calculated results are shown in the following table. Agreement with the above experimental value is good.

x_1	x_1^s	γ(Eq[I-78])($\frac{mN}{m}$)	γ(Eq[I-79])($\frac{mN}{m}$)
0.250	0.314	35.62	35.73
0.500	0.579	34.72	34.86

Example I-6. Surface Tension of Regular Solution (3)

Assume that diethyl ether (1) and acetone (2) form regular solutions at 15°C where γ_1 = 17.6 and γ_2 = 23.7 mN/m. The interchange energy parameter α may be taken as 450 cal/g-mol and the area per molecule a as 0.30 nm^2. Calculate the surface tension and surface composition of a liquid mixture having x_2 = 0.7. Make two calculations, one for a simple cubic lattice (ℓ = 2/3, m = 1/6) and one for a close packed lattice (ℓ = 1/2, m = 1/4). Compare your results with the predictions of ideal solution theory and with the experimental value of 20.7 mN/m.

Solution

From Eq. [I-83] we find by trial that

$$x_1^s = 0.46, \quad \gamma = 21.0 \text{ mN/m for simple cubic lattice}$$

$$x_1^s = 0.45, \quad \gamma = 20.8 \text{ mN/m for close packed lattice}$$

From Eqs. [I-75] and [I-77] for an ideal solution the corresponding values are

$$x_1^s = 0.40, \quad \gamma = 21.6 \text{ mN/m}$$

Clearly the calculations based on regular solution theory are in better agreement with the experimental measurement.

9. SURFACTANTS

In the preceding section we saw that the surface of a binary liquid mixture is enriched in the species with the lower surface tension. This result is qualitatively consistent with the Gibbs adsorption equation [I-25] although it can be shown that agreement is not quantitative owing to limitations of modeling the surface as a single layer of uniform composition x_1^s (3).

Some materials are so highly enriched at the surface that they are called surface-active agents or "surfactants". These materials are discussed extensively in Chapter IV. Some aspects of Chapters II and III will be clearer, however, if a brief account of their two key properties is given here.

Water, a highly polar material, resists incorporation into its structure of nonpolar entities such as hydrocarbon chains. One familiar consequence is that pure hydrocarbons are rather insoluble in water. Polar materials, in contrast, have considerable solubility in water. We might ask what happens to a material such as sodium laurate ($C_{12}H_{25}COONa$), a typical soap which has both a hydrocarbon chain and a polar group, when it is added to water. It seems clear that the most favorable configuration would be one where the hydrocarbon chain of each molecule is removed from the water but the polar group remains in the water. As shown in Figure I-13, this ideal condition can be achieved if the molecules adsorb at the interface between the water and an adjacent fluid as a monomolecular layer or "monolayer." We conclude that sodium laurate and similar "amphiphilic" compounds which have separate nonpolar and polar regions are surfactants.

For most situations where two bulk phases are in equilibrium, the amount of a surface-active contaminant required to form an interfacial monolayer is quite small. For instance, if each contaminant molecule occupies 0.4 nm^2, a simple calculation shows that 2.5×10^{14} molecules are required to cover an area of 1 cm^2. But this number of molecules corresponds to a bulk concentration of only about 4×10^{-7} molar when dissolved in 1 cm^3 of a solvent. Because of its relatively high surface tension, water is particularly susceptible to surface contamination. Indeed, the surface tension of water is an important indicator of its purity. The possibility of surface contamination and the associated requirement for high purity must be kept in mind when planning, performing, and evaluating experiments involving interfacial phenomena.

Surfactants have another important property. It too is a consequence of the reluctance of water to incorporate hydrocarbon chains. At very low concentrations dissolved surfactant exists as individual molecules, as with most other solutes. But at a particular concentration it becomes more favorable for the surfactant molecules to form aggregates called micelles. As illustrated in Figure I-13, the polar groups are all in contact with water while the hydrocarbon chains form the interior of the micelle. For typical water-soluble surfactants such as sodium laurate, the

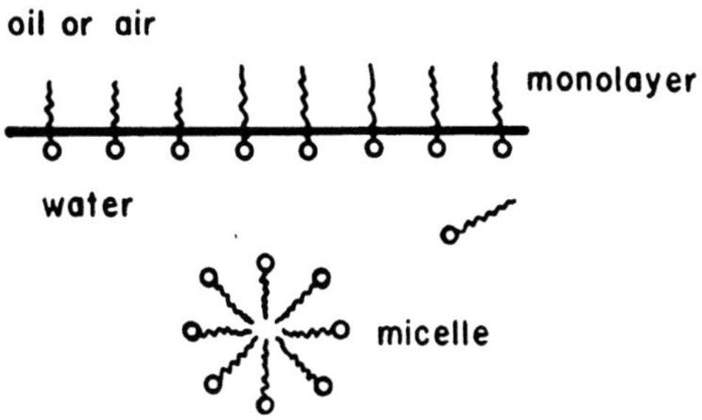

Figure I-13. Idealized diagram showing surfactant behavior.

micelles contain perhaps a hundred molecules and are approximately spherical in shape.

Because many types of polar groups exist and because the hydrocarbon portion of a surfactant may take different forms, many types of surfactants exist. Some simple polar groups are carboxylic acid (COOH), carboxylate ion (COO^-), alcohol (OH), sulfate (SO_4^-), sulfonate (SO_3^-), and amine (NH_3^+). Hydrocarbon chains may be straight or branched, saturated or unsaturated. Aromatic and naphthenic rings may also be present in the hydrocarbon region. It is evident that, given suitable methods of synthesis, practically an infinite variety of surfactants could be made.

REFERENCES

General Texts on Interfacial Phenomena

Adam, N.K. (1941) *The Physics and Chemistry of Surfaces*, 3rd ed., Oxford University Press.

Adamson, A.W. (1982) *Physical Chemistry of Surfaces*, 4th ed., Wiley, New York.

Aveyard, R. and Haydon, D.A. (1973) *An Introduction to the Principles of Surface Chemistry*. Cambridge University Rress.

Davies, J.T. and Rideal, E.K. (1963) *Interfacial Phenomena*, 2nd ed., Academic Press, New York.

Defay, R., Prigogine, I., Bellemans, A., and Everett, D.H. (1966) *Surface Tension and Adsorption*, Longmans, London.

Harkins, W.D. (1952) *The Physical Chemistry of Surface Films*, Reinhold, New York.

Jaycock, M.J. and Parfitt, G.D. (1980) *Chemistry of Interfaces*, Ellis Horwood, New York.

Osipow, L.I. (1962) *Surface Chemistry*, van Nostrand-Reinhold, New York.

Rowlinson, J.S. and Widom, B. (1982) *Molecular Theory of Capillarity*, Clarendon Press, Oxford.

Vold, R.D. and Vold, M.J. (1983) *Colloid and Interface Chemistry*, Addison-Wesley, Reading, Mass.

Textual References

1. Hayward, A.T.J. (1971) *Am. Scientist* **59**, 434.
2. Gibbs, J.W. (1878) *Trans. Conn. Acad.* **3**, 108, 343. Reprinted in *The Scientific Papers of J. Willard Gibbs*, **Vol 1**, New York, Dover.
3. Defay, R., Prigogine, I., Bellemans, A., and Everett, D.H. (1966) *Surface Tension and Adsorption*, Longmans, London.
4. Weatherburn, C.E. (1955) *Differential Geometry of Three Dimensions*, **Vol. 1**, Cambridge University Press.
5. Buff, F.P. (1956) *J. Chem. Phys*. **25**, 146.
6. Cahn, J.W. and Hilliard, J.E. (1958) *J. Chem. Phys*. **28**, 258.
7. Courant, R. and Hilbert, D. (1953) *Methods of Mathematical Physics*, **Vol. 1**. Wiley, New York. Chapter 4.
8. Davis, H.T. and Scriven, L.E. (1981) "Stress and Structure in Fluid Interfaces." *Adv. Chem. Phys*. **49**, 357.
9. Rowlinson, J.S. and Widom, B. (1982) *Molecular Theory of Capillarity*, Clarendon Press, Oxford.
10. Adamson, A.W. (1982) *Physical Chemistry of Surfaces*, 4th ed., Wiley, New York.
11. Bashforth, F. and Adams, J.C. (1893) *An Attempt to Test the Theory of Capillary Action*, Cambridge University Press.
12. Padday, J.F. (1969) "Theory of Surface Tension," in *Surface and Colloid Science*, E. Matijevic (ed.) **Vol. 1**, Wiley, New York, p. 39, 101, 151.
13. Ambwani, D.S. and Fort, Jr., T. (1979) "Pendant Drop Technique for Measuring Liquid Boundary Tensions," in *Surface and Colloid Science*, R.J. Good and R.R. Stromberg (ed.), **Vol. 11**, Plenum, New York. p. 93.
14. Good, R.J. (1976) *Pure Appl. Chem*. **48**, 427.
15. Murphy, C.L. (1966) *Thermodynamics of Low Tension and Highly Curved Surfaces*, Ph.D. Thesis, University of Minnesota.
16. Ono, S. and Kondo, S. (1960) *Encyclopedia of Physics*, S. Flugge (ed.), **Vol. 10**, Springer, Berlin. p. 134.
17. Nishioka, G. and Ross, S. (1981) *J. Colloid Interface Sci*. **81**, 1.

18. Lando, J.L. and Oakley, H.T. (1967) *J. Colloid Interface Sci*. **25**, 526.

19. Sugden, S. (1921) *J. Chem. Soc*., 1483.

PROBLEMS

I-1. (a) Starting with Eq. [I-12], show that

$$(\partial S^s/\partial A)_{T,n_i^s} = -(\partial \gamma/\partial T)_{A,n_i^s}$$

(b) Use this result to show that

$$\left(\frac{\partial U^s}{\partial A}\right)_{T,n_i^s} = \gamma - T\left(\frac{\partial \gamma}{\partial T}\right)_{A,n_i^s}$$

(c) Consider a single-component system with the reference surface chosen to make $n_1^s = 0$. In this case $\gamma = (F^s/A)$, according to Eq [I-29]. Also

$$\left(\frac{\partial U^s}{\partial A}\right)_T = \frac{U^s}{A} \quad , \quad \left(\frac{\partial S^s}{\partial A}\right)_T = \frac{S^s}{A}$$

Except near the critical point, γ is often found to decrease linearly with T in this case. Assuming such behavior to apply, sketch the expected behavior of U^s, F^s, and S^s as a function of temperature.

I-2. Is the vapor pressure p_v of a liquid inside a small capillary tube which it completely wets greater or smaller than the vapor pressure p_v^o of a large pool of the same liquid at the same temperature? Calculate (p_v/p_v^o) for water at 100°C inside a capillary tube with a radius of 1 μm . Use the values of physical properties given in Example I-1. The change in vapor pressure for a liquid in a capillary tube or, more generally, a porous medium is usually called the capillary condensation effect.

I-3. For a single component liquid-vapor interface it is not possible to vary temperature and pressure independently while maintaining the phases in equilibrium. However, measurement of interfacial tension variation with pressure at constant temperature is possible in a binary system. Good (14) has shown that such data can be used to obtain useful information about interfacial characteristics in this case.

(a) Use the Gibbs adsorption equation to derive formulas for Γ_2 when $\Gamma_1 = 0$ and Γ_1 when $\Gamma_2 = 0$ in terms of the pressure coefficient of interfacial tension at constant temperature. For simplicity assume a liquid liquid system where $x_2 \ll 1$ in phase 1 and $x_1 \ll 1$ in phase 2.

(b) Let Λ be the distance between the reference surfaces corresponding to $\Gamma_1 = 0$ and $\Gamma_2 = 0$. Show that $\Lambda = -(\partial\gamma/\partial p)_T$ and hence that the pressure variation of interfacial tension provides some information on interfacial thickness.

(c) In the benzene-water system at room temperature $(\partial\gamma/\partial p)_T$ has been found to be -7×10^{-4} dyne/cm atm. Calculate Λ and Γ for each component at the reference surface where Γ for the other component vanishes.

I-4. For interfaces with very low interfacial tensions or with small radii of curvature the assumption made in writing Eq. [I-9] is not valid. Consider the special case of a spherical interface and rewrite Eq. [I-9], adding a term $(C d(2/a))$ where a is the radius of the sphere and C is defined by

$$C = \left(\frac{\partial U^s}{\partial(2/a)} \right)_{S^s, A, n_i^s}$$

(a) Repeat the derivation of Section 3 to obtain the following generalization of the Young-Laplace equation for a spherical interface

$$p_A - p_B = \frac{2\gamma}{a} - \frac{2(C/A)}{a^2} \qquad \text{(i)}$$

This equation can be generalized (5,15) to include nonspherical interfaces, the result being

$$p_A - p_B = -2 H \gamma - (4H^2 - 2K) (C/A) \qquad \text{(ii)}$$

where K is the Gaussian curvature of the reference surface defined as the product of $(1/r_1)$ and $(1/r_2)$. It can also be shown that the quantity (C/A) has a mechanical interpretation:

$$\frac{C}{A} = \int_{\lambda_A}^{\lambda_B} \Delta p_T \, (\lambda - \lambda_s) \, d\lambda \qquad \text{(iii)}$$

where Δp_T is as defined at Eq. [I-33] and λ_s is the coordinate of the reference surface. Murphy (15) termed (-C/A) the "bending stress" because of this relation to the first moment of the tangential stress distribution. His general derivation yields yet another term in (i) and (ii). In the former equation it is proportional to $(1/a^3)$. Finally, we note that under conditions where (i) or (ii) holds, Eq. [I-39] must be corrected for curvature effects as well:

$$\gamma = -\int_{\lambda_A}^{\lambda_B} \Delta p_T(1 - 2H\ (\lambda-\lambda_s))\ d\ \lambda \qquad \text{(iv)}$$

(b) Consider the situation of the adjacent diagram where the three spherical surfaces are parallel and occupy a solid angle ω. With the reference surface maintained at

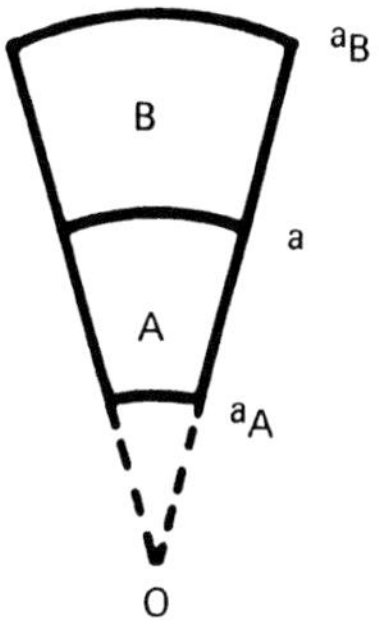

$r = a$ but with a_A and a_B allowed to change calculate the change in Helmholtz free energy for the closed system at constant temperature and volume when it is deformed so that it occupies a solid angle $(\omega + d\omega)$ and show that the interfacial tension is given by

$$\gamma = (p_A - p_B)\frac{a}{3} + \frac{Q}{a^2} \qquad \text{(v)}$$

where Q is a constant. By differentiating this expression show that (i) may be written in the following form given by Ono and Kondo (16)

$$p_A - p_B = \frac{2\gamma}{a} + \frac{d\gamma}{da} \qquad \text{(vi)}$$

(c) The reference surface where $(d\gamma/da)$ vanishes, i.e., where $C = 0$, is usually called the "surface of tension," an expression coined by Gibbs. Rewrite (v) in terms of the radius a_s of the surface of tension and the corresponding interfacial tension γ_s and hence show that

$$\gamma = (\gamma_s/3)\ [2\ (a/a_s) + (a_s/a)^2]$$

Note that this expression implies that γ attains a minimum value γ_s at the surface of tension.

I-5. A drop of a perfectly wetting liquid is placed between two parallel plates as shown in the accompanying figure. As the plates are moved together, the diameter of the wetted region increases. For the case D >> h:

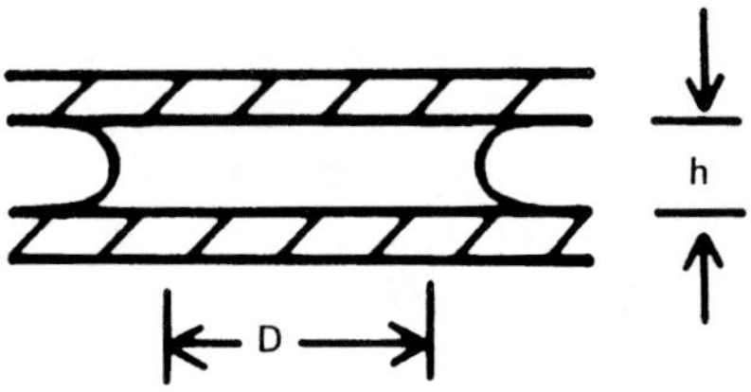

(a) Derive an equation for net force acting on each plate

(b) Calculate the value of the force for water (γ = 72 mN/m) when drop volume is 2 cm^3 and h = 1 mm. This force is a major factor contributing to the adhesion of surfaces.

I-6. The surface divergence theorem was stated at Eq. [I-36] without proof. Except for the last term of the integrand, it is basically what one would expect in view of the more familiar three-dimensional relationship.

$$\int_S \mathbf{n}\, f\, d\mathcal{S} = \int_V \nabla f\, dV$$

Some insight into the origin of the last term in Eq. [I-36] may be gained by considering a special case where (i) f is uniform so $\nabla_s f$ vanishes and (ii) the surface is cylindrical, having a radius R and a length L and enclosing an angle of $2\theta_1$. Show that Eq. [I-36] is satisfied for this case.

I-7. Nishioka and Ross (17) have proposed the following method for measuring the decrease with time in the surface area of a foam, which provides a good indication of foam stability. Foam is generated in a chamber of fixed volume which is maintained at constant temperature. The pressure in the container but outside the foam itself is monitored as a function of time as the foam decays. The pressure p_∞ corresponding to complete decay of the foam is obtained by injecting a defoaming agent at the end of the experiment.

(a) Use the Young-Laplace equation [I-22] and the ideal gas equation to derive the following equation of state for a foam

$$n_f RT = p_e V_f + \frac{2\gamma A}{3}$$

where n_f is the number of moles of gas in the foam, p_e is the pressure external to the foam, A and V_f are foam area and volume, and γ is the surface tension of the foam lamellae.

(b) Assuming constant γ and T and imposing requirements that the total number of moles of gas in the vessel (both internal and external to the foam) and the total volume V_g of this gas remain constant, show that for times t_1 and t_2

$$3\, V_g\, (p_{e2} - p_{e1}) + 2\gamma\, (A_2 - A_1) = 0$$

(c) Use the information that A = 0 for $p_e = p_\infty$ to show that

$$A = \frac{3V_g}{2\gamma}\, (p_\infty - p_e)$$

I-8. At small saturations, the wetting fluid in a porous medium forms "pendular rings" near the points of contact of individual grains. To simplify the calculations we will consider the junction between two planes as sketched instead of between two spheres. The fluid interface between A and B may be taken to be a portion of a circle which, assuming zero contact angle, is tangent to the planes at A and B. The angle between the planes is 2α. Calculate the area of region OAB occupied by the wetting fluid and the pressure difference between the fluids for the following four cases:

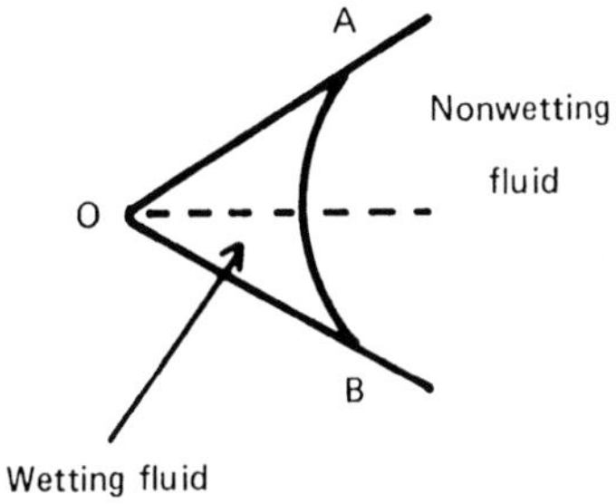

(a) $\alpha = 10°$, $\overline{OA} = 10\ \mu m$
(b) $\alpha = 10°$, $\overline{OA} = 1\ \mu m$
(c) $\alpha = 30°$, $\overline{OA} = 10\ \mu m$
(d) $\alpha = 30°$, $\overline{OA} = 1\ \mu m$

I-9. A very long solid body having a square cross section floats at an interface with the configuration sketched. Determine h and α at equilibrium in terms of a, densities ρ_A and ρ_B, and interfacial tension γ_{AB}. (It is not necessary to solve explicitly for h and α from the final equations you obtain.)

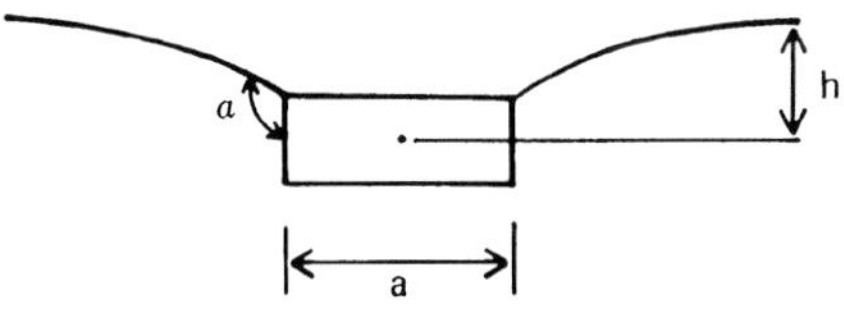

Hint: The differential equation of interfacial statics need not be solved if one is interested only in finding h and α (and not details of the interfacial profile). One can instead write a horizontal force balance for the entire fluid interface on one side of the solid.

That surface tension can keep a small solid body from sinking is important to certain insects and, from a more practical point of view, to people interested in using flotation to separate valuable ores from material containing no minerals. Note that the basic method of solution can be used even if the fluid surface does not contact the solid along an edge; for example, if one had a long cylinder floating at the surface. Note also that the surface tension can't help keep the solid at the surface unless the equilibrium contact angle (on a plane surface) differs from zero; in other words, the mechanism is ineffective if the liquid spreads spontaneously on the solid.

I-10. As indicated in Section 7, one method of measuring interfacial tension between two fluids is the spinning drop device. A drop of the less dense fluid A is placed in the denser fluid B, and both fluids are rotated at a constant angular velocity ω (Figure I-9).

(a) Neglecting gravity and assuming an axisymmetric drop, derive a differential equation for the drop profile z as a function of radial coordinate r, interfacial tension γ, fluid densities ρ_A and ρ_B, and the angular velocity ω. Recall that the principal radii of curvature are given by Eqs. [I-52] and [I-53]. Your final equation should be in terms of *dimensionless* variables and should contain only a single dimensionless group involving the various parameters.

(b) Write the boundary conditions needed to solve your differential equation.

(c) For high angular velocities the drop becomes long and thin, i.e., $z_m \gg r_m$, where z_m and r_m are coordinates at the point of maximum drop diameter (see sketch). Under these conditions it can be shown that $(r_m/r_0) = 1.5$, where r_0 is the radius of curvature at the origin 0. Use this information to derive Eq. [I-58] (Note that you do *not* need to solve the differential equation to do this.)

I-11. Eq. [I-65] lists variables upon which the interfacial tension γ is thought to depend when the drop volume method is used. Assume that the dependence takes the form

$$\gamma = B\,V^{a}(\rho_A - \rho_B)^{b}\,g^{c}\,r^{d}$$

where B is a dimensionless constant. By requiring dimensional consistency of this equation, show that

$$\frac{\gamma}{(\rho_A - \rho_B) g\, r^2} = B\left(\frac{r}{V^{1/3}}\right)^{-3a}$$

Thus, two dimensionless groups are obtained as stated in the text.

I-12. The drop volume method is being used to measure the interfacial tension between water (ρ_A = 1.00 gm/cm^3) and benzene (ρ_B = 0.88 gm/cm^3). For a capillary tube with an outside diameter of 4 mm, the volume of each drop is found to be 0.273 cm^3. Calculate the interfacial tension. Note that the usual formulation of the equation takes the form

$$\frac{\gamma r}{(\rho_A - \rho_B) g V} = \left(\frac{\gamma}{(\rho_A - \rho_B) g r^2}\right)\left(\frac{r}{V^{1/3}}\right)^{3} = F\left(\frac{r}{V^{1/3}}\right)$$

where the function F is obtained from experiment and tabulated (see Table III for selected values in the range of interest here). The reason for this formulation is that the first expression is related to the ratio of the upward force on the drop produced by interfacial tension to the effective weight of the suspended drop. As is clear from dimensional analysis, it is not necessary to consider this ratio explicitly in solving for γ. One need only determine the relationship between the two pertinent dimensionless variables. But since the function F as defined above is the quantity tabulated, it is easiest to follow the usual procedure.

I-13. As indicated in Section 7, it is necessary to correct for the nonspherical shape of the meniscus in accurate work using the capillary rise technique.

(a) Show that if h is the height to the bottom of the meniscus and b is the radius of curvature there,

$$\frac{2\gamma}{b} = (\rho_A - \rho_B)\, g\, h \qquad \text{(i)}$$

Hence show that

$$\frac{R}{(bh)^{1/2}} = \frac{R}{b}\left(\frac{\beta}{2}\right)^{1/2} \qquad \text{(ii)}$$

where R is the tube radius and β is as defined by Eq. [I-57].

Table III. Partial List of Correction Factors for Drop Volume Method

$r/V^{1/3}$	Correction Factor F
0.300	0.2166
0.301	0.2168
0.302	0.2169
0.303	0.2171
0.304	0.2173
0.305	0.2175
0.306	0.2176
0.307	0.2178
0.308	0.2180
0.309	0.2182
0.310	0.2183

Values from Lando and Oakley (18).

(b) The procedure commonly used to make the correction is due to Sugden (19). For any β , the numerical solution of Eq. [I-56] yields a value of (r/b) for $\phi = 90^0$. Sugden tabulated (r/b) for $\phi = 90^0$ as a function of $(r/b)(\beta/2)^{1/2}$. A portion of his table is given in Table IV. Show that the entry for (r/b) = 0.9631 in Table IV corresponds to a value for β of 0.249.

(c) Sugden's table is used to find γ in the following way. Because the contact angle is zero, r = R and $\phi = 90°$ at the point where the meniscus meets the tube. A successive approximation scheme is used. First γ is calculated from (i) with b = R (spherical interface approximation). This value of γ is then used to calculate $(R/b)(\beta/2)^{1/2}$. The table then gives a value of (R/b) from which an improved value of b can be obtained. This improved b is substituted into (i) to get an improved γ, and the process is repeated until the same value of γ is obtained on successive iterations. Calculate the surface tension of toluene (ρ = 0.866 gm/cm^3) if h = 0.838 cm in a capillary tube of radius 0.0777 cm.

I-14. Derive Eq. [I-71] starting with the known relationship between the Helmholtz and Gibbs free energies F and G and the well known equation for an ideal solution

$$\mu_i = \mu_i^o (T,p) + RT \ln x_i$$

Table IV. Selected Values from Sugden's Table of Corrections for Capillary Rise Method (19)

$(r/b)\ (\beta/2)^{1/2}$	(r/b)
0.30	0.9710
0.31	0.9691
0.32	0.9672
0.33	0.9652
0.34	0.9631
0.35	0.9610
0.36	0.9589
0.37	0.9567
0.38	0.9545
0.39	0.9522
0.40	0.9498

where μ_i^0 is the chemical potential of pure i at T and p. Remember that μ_i is the partial molar Gibbs free energy $\bar{G}_i$.

I-15. Show that for an ideal dilute solution the surface tension is given approximately by

$$\gamma_1 - \gamma = \frac{RT}{a} x_2 \left[H \exp\left(\frac{\gamma a}{RT}\right) - 1\right]$$

where γ_1 is the surface tension of the pure solvent, a is the area of solute and solvent molecules and H is a constant which must be determined experimentally. Recall that the following expression applies in a dilute ideal solution in a bulk phase:

$$\mu_2 = \mu_2^\infty (T,p) + RT \ln x_2 \qquad (x_2 \ll 1)$$

Here μ_2^∞ is *not* the chemical potential of pure 2 at T and p. The chemical potential μ_1 of the solvent in such a solution is still given by Eq. [I-16]. Assume that the surface is an ideal dilute solution in your derivation.

II

Fundamentals of Wetting and Contact Angles

1. INTRODUCTION

One is familiar with the ways in which small amounts of liquid on a solid surface behave, not only from laboratory studies but also from everyday observations made of our immediate surroundings. One knows that drops of water on inclined solid surfaces can remain stationary, or can move under gravity. If a waxed paper is withdrawn from water, the thin sheet of water formed initially will break up into beads. On the other hand, if the waxed paper is withdrawn from oil, the sheet of oil formed will drain and grow thinner but will not form beads. Leaving aside the effects of gravity, two types of behavior are observed on a given solid surface, one where the liquid forms beads and the other where it forms a film. It is now easy to assume that the liquids which have strong affinities for a solid will form films such that the liquid-solid contact is maximized while those which have weaker affinities will collect themselves into beads.

The above affinity is referred to as the wettability. It is important for phenomena ranging from flow in packed columns, underground oil reservoirs, and other porous media to condensation of liquids on solid surfaces to flotation schemes for concentrating mineral ores.

It is apparent from the above discussion that the success of adhesives and coatings depends strongly on the wettabilities of these materials on the adherent or solid substrate. If the coating material does not wet the substrate, not only would it be difficult to apply, it would also eventually peel away. Adhesives are not allowed to behave in this fashion even under severe conditions. The analysis of adhesive behavior and determination of adhesive bond strength require knowledge of the factors which influence the energy of an interface between two materials.

2. YOUNG'S EQUATION

Since the configurations of a wetting liquid and a non-wetting liquid on a solid surface differ substantially in their interfacial areas, any quantitative description of wettability must involve the interfacial energies in a major way. Consider the axisymmetric bead shown in Figure II-1. There are three interfaces: the solid (S) - liquid (L), the liquid-vapor (V) and the solid-vapor, and consequently there are three interfacial tensions γ_{SL}, γ_{SV}, and γ_{LV}. The line common to the three phases (the basal circle of the drop) is called the common line or the contact line. The angle that the vapor-liquid interface makes with the solid surface is called the contact angle. If it is assumed that the interfacial tensions can be taken as forces even for the solid-fluid interfaces, the force balance at point 0 on the contact line in the horizontal direction is

$$\gamma_{SV} = \gamma_{SL} + \gamma_{LV} \cos \lambda$$

which is rewritten in the form

$$\gamma_{LV} \cos \lambda = \gamma_{SV} - \gamma_{SL} \qquad [II-1]$$

This is Young's equation.

Alternately, interfacial thermodynamics can be used to develop a more rigorous derivation. Small changes in Helmholtz's free energy can be written by generalizing Eq. [I-14] for multiple interfaces. If temperature and the individual phase volumes are constant, the result is

$$\begin{aligned} dF = & \sum_i \mu_i (dn_i^{LV} + dn_i^{SV} + dn_i^{SL}) \\ & + \sum_i \mu_i (dn_i^{S} + dn_i^{L} + dn_i^{V}) \\ & + \gamma_{SV} dA_{SV} + \gamma_{LV} dA_{LV} + \gamma_{SL} dA_{SL} \end{aligned} \qquad [II-2]$$

As the number of moles remains unchanged,

$$(dn_i^{LV} + dn_i^{SV} + dn_i^{SL}) + (dn_i^{S} + dn_i^{L} + dn_i^{V}) = dn_{iT} = 0,$$

and one has

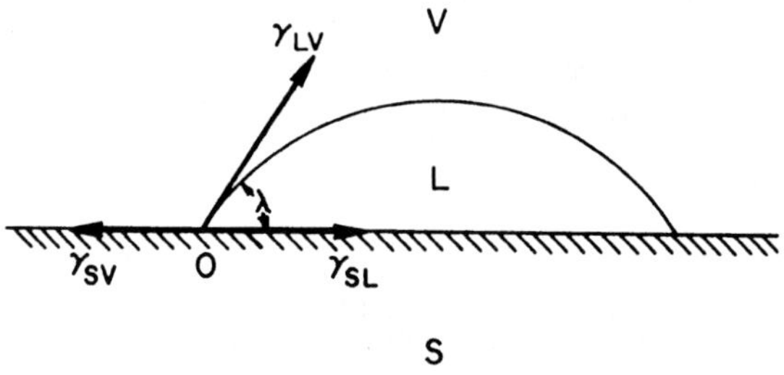

Figure II-1. Configuration analyzed in Section 2. The phases are vapor (V), liquid (L) and solid (S). The contact line is along O.

$$dF = \gamma_{SV} dA_{SV} + \gamma_{LV} dA_{LV} + \gamma_{SL} dA_{SL} \qquad [II\text{-}3]$$

Further, geometry dictates that

$$dA_{SV} = - dA_{SL} \qquad [II\text{-}4]$$

If the drop is a spherical cap, it can be shown that

$$dA_{LV} = \cos \lambda \, dA_{SL} \qquad [II\text{-}5]$$

Combining Eqs. [II-3] to [II-5], one has at constant volume of the drop:

$$\left(\frac{\partial F}{\partial A_{SL}} \right) = \gamma_{SL} - \gamma_{SV} + \gamma_{LV} \cos \lambda \qquad [II\text{-}6]$$

At equilibrium F has to be a minimum with respect to A_{SL}, i.e., the left-hand side of Eq. [II-6] is zero and Eq. [II-1] is obtained. Note that this derivation uses the principle of minimizing the free energy. It is known that on minimization of total energy independent sets of force balances are obtained and consequently the fact that Young's equation can be derived in this way is appropriate.

More detailed derivations of Eq. [II-1] are available (1,2). They uncover some basic limitations of Young's equation which are well worth noting. These are that, (a) Eq. [II-1] is valid when there is no adsorption at the interfaces. Adsorption and some of its effects on contact angles are discussed later. (b) The interfacial tensions that appear in Eq. [II-1] are those which are evaluated far from the contact line. The interfacial tensions at the contact line may have different values, however, and indeed may not be well defined owing to overlap of portions of the three interfacial regions. All such deviations near the contact line can be lumped together in the form of the additional effect of *line tension.* (Note that in Chapter I it has been shown that the interfacial tension is also a lumped quantity.) Gibbs (3) anticipated this effect when he argued that just as a surface separating two bulk media is in a state of tension (interfacial tension), so is a *line* separating two interfaces (line tension). Thus in Figure II-1, the common line will be in a state of tension. It is noteworthy that the measured values of line tensions show that they are extremely small (4). Thus the contact angle that is predicted by Young's equation is quite likely the one that is measured.

Some interesting conclusions are reached from Young's equation. Since the interfacial tensions in Eq. [II-1] are equilibrium properties, the λ there is also an equilibrium property and is consequently referred to as the *equilibrium* contact angle. For Eq. [II-1] to be meaningful, $|\cos \lambda| = |(\gamma_{SV} - \gamma_{SL})/\gamma_{LV}| < 1$, which makes it necessary to examine the physical implication of this constraint. Assume that the drop sizes are small such that gravity forces can be neglected. In that case the shape of the vapor-liquid interface is that of a spherical cap. For drops of constant volume, as λ decreases to zero (or $\cos \lambda$ increases to + 1), the drops approach a thin film configuration. As λ increases to 180° (or $\cos \lambda$ decreases to -1), the drops become spherical, having only one point of contact with the solid at $\lambda = 180°$. Thus an ordering is established in the spectrum of available λ . For $\lambda = 0°$, the liquid wets the solid surface completely, i.e., it is a wetting liquid. As λ increases beyond 90°, the partially wetting liquid becomes non-wetting, and it ultimately becomes completely non-wetting for $\lambda = 180°$.

While this concludes the discussion of the nature of non-wetting liquids, more remains to be said about wetting liquids. These completely spread out on the solid surface and consequently the process must be accompanied by a decrease in the free energy. This means that for wetting liquids the left hand side of Eq. [II-6] is negative. It is also seen in experiments that when a wetting liquid spreads on a solid surface, a zero contact angle is maintained (5-7). A spreading coefficient is defined as

$$S_{L/S} = \gamma_{SV} - \gamma_{LV} - \gamma_{SL} \qquad [II-7]$$

Substituting Eq. [II-7] into Eq. [II-6] with $\cos \lambda = 1$, one has $(\partial F/\partial A_{SL}) = -S_{L/S}$, that is, the spreading coefficient for a wetting liquid is *positive.* For a non-wetting liquid one finds on substituting Eq. [II-1] into Eq. [II-7] that $S_{L/S} = \gamma_{LV} (\cos\lambda - 1)$. Thus the spreading coefficient is negative here. Consequently the knowledge of interfacial tensions alone is sufficient to predict the wettability of a liquid on a solid surface using Eq. [II-7]. Unfortunately, the solid-fluid interfacial tensions γ_{SV} and γ_{SL} cannot ordinarily be measured although one method of obtaining γ_{SV} for nonpolar solids is discussed in Section 4 below.

3. WORK OF ADHESION AND WORK OF COHESION

Consider a cylindrical column of two materials A and B as shown in Figure II-2a. If the column is torn apart at the A-B interface, (Fig. II-2b), then the work done per unit interfacial area is

$$W_{AB} = \gamma_A + \gamma_B - \gamma_{AB} \qquad \text{[II-8]}$$

This is the *work of adhesion.* As the name implies it forms an important concept in the theory of adhesive materials where it is necessary to have a measure of the strength of the adhesive to the adherend. Eq. [II-8] follows from the definition of interfacial tension as the work done in creating a unit area.

If the column is made entirely of the same material A, then the work done per unit area in tearing the column is

$$W_{AA} = 2\gamma_A \qquad \text{[II-9]}$$

which is the *work of cohesion.* It represents the condition of failure in the bulk of a material.

If A is a liquid (L) and B is a solid (S) then if $W_{AA} > W_{AB}$, one has from Eqs. [II-8] and [II-9], $\gamma_L > (\gamma_S - \gamma_{SL})$. This condition is satisfied only if the contact angle is non-zero (Eq. [II-1] is used). Consequently one may conclude that if an adhesive does not spread spontaneously, it can come unstuck ("adhesive failure"), before the failure of the material itself ("cohesive failure").

To return to the definitions of the work of adhesion and the work of cohesion, one notes an interesting feature, viz., that the work done in moving apart molecules can be related to interfacial tensions. It is noteworthy that the important energy that is dependent on the intermolecular distances is the intermolecular potential. Thus on assuming that the work done is solely the change in intermolecular potential on tearing apart a column of material, one finds a means for relating interfacial tensions to molecular effects. A simple form of intermolecular potential between molecules i and j is

$$\varphi'_{ij} = \begin{cases} -\varepsilon_{ij}\left\{\dfrac{\sigma_{ij}}{|\mathbf{r}_{ij}|}\right\}^6 & \text{for } |\mathbf{r}_{ij}| > \sigma_{ij} \\ +\infty & \text{for } |\mathbf{r}_{ij}| \leq \sigma_{ij} \end{cases} \qquad \text{[II-10]}$$

where ε_{ij} is a characteristic energy, σ_{ij} is a characteristic length, and $|\mathbf{r}_{ij}|$ is the distance between the centers of mass of the molecules i and j. σ_{ij} is the center-to-center distance at contact between molecules i and j.

The first part of the potential which is negative and varies inversely with the sixth power of the intermolecular distance is called the van der Waals or the London dispersion potential. It represents the attraction between the two molecules. The second part is positive and represents repulsion. The fact that it is infinite implies that the molecules are hard spheres, and cannot penetrate each other. A sketch of the dimensions involved is shown in Figure II-3, as well as the form of Eq. [II-10]. For $i = j$ the distance of closest approach (when the spheres touch one another) is σ_{ii} which is also the diameter of the molecule. When $i \neq j$ one often takes $\sigma_{ij} = \frac{1}{2}(\sigma_{ii} + \sigma_{jj})$. However, sometimes it is more convenient to define $\sigma_{ij}^3 = \frac{1}{2}(\sigma_{ii}^3 + \sigma_{jj}^3)$. Both forms, it should be noted, are approximate. For the dispersion energy ε_{ij}, a simple rule for mixing is $\varepsilon_{ij} = (\varepsilon_{ii}\varepsilon_{jj})^{1/2}$ (8).

Eq. [II-10] is confined to non-polar molecules. More detail on intermolecular potentials is given elsewhere (8). Eq. [II-10] can be written in a more convenient form:

$$\varphi_{ij} = \begin{cases} -\dfrac{n_i n_j \beta_{ij}}{|\mathbf{r}_{ij}|^6} & \text{for } |\mathbf{r}_{ij}| > \sigma_{ij} \\ +\infty & \text{for } |\mathbf{r}_{ij}| < \sigma_{ij} \end{cases} \qquad \text{[II-11]}$$

where n_i and n_j are moles per unit volume of the i and j species. β_{ij} is a new constant, and ϕ_{ij} has the dimensions of energy/(volume of i)/(volume of j).

With this form of interaction energies, Hamaker (9) determined that the attraction between A & B phases per unit area of a planar interface is given by $(-\pi n_A n_B \beta_{AB})/(12\,\sigma_{AB}^2)$. The derivation of a more general form,

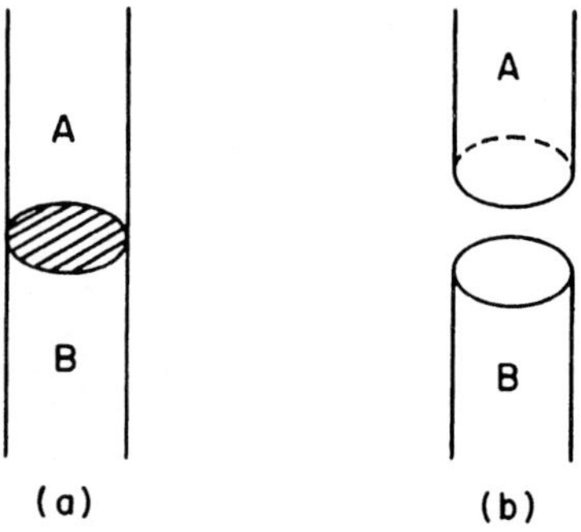

Figure II-2. Configuration analyzed in Section 3. (a) Before separation of A and B. (b) After separation.

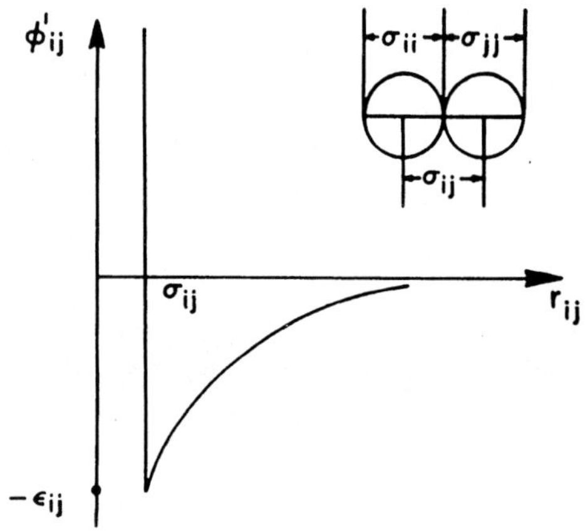

Figure II-3. A model for the intermolecular potential.

where A and B are separated by a distance h is given in Chapter III. The potential energy in that case is given by Eq. [III-15], and can be written here as $(-\pi n_A n_B \beta_{AB})/(12h^2)$. The smallest value of h is σ_{AB}, the smallest distance of approach of the first lines of the A and B molecules which are considered to be hard spheres.

If A and B are the same, then the work done in tearing apart the system at a plane A'B' is

$W^d_{AA} = 2\gamma_A{}^d$ = (interaction energy when the two pieces are separated by an infinite distance)
- (interaction energy when the two pieces are continuous).

Since there is no interaction at infinite distance ($h = \infty$),

$$W_{AA}^{d} = 2\gamma_A{}^{d} = \frac{\pi n_A n_A \beta_{AA}}{12\sigma_{AA}{}^2} \qquad \text{[II-12]}$$

The superscript d indicates that only the contribution of the dispersion potential to the interfacial tenison and work of cohesion has been included. If the molecules are non-polar, $W_{AA} = W_{AA}{}^{d}$ and $\gamma_A = \gamma_A{}^{d}$. Padday (10) has examined the above relation, after adjustments have been made for the improvement of the intermolecular potentials, improved liquid theories, etc. He found the agreement with measured values of liquid-vapor interfacial tensions to be excellent.

Along the same lines one obtains the work of adhesion as

$$W_{AB}^{d} = \frac{\pi n_A n_B \beta_{AB}}{12\ \sigma_{AB}^2} \qquad \text{[II-13]}$$

One of the consequences of the above equation is that if $\sigma_{AA} \simeq \sigma_{BB} = \sigma$, i.e., if the molecules are of comparable size, then $W_{AB}^{d} \cong 2(\gamma_A^d\gamma_B^d)^{1/2}$, as $\sigma_{AB} \simeq \sigma$. If A and B interact through dispersion alone, then $W_{AB} = W_{AB}^{d} \cong 2(\gamma_A^d\gamma_B^d)^{1/2}$. Substituting this result into Eq. [II-8], one has

$$\gamma_{AB} = \gamma_A + \gamma_B - 2(\gamma_A^d\gamma_B^d)^{1/2} \qquad \text{[II-14]}$$

This equation is due to Fowkes (11) and constitutes a special case of the more general equation of Girifalco and Good (12). It is noteworthy that Eq. [II-14] has been derived when the interaction between A and B occurs through dispersion only. This is true when both A and B are non-polar, i.e., when $\gamma_A = \gamma_A^d$ and $\gamma_B = \gamma_B^d$, and Eq. [II-14] becomes

$$\gamma_{AB} = \gamma_A + \gamma_B - 2(\gamma_A\gamma_B)^{1/2} \qquad \text{[II-15]}$$

However, Eq. [II-14] also constitutes a very good approximation for the case where either A or B is non-polar: the interaction in that case is almost solely through dispersion. In fact Eq. [II-14] even works when A and B are both polar, but A is a solid. In this case the *primary polar* interaction energy is zero and the dispersion interaction dominates.

4. PHENOMENOLOGICAL THEORIES OF EQUILIBRIUM CONTACT ANGLES.

If Eq. [II-14] is substituted into Eq. [II-1], one obtains

$$\cos \lambda = -1 + 2 (\gamma_S^d \gamma_L^d)^{1/2} / \gamma_L \qquad \text{[II-16]}$$

For the purposes of investigating Eq. [II-16], it is assumed here that the same solid is used, i.e., γ_S^d is a constant, and the contact angles of a number of nonpolar liquids, i.e., $\gamma_L^d = \gamma_L$, are measured. Eq. [II-16] becomes

$$\cos \lambda = -1 + 2(\gamma_S^d / \gamma_L)^{1/2} \qquad \text{[II-17]}$$

One notes that as $\gamma_L \to \infty$, $\cos \lambda \to -1$, or that such a liquid is completely non-wetting. As γ_L is decreased, one finds that for $\gamma_L = \gamma_S^d$, $\cos \lambda = +1$. Liquids with surface tensions smaller than this value completely wet the given solid surface. The fact that $\gamma_L < \gamma_S^d$ represents wetting liquids can be verified on combining Eq. [II-14] with Eq. [II-7], whence

$$S_{L/S} = -2\gamma_L + 2(\gamma_S^d \gamma_L)^{1/2} \qquad \text{[II-18]}$$

for $\gamma_L = \gamma_L^d$. Obviously $S_{L/S} = -2\gamma_L[1 - (\gamma_S^d / \gamma_L)^{1/2}]$, is negative if $\gamma_L > \gamma_S^d$ (non-wetting) and positive (wetting) if $\gamma_L < \gamma_S^d$. Thus Eq. [II-17] needs to be rewritten as

$$\cos \lambda = \begin{cases} -1 + 2(\gamma_S^d / \gamma_L)^{1/2}, & \gamma_L > \gamma_S^d \\ +1, & \gamma_L < \gamma_S^d \end{cases} \qquad \text{[II-19]}$$

In the second part of Eq. [II-19], the fact that wetting liquids always have λ equal to zero (5-7) has been used.

If it is assumed that $|\gamma_L - \gamma_S^d|$ is small, then Eq. [II-19] can be expanded in a Taylor series about $\gamma_L = \gamma_S^d$ and only the first term retained. One has

$$\cos \lambda = \begin{cases} 1 - [(\gamma_L - \gamma_S{}^d) / \gamma_S{}^d], & \gamma_L > \gamma_S{}^d \\ 1, & \gamma_L < \gamma_S{}^d \end{cases} \qquad \text{[II-20]}$$

Zisman and co-workers (13) measured the values of the equilibrium contact angles for various liquids on a given solid surface. The results of Fox and Zisman (5) for hydrocarbons on teflon are shown in Figure II-4.

The plot is a straight line with a constant slope, in agreement with Eq. [II-20]. The value of γ_L at which the line intersects the horizontal line cos γ = 1, is known as the critical surface tension of the solid. From Eq. [II-20], this is equal to γ_S^d. For solids which are non-polar, $\gamma_S^d = \gamma_S$ and consequently the Zisman plot constitutes one of the very few ways of determining γ_S. In all cases the critical surface tension is a very good way of arriving at an estimate for γ_S.

Liquid surface tensions γ_L are typically in the range of 20 - 70 dynes/cm. Solids surfaces may be divided into high energy surfaces (~ 500 dynes/cm and more, e.g., glass, metals, etc.) and low energy surfaces (~ 20 - 40 dynes/cm, e.g., hydrocarbons, polymers, etc.). From these values and the above results, one may conclude that high energy surfaces are almost always wettable. Such a generalization cannot be made for the low energy surfaces. The latter are usually not wet by water since water has a high surface tension of 72 dynes/cm. It is noteworthy that fluorocarbon surfaces have very low critical surface tensions. For instance, Teflon (polytetrafluorethylene) has a critical surface tension of about 18 dynes/cm and is consequently wet by very few liquids. Polyethylene, in

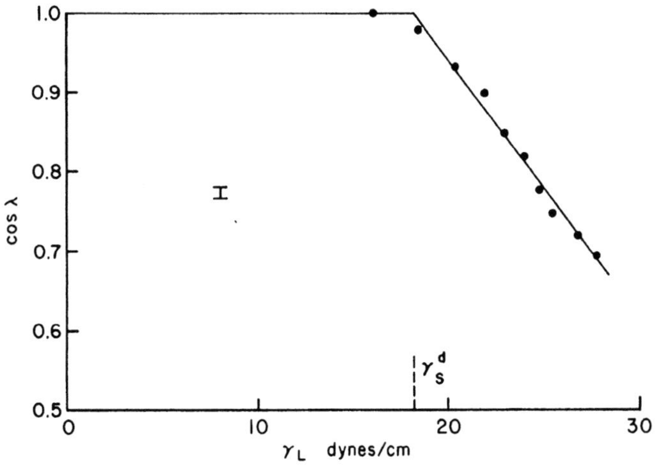

Figure II-4. The Zisman plot for n-alkanes on teflon (5). The liquids with cos λ = 1 are wetting liquids. The error bar is shown in the inset.

contrast, has a critical surface tension of about 31 dynes/cm and is wet by numerous organic liquids. Extensive compilations of contact angles and critical surface tensions have been made by Adamson (14).

To put matters in perspective, we note that Eq. [II-16] could be derived only because the work of cohesion was obtained independently, viz., through consideration of intermolecular energies of interaction. This work was expressed in terms of interfacial tensions, and Young's equation, Eq. [II-1], was thus simplified to Eq. [II-16]. An alternate method exists for relating the equilibrium contact angle λ to molecular effects (15,16). Consider Figure II-5 where a gas (G) - liquid (L) - solid (subdivided into S_A and S_B) three phase region is shown and where 0 is the contact line. The liquid is wedge-shaped and has a plane gas-liquid interface. Q is a point at the gas-liquid interface. ϕ is the potential energy of interaction at Q, per unit volume of liquid, with the gas phase G, the liquid phase L and the solid phase S_A and S_B. Now a datum potential ϕ_F is defined in the following way. If the contact line region were not present, there would have been only the gas and the liquid phases with the two separated by the original G-L interface and its extrapolation as shown by the dashed line in the figure. Thus G + S_A would then be the gas phase, and L + S_B would be the liquid phase. ϕ_F for this situation is obviously a constant related to the energy at a plane gas-liquid interface. The energy of interaction with the gas phase can be neglected for there are too few molecules in the gas phase. Hence,

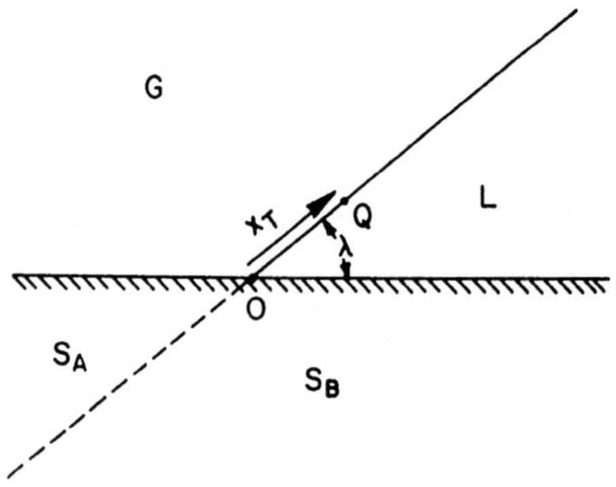

Figure II-5. The construction used for determining the molecular potential at the point Q. The solid has been partitioned into regions S_A and S_B.

$\phi-\phi_F$ = (interaction of Q with S_B as a solid - that with S_B as a liquid)
+ (interaction of Q with S_A as a solid)

With the form of interaction energy given by Eq. [II-11], the above relation can be expressed in terms of integrals over the respective volumes, i.e.,

$$\phi-\phi_F = \int_{V_{S_B}} \frac{n_L(n_L\beta_{LL} - n_S\beta_{SL})}{|\mathbf{r}|^6} dV - \int_{V_{S_A}} \frac{n_L n_S \beta_{SL}}{|\mathbf{r}|^6} dV \qquad \text{[II-21]}$$

where $|\mathbf{r}|$ is the radial distance from the point Q. The volume integrations are performed for the profile shown in Fig. II-5 in a cylindrical coordinate system with O on the axis of the cylinder. The result is

$$\phi-\phi_F = \frac{\pi}{12x_T^3}(n_L^2\beta_{LL} - n_L n_S \beta_{SL})\, G(\lambda) - \frac{\pi}{12x_T^3} n_L n_S \beta_{SL}\, G(\pi-\lambda) \qquad \text{[II-22]}$$

where the function G is

$$G(\alpha) = \operatorname{cosec}^3 \alpha + \cot^3 \alpha + \frac{3}{2} \cot \alpha \qquad \text{[II-23].}$$

α is the angle of wedge. x_T is the distance of the point Q from O. If the system is at equilibrium, then the hydrodynamic force $[-d(\phi-\phi_F)/dx_T]$ along the gas-liquid interface (locus of Q) must be zero, leading to the result

$$\frac{G(\pi-\lambda)}{G(\lambda)} + 1 = \frac{n_L^2\beta_{LL}}{n_L n_S \beta_{SL}} \qquad \text{[II-24]}$$

Eq. [II-24] defines the equilibrium contact angle λ. From Eqs. [II-12] and [II-13] with $\sigma_{SS} \sim \sigma_{LL}$, the right hand side of Eq. [II-24] is equal to W_{LL}^d/W_{SL}^d, i.e., the work of cohesion divided by the work of adhesion. Thus for non-polar systems the right hand side may be obtained from Young's equation, Eq. [II-1] as $2/(1 + \cos\lambda)$. If the above derivation is correct (including the determination of the work of adhesion and the work of cohesion), one should have an *identity*

$$\frac{G(\pi-\lambda)}{G(\lambda)} + 1 = \frac{2}{1+\cos\lambda} \qquad \text{[II-25]}$$

Eq. [II-25] is not an identity. The left hand side and the right hand side

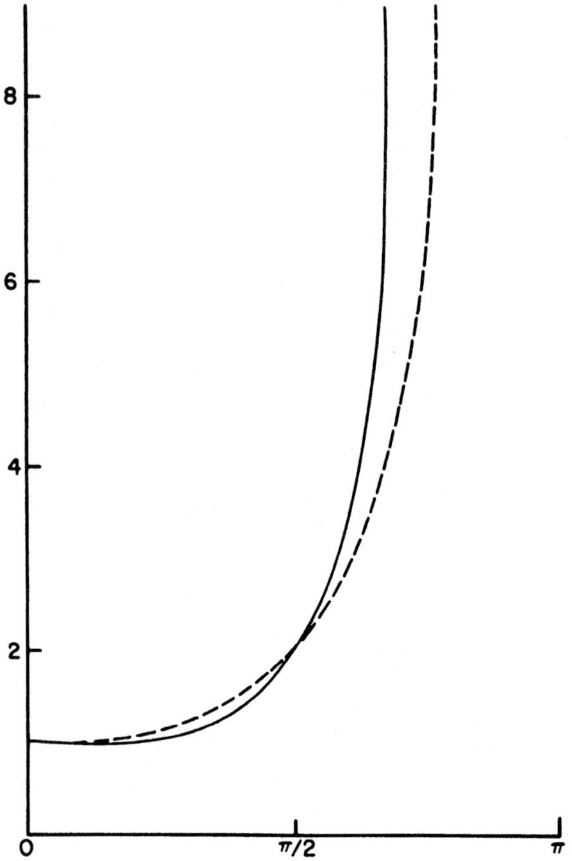

Figure II-6. The left hand side and the right hand side in Eq. [II-25] have been shown with bold and dashed lines respectively for different values of λ .

of the equation have been shown in Figure II-6 in bold and dashed lines respectively; the agreement is good. The plots form mirror images about $\lambda = \pi$. Eq. [II-24] has been derived by Miller and Ruckenstein (15) and in a somewhat different way by Jameson and del Cerro (16).

Thus, Eq. [II-24] provides an approximation to Young's equation. The important contribution of the Miller-Ruckenstein equation, Eq. [II-22], is that it is valid (within its degree of approximation) in non-equilibrium situations as well and provides a means for obtaining an appropriate hydrodynamic driving force in the liquid for those cases where λ in Figure II-5 differs from the equilibrium contact angle.

5. CONTACT ANGLE HYSTERESIS

Experimental measurements of equilibrium contact angles may be accomplished photographically. There are no formulas to apply. In some cases, notably the sessile drop, the equipment used to measure the interfacial tensions also serves to measure the contact angles, provided that in such systems well defined contact lines exist. If interfacial tensions are known from separate measurements, the Wilhelmy plate and capillary rise methods may be used to obtain contact angles. Although the basic arrangements are simple, a great many precautions are necessary for sufficiently accurate results [see Refs. (17,18)]. A particular method exists where a flat plate is partially immersed in a liquid and tilted until the clinging meniscus disappears. The angle of the tilt describes the contact angle. The method has been sketched in Figure II-7(a) and (b).

The first experimental data on contact angles were quite confusing. If they were at all reproducible, they could not be explained. Although in general they were supportive of the theory, the data did a great many things other than those anticipated by the theory. In what followed, virtually everything was put under doubt as a matter of principle. It was argued that some of the conditions under which the experiments were performed could have led to spurious results, e.g., vibrations in the system. Alternately it was said that the theory was incomplete and a list of a dozen physical effects was advanced which, it was claimed, should be incorporated into the theory. Obviously, in such a case, both arguments could be true, a point that appeared to be quite unclear given the polemics of those times. Thanks to the meticulous and very provocative work of Zisman, Johnson, Dettre, and others, the reasons behind the anomalous behavior are now well known. The initial inability to explain contact angle hysteresis stemmed from the usual simplifying assumption that solid surfaces are perfectly smooth, impermeable, pure, homogeneous, etc. Obviously, these terms apply only to an idealized solid surface, reasonable for most purposes, but quite a naive description in the science of interfacial phenomena.

A description of an experiment is necessary first to delineate contact angle hysteresis. A flat plate is shown immersed in a pool of liquid in Figure II-7(a). A meniscus forms near the contact line. The plate is rotated till the meniscus disappears; the inclination of the plate to the horizontal liquid surface is the equilibrium contact angle λ as shown in (b). The contact line is at point P on the solid. If the plate is dipped

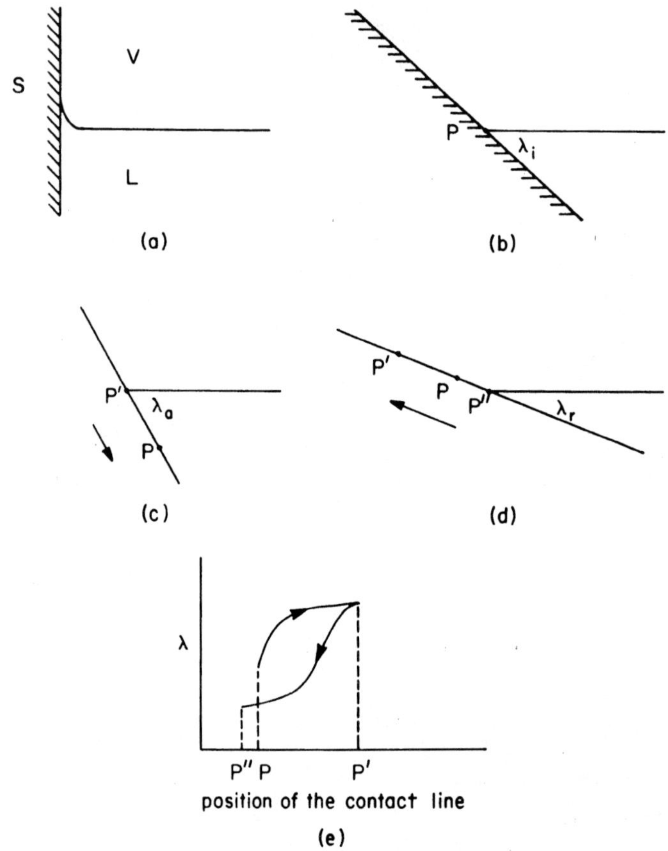

Figure II-7. The solid surface in (a) is turned to obtain a horizontal vapor (V) - liquid (L) interface in (b). In (c) the advancing contact angle λ_a is shown, and (d) shows the receding contact angle λ_r. In (e), the measured values of λ are shown to illustrate the "hysteresis".

some more into the liquid till the point P' is reached on the solid, the *new* contact angle there is seen to be λ_a. This is called the *advancing* contact angle since the contact line has advanced to a new position over a dry solid surface (c). The plate is withdrawn to a point where the contact line is at P". The contact angle there is found to be λ_r (d). This is called the *receding* contact angle since the contact line has receded into the region previously occupied by the liquid. On repeated dipping and withdrawing the plate λ_a and λ_r may reach steady values but are seen not to be equal to one another (e). Johnson and Dettre (19) have discussed why the term hysteresis came to be applied. It is only of historical interest. However, the understanding of as to why λ_a is different from λ_r is of importance. The main reasons are discussed below.

5.a Impurities on the Surface

It has been argued previously that a high energy glass surface is completely wet by water. A casual experiment will show this as *not* the case. Since glass has a high surface energy, it is wet by almost all substances and consequently small amounts of grease (and dirt) wet the surface and adhere to it. The waxy material brings down the interfacial tension to sufficiently low values that it is *not* wet by water. Glass has to be cleaned with strong acids and alkalis, and finally with chromic acid to obtain a surface sufficiently clean to be wet with water. This simple example serves to show the importance of contamination and how difficult it is to avoid.

Surface contamination, which gives rise to a patchy or heterogeneous surface, can also cause contact angle hysteresis. In Figure II-8, the advancing and the receding contact angles are shown for water on a glass surface (titania coated glass treated with waxy trimethyloctadecyl ammonium chloride) as a function of the number of coating treatments with 1% polydibutyl titanate (20). It is seen that about 50 coating treatments are required to remove or recoat the contaminant and make the surface sufficiently homogeneous that there is no hysteresis.

5.b Effect of Adsorption

It is noteworthy that the impurities discussed in Section 5a are in adsorbed form. Here a special type of adsorption effect is analyzed, viz., it is postulated that due to adsorption γ_S for determining the advancing contact angle is different from that of the receding contact angle. Zisman and co-workers (22,23) coated the surface of platinum with alkyl amines by (a) adsorbing them from aqueous solution, (b) adsorbing them from melts, and (c) adsorbing them from a cetane solution. The advancing and receding contact angles for water on the above three surfaces are shown in Figure II-9, plotted against the carbon number of the n-alkyl amines. A striking feature is that the advancing and receding contact angles for water in case (a) are the same. It is also seen that the advancing and receding contact

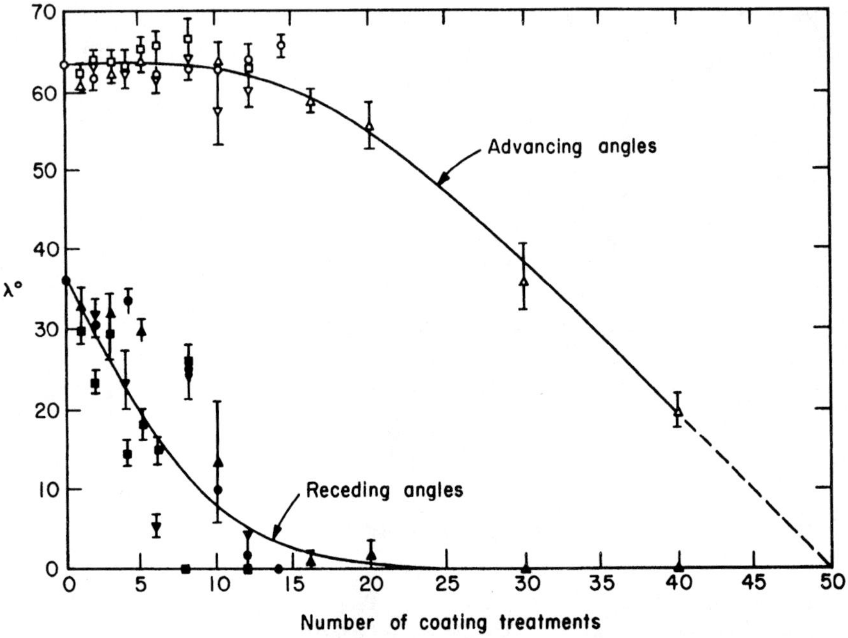

Figure II-8. The advancing and receding contact angles of water on titania coated glass after treatment with trimethyloctadecylammonium chloride have been shown as a function of coating treatments with 1.1% polydibutyl titanate. Reprinted with permission from Ref. (20). Copyright 1965 American Chemical Society.

angles for cases (b) and (c) fall on only two curves, the "advancing" curve and the "receding" curve labeled in the figure. Case (a) falls on the "receding" curve. Interesting conclusions can be drawn.

It is known that the amine group adsorbs onto the platinum surface. Thus the hydrocarbon tails are directed away from the platinum suface as shown in Figure II-10. Under the bulk water, the water molecules penetrate the tail region, and the nature of the advancing contact angle is as shown in Figure II-10(a). Now, for the receding contact angle, the situation is as shown in Figure II-10(b). The "solid surface" has alkyl amines adsorbed on it as in Figure II-10(a), *but* with water molecules among the hydrocarbon tails. Obviously, the appropriate values of γ_S in Figures II-10(a) and II-10(b) are different, and the advancing and the receding contact angles differ. However, if the amine is adsorbed from aqueous solution, water molecules will be interdispersed in the hydrocarbon tail region everywhere. Consequently, in this case the advancing and receding contact

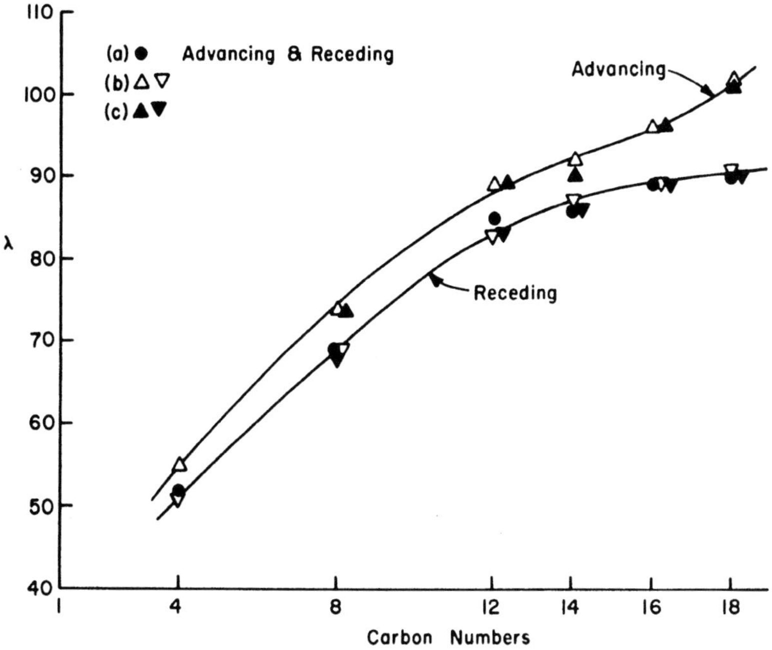

Figure II-9. The advancing and receding contact angles of water are plotted against the carbon number of n-alkyl amine coated platinum, for amines adsorbed (a) from aqueous solutions (b) from melts and (c) from cetane solutions. Data from (22,23).

angles will be the same (same γ_S) and will be equal to the receding contact angles obtained in the other cases [Figure II-10(b)].

One attempts to generalize the above effect by rewriting Eq. [II-1] as

$$\gamma_{LV} \cos \lambda = (\gamma_{SV} - \pi_e) - \gamma_{SL} \qquad \text{[II-26]}$$

where π_e is the surface pressure of the adsorbed material on the solid surface. The adsorbed material changes the interfacial tension of the solid from γ_{SV} to $\gamma_{SV} - \pi_e$. The π_e term is usually unimportant for non-wetting liquids, but becomes very important as the wettability increases (7). In this connection a special class of systems where γ_L is close to the critical surface tension of the solid needs to be mentioned. It is found that in such cases the liquid wets the dry solid but will not wet the solid surface which has the vapor of the liquid already adsorbed on it

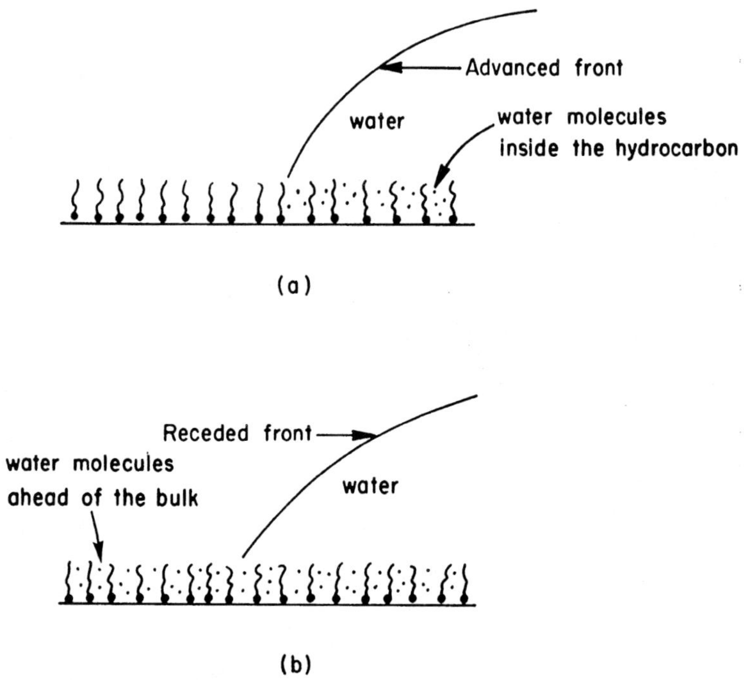

Figure II-10. Schematic diagrams explaining the nature of results shown in Figure II-9.

(24). This phenomenon goes under the name of autophobicity. The contact angles so formed are usually very small.

5.c Surface Roughness

Vibrations in contact angle measuring systems are known to affect the experimental results (25). Investigators found that contact angle hysteresis is made more acute as steps were taken to eliminate vibrations, i.e., the data became more irreproducible. In contrast, the problem was minimized by introducing controlled amounts of vibration in the system (25-27). Eventually, Johnson and Dettre (19,20) showed that vibrations or kinetic energy of the drop very seriously affect the measured contact angles in experiments on rough solid surfaces. The degree of roughness in most cases is small but sufficient to give rise to hysteresis.

They calculated the surface free energy of a drop lying on a rough surface. The roughness was idealized as concentric sinusoidal grooves.

The liquid-vapor interface was taken to be a spherical cap. The scheme is shown in Figure II-11. It is seen that there is an apparent equilibrium contact angle ϕ, which is the angle that the drop makes with the horizontal direction. Its significance lies in the fact that although the actual equilibrium contact angle λ remains the same, ϕ varies depending on the position of the contact line. It is also this angle ϕ that is measured experimentally by extrapolating the observed shape of the liquid-vapor interface to meet the horizontal direction.

They were able to show that of all possible positions of the foot of the drop inside a given groove, there was only one where the free energy was a minimum. The minima in two neighboring grooves were separated by an energy barrier. For a drop of a given volume, the values of the free energy minima for different grooves were different and the system had one absolute minimum.

They concluded that although the drop had only one position favored energetically, i.e., where the position of the foot gave rise to the overall free energy minimum, in practice, the foot could get trapped in any groove between two energy barriers. How and where such an entrapment would occur would depend on the vibrations in the system, i.e., on the kinetic energy of the drop available to overcome the energy barriers between adjacent positions of metastable equilibrium (see Figure II-12). It is noteworthy that a controlled amount of vibration in the system could lead the foot of the drop eventually to the position of the overall minimum energy. ϕ may still differ from λ, but it is at least reprodicible and

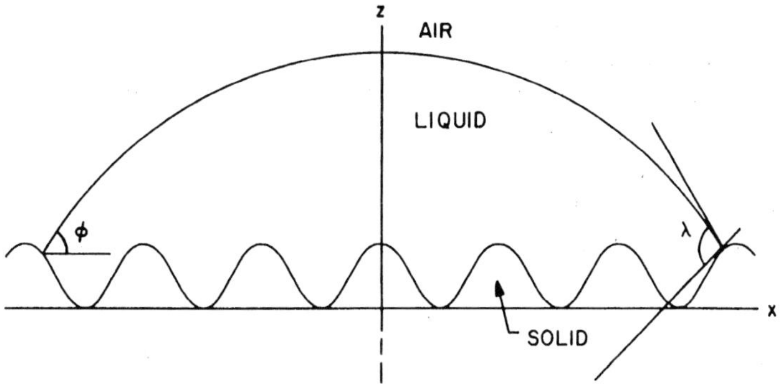

Figure II-11. A drop on an idealized rough surface. Reprinted with permission from Ref. (19). Copyright 1964 American Chemical Society.

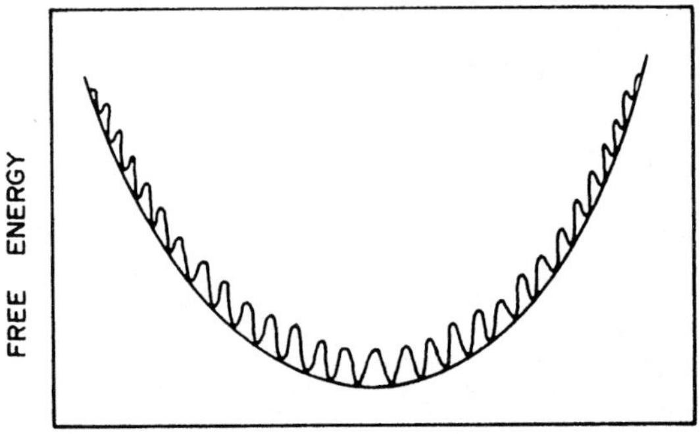

Figure II-12. In Figure II-11 it has been shown that the drop can have many configurations. Being symmetric, each configuration has a fixed apparent contact angle ϕ and a total free energy. The latter has been plotted here as a function of ϕ. It is seen that for the foot of the drop lying inside any groove there is a minimum. Two neighboring minima (meta-stable equilibrium positions) are separated by an energy barrier which is highest near the overall minimum. Consequently, this overall minimum is the most difficult to attain without sufficient vibrations. Reprinted with permission from Ref. (24).

connected to the thermodynamics of the system. A sketch of the energy barriers near the region of the contact line is shown in Figure II-12. Note that the energy barriers are largest near the equilibrium position. For a simple treatment showing that the roughness will affect the contact angle see Problem II-5.

The contact angle hysteresis of water on a rough wax surface is shown in Figure II-13 as the surface is made smoother by annealing. Beyond the seventh annealing treatment the difference between the advancing and receding contact angles progressively vanishes. Prior to the seventh treatment, the surface is so rough that solid-liquid contact is broken at numerous locations by air gaps, and the above arguments must be modified.

Besides the above three important features, there are many others, some of which are mentioned below in brief. It is often questioned that if the horizontal force balance at the contact line leads to Young's equation, what happens to the normal component $\gamma_{LV} \sin \lambda$? It appears that such a component does exist and tends to distort the solid surface. Since in most cases the magnitude of this force is low and the modulii of elasticity of the solids are high, no significant distortion can be observed. [But Bailey (28) has actually observed the distortion of the solid near the contact line region of a mercury drop on a mica sheet 1 μm thick.] Liquid

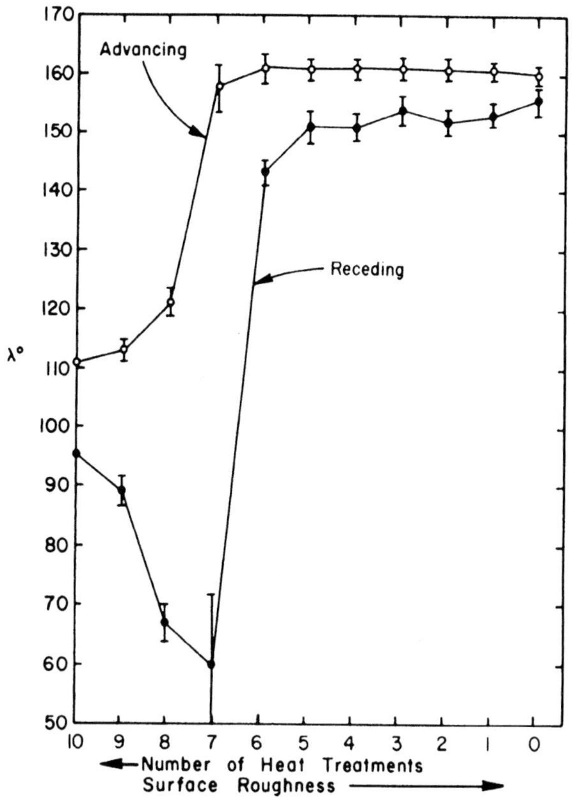

Figure II-13. Water contact angles on TFE-methanol telomer wax surface as a function of roughness. Reprinted from (19) with permission. Copyright 1964 American Chemical Society.

imbibition into the solid substrate can also alter the results. Fort (29) reports such effects on nylon polymers.

It appears that although one knows why hysteresis can occur, or even why it occurs in a given system, it is a formidable job to get rid of it entirely. Extremely smooth and homogeneous surfaces can rarely be prepared. Certainly the means or methods are not available to investigators outside this select field, nor employed by say a fluid mechanist who also works on systems containing contact lines. However, contact angle hysteresis *is* tractable. Consequently, one may say that in any *one* experiment in a given set of identical experiments, the contact angle can be treated as a constant. However, its values in different experiments may differ, due to a difference in histories of the processes or the materials. Similarly one may say that for a given solid-liquid pair different investigators will record different advancing (say) contact angles, for the history or the quality of the materials may differ as well as the experimental conditions. It is only here that one finds out the fact the contact angle is a somewhat robust constant, so that with care it can generally be reproduced to within ± 2° and certainly within ± 5° among investigators.

6. ADSORPTION

Previously the influence of adsorbed material on contact angles has been seen to be important. In general, the study of adsorption has been prompted by its importance in catalysis and rections at interfaces. The Langmuir-Hinshelwood model for the kinetics of such reactions is both popular and dependent on the adsorption-desorption rates (30). The nature of adsorption also needs to be studied to determine the viability of chromatographic separation processes for a given system. Its attraction is that it is one of the very few means for separating similar components, such as isomers, azeotropes, etc., that are traditionally "difficult" (31).

As noted previously, the adsorbed amount Γ is the surface excess. The concentration of a species changes smoothly across an interface attaining bulk values in the interiors of the two phases. If for simplicity the bulk concentrations are extrapolated to the dividing surface, it leaves one with an excess material which is assumed to reside at the dividing surface. This is the surface excess as seen in Chapter I. Consider a gas-solid interface where the gas is ideal and the solid is impermeable. Now, the gas molecules are attracted towards the solid because of the intermolecular forces. However, these forces are not felt in the bulk of the gas far away from the interface. At equilibrium the chemical potential of the gas in the bulk is $\mu^0 + RT\ell n\rho_\infty$ and that near the interface is $\mu^0 + RT\ell n\rho + \Phi$. Here ρ is the gas density and ρ_∞ is the density in the bulk. μ^0 is the standard state chemical potential. Φ is the attraction potential between the gas and the solid and $\Phi(z) \to 0$, $\rho(z) \to \rho_\infty$ as $z \to \infty$, where z is the perpendicular distance from the interface. At equilibrium, the two chemical potentials are equal, leading to $\rho = \rho_\infty \exp(-\Phi/kT)$. Following the definition of Γ one has

$$\Gamma = \int_0^\infty (\rho-\rho_\infty)\, dz$$

or

$$\Gamma = \rho_\infty \int_0^\infty [\exp(-\Phi/RT) - 1]\, dz \qquad [II-27]$$

As the integral is a constant, Eq. [II-27] reduces to Henry's law

$$\Gamma = Hp \qquad [II-28]$$

where H is the Henry's law constant given by

$$H = \frac{1}{RT} \int_0^\infty [\exp(-\Phi/RT) - 1]\, dz$$

and the ideal gas law has been used to replace ρ_∞ with the pressure p. It is found that Henry's law is valid for moderately small pressures.

It is seen that ρ deviates most from ρ_∞ at the interface. Thus the gas molecules that make up the surface excess concentration Γ reside very close to the interface. One may now assume a model where the adsorbed gas molecules are supposed to be stuck to the solid surface. A quantity θ, the fractional surface coverage, is thus defined as

$$\theta = \sigma^0 \Gamma \qquad [II-29]$$

where σ^0 is the projected area of a molecule adsorbed at the interface.

6.a Langmuir Adsorption Isotherm

When adsorption is such that $\theta < 1$, the rate of adsorption is proportional to $p(1-\theta)$, where $(1-\theta)$ is the fractional available space for adsorption and p is the gas pressure. The model is similar to that describing the kinetics of a bimolecular reaction. The rate of desorption is proportional to θ or the concentration of the adsorbed species. At equilibrium

$$k_a\, p(1-\theta) = k_d \theta$$

where k_a and k_d are the adsorption and desorption rate constants. Rearranging the above, one has

$$\theta = \frac{(k_a/k_d)\, p}{1+(k_a/k_d)p} \qquad [II-30]$$

or

$$\Gamma = \frac{[k_a/(k_d\sigma^0)]p}{1+(k_a/k_d)p} \qquad [II\text{-}31]$$

which is the Langmuir adsorption isotherm. At small pressures, Henry's law

$$\Gamma = \left(\frac{k_a}{k_d\sigma^0}\right) p \qquad [II\text{-}32]$$

is obtained. Note that from Eq. [II-30], $\theta \leq 1$.

6.b BET (Brunauer-Emmett-Teller) Isotherm

It is sometimes seen that as $p \to p^0$, the saturation pressure, the adsorbed amount becomes infinite. This seems to imply that adsorption turns into condensation at some stage. In the BET adsorption isotherm, it is assumed that the first layer is an adsorption layer. Multiple layers are built on it due to condensation. The activation energy for the first layer is the heat of adsorption Q. The activation energy for adsorbing a layer on an n-stack layer (n > 1) is the heat of condensation Q_v.

Thus the equilibrium for the first layer is given by

$$ap\theta_0 = b\theta_1 e^{-Q/RT} \qquad [II\text{-}33]$$

and for others by

$$ap\theta_{n-1} = b\theta_n e^{-Q_v/RT} \quad , \quad n > 1 \qquad [II\text{-}34]$$

In Eqs. [II-33] and [II-34], the left hand sides denote the rates of adsorption which are proportional to the pressure and the unoccupied surface. The right hand sides denote the rates of desorption which are proportional to the occupied surface. The rate constants are a for adsorption and [b exp(-Q/RT)] and [b exp($-Q_v$/RT)] for desorption in the two cases. θ_n is the fractional area covered by the n-stack layer. Obviously the volume adsorbed per unit area is proportional to $\sum_{n=1}^{\infty} n\theta_n$.

Thus

$$\frac{v}{v_m} = \frac{\sum_{n=0}^{\infty} n\theta_n}{\sum_{n=0}^{\infty} \theta_n} \qquad [II\text{-}35]$$

where v_m is a reference quantity. The summation can be carried out using Eqs. [II-33] and [II-34], to yield

$$\frac{v}{v_m} = \frac{Ky}{(1-y)[1+(K-1)y]} \qquad [II-36]$$

where $K = \exp[(Q-Q_v)/RT]$ and $y = (a/b)\, p \exp(Q_v/RT)$ One notes in Eq. [II-36] that at $y = 1$, $v = \infty$. This is precisely the condition necessary to have complete condensation at $p = p^o$. Thus at $p = p^o$, $y = 1$, or

$$\frac{a}{b} p^o e^{Q_v/RT} = 1 \text{ and } y = p/p^o$$

Eq. [II-36] is the BET isotherm (14c). For small y, i.e., $1 - y \sim 1$ and large K, i.e., $K-1 \sim K$, the Langmuir isotherm is obtained. Henry's law, Eq. [II-28], the Langmuir isotherm, Eq. [II-30], and the BET isotherm, Eq. [II-36] are plotted in Figure II-14, with $\sigma^o H p^o = p^o k_a/k_d = K$. Among them, they describe wide ranges of adsorption behavior.

Experimentally v is measured in volume at STP. If the adsorption is only a saturated monolayer, then $\theta_1 = 1$ and all other θ_i are zero. In that case one has $v = v_m$, allowing one to interpret v_m as the volume of gas at STP in cc required to cover the surface of the adsorbent completely and with no excess material. Thus

$$v_m = \frac{Sv^o}{N_A \sigma^o} \qquad [II-37]$$

where $v^o = 22400$ cc/mole and N_A is Avogadro's number. S is the surface area of the adsorbent, and σ^o is the projected area of one molecule of the adsorbed gas. Hence from the experimental data, v_m and K can be determined (see Problem II-16 for a useful method). Then S can be found from Eq. [II-37] if a good estimate of σ^o is available. Consequently the BET equation has been widely used to estimate S, the knowledge of which is invaluable in catalysis.

Adsorption of a component from a liquid follows similar behavior. In fact, similar adsorption equations are used. An important case is where a component is adsorbed from the bulk liquid onto a solid surface. The previous equations are altered by replacing the pressure p with c, the concentration of the adsorbed species in the bulk phase.

The spreading pressure π is defined as

$$\pi = \gamma^o - \gamma \qquad [II-38]$$

where γ^o is the interfacial tension of the pure substance and γ is that

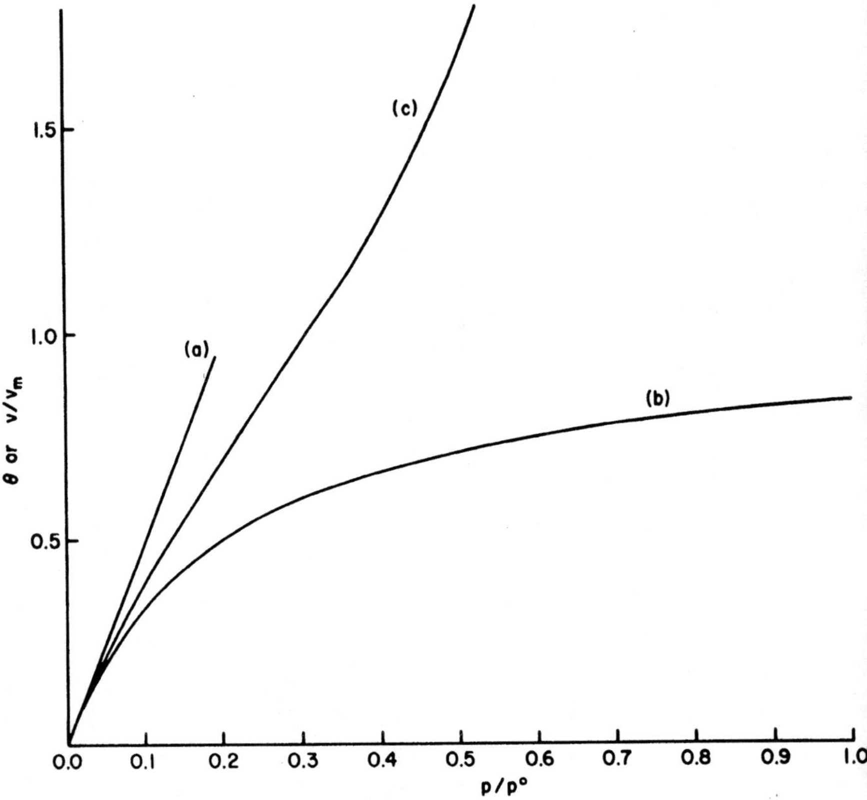

Figure II-14. Henry's law (a), Langmuir adsorption isotherm (b), and BET isotherm (c), plotted with parameters such that (b) and (c) reduce to (a) at low p/p°.

after adsorption. For liquid-fluid interfaces π is particularly significant as it can be measured directly (see Chapter IV). Using Gibbs' adsorption equation, Eq. [I-24], one has

$$\pi = \int_0^c \Gamma d\mu \qquad [II\text{-}39]$$

Assuming that $\mu = \mu^0 + RT\ln c$ and that Henry's law applies in the form $\Gamma = Hc$, one has

$$\pi = RTHc = RT\,\Gamma$$

which on rearrangement yields

$$\pi\sigma = RT \qquad [II-40]$$

where $\sigma = \Gamma^{-1}$ is the interfacial area per adsorbed molecule. Eq. [II-40] is reminiscent of the ideal gas law pv = RT, where π is the two-dimensional analog of the pressure p and σ is the analog of specific volume v. Consequently models are built in more complicated systems along the lines of appropriate equations of state in three dimensions. It is noteworthy that the analogy is not exact. For instance, negative values of π, unlike negative pressures p, are physically meaningful and are known to occur in strong electrolyte solutions. Adamson (14) provides a more comprehensive discussion on the subject of adsorption.

7. DENSITY PROFILES IN LIQUID FILMS ON SOLIDS

The density profile at a vapor-liquid interface was considered briefly in Section 5 of Chapter I. Prominent in that discussion was the positive contribution to the local free energy density produced by sharp gradients in density (cf. Eq. [I-41]). The buildup of a thin film on a solid surface is also influenced by this gradient energy (34,35). As we shall see, gradient effects sometimes prevent the continuous film buildup predicted by conventional adsorption isotherms as the bulk density of the vapor approaches its saturation value. Moreover, when both liquid and vapor are present, the liquid spreads spontaneously on the solid for temperatures sufficiently near the critical temperature but has a finite equilibrium contact angle at lower temperatures.

Suppose that a solid surface is placed in a bulk vapor phase having a pressure p_B and molecular density n_B below the saturation values at the existing temperature. Attraction between the solid and molecules of the vapor favors buildup of an adsorbed film. However, the additional energy associated with the density gradient between the film and bulk vapor phase opposes film formation, an effect not considered in developing the BET adsorption isotherm of the preceding section.

The local Helmholtz free energy density corresponding to Eq. [I-41] in such a film is

$$f = f_o(n) + k\left(\frac{dn}{dz}\right)^2 + n\,\Phi(z) \qquad [II-41]$$

where $\Phi(z)$ is the interaction potential between solid and gas as in Section 6. Since the temperature T and the external pressure p_B are fixed, the Gibbs free energy G of the system must be minimized at equilibrium. Evaluating the local Gibbs free energy at each position relative to that of the same number of molecules at T and p_B, we obtain

$$G = \int_0^\infty [\Delta f + k \left(\frac{dn}{dz}\right)^2 + n\,\Phi(z)]\,dz \qquad [II\text{-}42]$$

$$\Delta f = f_o(n) + p_B - \frac{n}{n_B}\,(f_o(n_B) + p_B) \qquad [II\text{-}43]$$

Invoking the calculus of variations, we find that the density profile which minimizes G must satisfy the following condition:

$$2k\,\frac{d^2n}{dz^2} - \frac{\partial \Delta f}{\partial n} - \Phi(z) = 0 \qquad [II\text{-}44]$$

Finally, we substitute Eq. [II-43] into this equation, recognizing that $[(f_o(n_B) + p_B)/n_B]$ is the chemical potential μ_B:

$$2k\,\frac{d^2n}{dz^2} - \frac{df_o(n)}{dn} - \Phi(z) + \mu_B = 0 \qquad [II\text{-}45]$$

Teletzke et al (35) used Eq. [II-45] and the Peng-Robinson equation of state (36) to calculate density profiles for various values of the temperature T and bulk density n_B. They found that below the saturation density n_B^o relatively thick films can form near the solid when the temperature is below but not too far from the critical temperature T_c of the fluid. Under these conditions the gradient energy contribution is relatively small, as would be expected in the neighborhood of the critical point. For densities above n_B^o where the bulk phase is a liquid-vapor mixture, the liquid completely wets the solid, i.e., there is no equilibrium contact angle.

Below a "surface critical temperature" T_{cs} which is less than T_c, two profiles with the same total free energies are possible at certain bulk densities slightly less than n_B^o. One is a relatively thin film with but a small enhancement in density near the solid. It has both a relatively low attractive interaction with the solid and a low gradient energy. The other solution is a thicker film having a greater density near the solid and consequently both a greater attractive interaction and a larger gradient energy. In principle, films with these two profiles could coexist on the same surface. Again the liquid wets the solid for bulk densities exceeding the saturation value n_B^o.

Below a still lower temperature T_{cw} called the "critical wetting temperature" only a thin adsorbed film exists for $n_B < n_B^o$ because the gradient energy is so high that development of thicker films is energetically unfavorable. For densities above n_B^o a finite contact angle exists where liquid and vapor phases meet the surface. Thus, T_{cw} marks the boundary between the domains where the liquid has zero and nonzero contact angles with the solid. It can be shown that similar behavior exists in liquid-liquid systems, i.e., one liquid is completely wetting when conditions are sufficiently close to those corresponding to liquid-liquid criticality.

REFERENCES

General Reference

Gould, R.F. ed. (1964) *Contact Angle, Wettability and Adhesion*, Adv. in Chem. Ser., **V. 43**, American Chemical Society, Washington, D.C. (in addition to general references listed in Chapter I).

Textual References

1. Johnson, R.E., Jr. (1959) *J. Phys. Chem.* **63**, 1655.
2. Buff, F.P., and Saltzburg, H. (1957) *J. Chem. Phys.* **26**, 23.
3. Gibbs, J.W. (1961) *The Scientific Papers*, **V. 1**, p. 288 Dover, N.Y.
4. Platikanov, D., Nedyalkov, M. and Scheludko, A. (1980) *J. Colloid Interface Sci.* **75**, 612, and Platikanov, D., Nedyalkov, M. and Nasteva, V. (1980) *J. Colloid Interface Sci.* **75**, 620.
5. Fox, H.W. and Zisman, W.A. (1950) *J. Colloid Sci.* 5, 514.
6. Bascom, W.D., Cottington, R.L., and Singleterry, C.R. (1964) in *Contact Angle, Wettability, and Adhesion*, Adv. in Chem. Ser. **V. 43**, Am. Chem. Soc., Washington, D.C., p. 355.
7. Johnson, Jr., R.E. and Dettre, R.H. (1966) *J. Colloid Interface Sci.* **21**, 610.
8. Prausnitz, J.M. (1969) *Molecular Thermodynamics of Fluid-Phase Equilibria*, Prentice-Hall, Inc., N.J., p. 52-88.
9. Hamaker, H.C. (1937) *Physica* **4**, 1058.

10. Padday, J.F. (1969) in *Surface and Colloid Science* **V. 1**, E. Matijevic, ed., Wiley-Interscience, N.Y., p. 44-59.

11. Fowkes, F.M. (1963) *J. Phys. Chem.* **67**, 2538.

12. Girifalco, L.A. and Good, R.J. (1957) *J. Phys. Chem.* **61**, 904.

13. Zisman, W.A. (1964) in *Contact Angle, Wettability and Adhesion*, Adv. in Chem. Ser., **V. 43**, Am. Chem. Soc., Washington, D.C., p. 1.

14. Adamson, A.W. (1976) *Physical Chemistry of Surfaces*, 3rd ed., Wiley-Interscience, N.Y., (a) p. 333-371, (b) p. 256-260, (c) p. 548-571.

15. Miller, C.A. and Ruckenstein, E. (1974) *J. Colloid Interface Sci.* **48**, 368.

16. Jameson, G.J. and del Cerro, M.C.G. (1976) *JCS Faraday I* **72**, 883.

17. Good, R.J. (1979) in *Surface and Colloid Science* **V. 11**, R.J. Good and R.R. Stromberg, eds., Plenum Press, N.Y., 1.

18. Neumann, A.W. and Good, R.J. (1979) ibid, p. 31.

19. Johnson, Jr., R.E. and Dettre, R.H. (1964) in *Contact Angles, Wettability, and Adhesion*, Adv. in Chem. Ser., **V.43**, *Am. Chem. Soc.*, Washington, D.C., 112, 136.

20. Dettre, R.H. and Johnson, Jr., R.E. (1965) *J. Phys. Chem.* **69**, 1507.

21. Shafrin, E.G. and Zisman, W.A. (1954) in *Monomolecular Layers*, Amer. Assoc. Advance Sci., Washington, D.C.

22. Baker, H.R., Shafrin, E.G. and Zisman, W.A. (1952) *J. Phys. Chem.* **56**, 405.

23. Hare, E.F. and Zisman, W.A. (1955) *J. Phys. Chem.* **59**, 335.

24. Johnson, Jr., R.E. and Dettre, R.H. (1969) in *Surface and Colloid Science* **V. 2**, E. Matijevic, ed., Wiley-Interscience, N.Y., p. 85.

25. Fowkes, F.M. and Harkins, W.D. (1940) *J. Am. Chem. Soc.* **62**, 3377.

26. Phillipoff, W., Cooke, S.R.B. and Caldwell, D.E. (1952) *Mining Eng.* **4**, 283.

27. del Giudice, G.R.M. (1936) *Eng. Mining J.* **137**, 291.

28. Bailey, A.I. (1957) in *Proc. 2nd International Cong. Surf. Activity* **V. 3**, Butterworths, London, p. 189.

29. Fort, Jr., T. (1965) in *Contact Angles, Wettability, and Adhesion*, Adv. in Chem. Ser., **V. 43**, Am. Chem. Soc., Washington, D.C., p. 302.

30. Smith, J.M. (1970) *Chemical Engineering Kinetics*, 2nd ed., McGraw-Hill, N.Y.

31. Holland, C.D. and Liapis, A.I. (1983) *Computer Methods for Solving Dynamic Separation Problems*, McGraw-Hill, N.Y.

32. Harkins, W.D. (1952) *The Physical Chemistry of Surface Films*, Reinhold, N.Y.

33. Gaines, G.L. (1966) *Insoluble Monolayers at Liquid-Gas Interfaces*, Wiley-Interscience, N.Y.
34. Cahn, J.W. (1977) *J. Chem. Phys.* **66**, 3667.
35. Teletzke, G.F., Scriven, L.E., and Davis, H.T. (1982) *J. Colloid Interface Sci.* **87**, 550.
36. Peng, D.Y. and Robinson, D.B. (1976) *Ind. Eng. Chem. Fundam.* **15**, 59.

PROBLEMS

II.1. The data on the following interfacial tensions have been compiled by Girifalco and Good (12):

hydrocarbons	γ_{hc} (hydrocarbon)	γ_{w-hc} (water-hydrocarbon)
n-hexane	18.4	51.1
n-heptane	20.4	50.2
n-octane	21.8	50.8
n-decane	23.9	51.2
n-tetradecane	25.6	50.2
cyclohexane	25.5	50.2
decalin	29.9	51.4

where γ is in dynes/cm. Predict the values of the water-hydrocarbon interfacial tensions using Fowkes' equation [II-14] after appropriate assumptions. Compare the results with the experimental data given in the table. The surface tension of water is 72.8 dynes/cm.

II-2. In the above case can a general result be quoted for water-hydrocarbon interfacial tensions? Perfluorodibutyl ether, $(C_4F_9)_2O$ and perfluorodibutyl amine, $(C_4F_9)_3N$ have surface tensions of 12.2 and 16.8 dynes/cm respectively. Their interfacial tensions with water are 51.9 and 25.6 dynes/cm. Make an appropriate assumption about the polar nature of the ether and calculate the interfacial tension at the ether- n-heptane interface. Compare with the experimental value (12) of 3.6 dynes/cm.

If a similar assumption is made for the amine, what will the interfacial tension be at the amine - n-heptane interface? Compare with the experimental value of 1.6 dynes/cm. What are the discrepancies due to?

II-3. Obtain an expression for the work of adhesion W_A for non-wetting liquids using the Zisman equation [II-20]. What will be its value W_A^* when $\gamma_L = \gamma_s^d$?

II-4. Consider a drop on a solid surface where the effects of line tension have to be accounted for. Change Eqs. [II-2] - [II-6] appropriately to accommodate the energy τL where τ is the line tension and L is the length of the contact line. If it is assumed that the drop has a profile of a spherical cap, show that the following equation results

$$\left(\frac{\partial F}{\partial A_{SL}}\right) = \gamma_{SL} - \gamma_{SV} + \gamma_{LV} \cos \hat{\lambda} + (\tau/r)$$

where r is the radius of the basal circle and $2\pi r = L$. Show that at equilibrium this contact angle $\hat{\lambda}$ is greater than λ of Eq. [II-1] if τ is positive.

II-5. It is well-known that roughness of the solid surface affects the equilibrium contact angle. Modify Eqs. [II-2] - [II-6] to include the effects of surface roughness. Note the following:

(i) The surface area of the solid-liquid interface is A_{SL} and equal to $r\,A_{SL}^*$, where r is the roughness factor and A_{SL}^* is the area of the solid-liquid interface projected on a horizontal plane.

(ii) The quantity r is a constant and $dA_{SL} = r dA_{SL}^*$.

(iii) One also has the geometric relation that $dA_{LV} = \cos \hat{\lambda}\, dA_{SL}^*$, where $\hat{\lambda}$ is the contact angle on a rough surface.

Show with a sketch the significance of $\hat{\lambda}$ and compare it to λ of Eq. [II-1]. For $r = \sec \lambda$, a phenomenon called wicking occurs. What is it from a physical point of view?

II-6. When a drop rests on a liquid surface, mutual solubilities, which are very often extremely small, can have significant roles. The following data (32) are available

Compound	γ(dynes/cm)
benzene	28.9
benzene saturated with water	28.8
water	72.8
water-benzene interface	35.0
water saturated with benzene	62.2

In an experiment a drop of dry benzene is put on the surface of pure water. First, the benzene becomes saturated with water, then the water becomes saturated with benzene. Calculate the spreading coefficients in all three stages and from the results describe the course of the physical process.

II-7. In deriving Eq. [II-39] it was necessary to express the chemical potential in terms of the concentration of the adsorbed species in the bulk solution.

(a) The solubility of n-pentane in water is so low that it cannot be measured with simple experiments. On the other hand the surface tension is significantly altered by adsorption of n-pentane. Consequently the surface tension of water is measured as a function of the pressure of pentane in the gas phase. Derive Γ as a function of the pressure in terms of the measured variables. Assume that the gas phase is ideal.

(b) Normal cetyl alcohol is non-volatile and insoluble in water. How would one determine or control its surface concentration on water? (see ref. 33).

II-8. Determine the condition for (stable) equilibrium when a fluid interface intersects a solid along an edge of the latter. Assume that the angles between the fluid interface and the solid are α in fluid A and β in fluid B (Figure II-15). Comment briefly on the cases $\alpha + \beta < \pi$, $\alpha + \beta = \pi$, and $\alpha + \beta > \pi$.

Find the range of stable α and β for an edge where $\alpha + \beta = 250°$ and α (equilibrium) $= \alpha_e = 40^0$ for a flat surface.

II-9. The π-σ relation at small values of σ is often given by the equation

$$\pi(\sigma - \sigma_o) = f(T) \qquad (1)$$

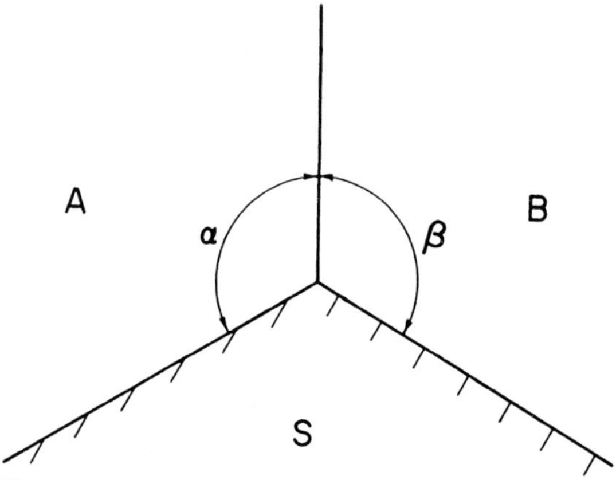

Figure II-15. Figure for Problem II-8.

where the right hand side denotes a known function of temperature. What does the constant σ_0 signify? From the answer to this question, is it possible to adapt the van der Waals equation of state to a two dimensional π-σ system? Argue that the system described by Eq. (1) is a liquid, whereas Eq. [II-40] describes a gas. Show that the adapted van der Waal equation encompasses both.

II-10. In the Gibbs adsorption equation for a two component system $d\gamma = -\Gamma_1 d\mu_1 - \Gamma_2 d\mu_2$, the left hand side is a total differential. Consequently, it is equivalent to $d\gamma = (\partial\gamma/\partial\mu_1)\, d\mu_1 + (\partial\gamma/\partial\mu_2) d\mu_2$, leading to the constraint that $(\partial\Gamma_1/\partial\mu_2) = (\partial\Gamma_2/\partial\mu_1)$. Modify or construct at least one adsorption isotherm for a two-component system such that the above constraint is satisfied.

II-11. If the Langmuir isotherm is written in the form $\Gamma = \Gamma_0 c/\ (c + b)$, show that the equation of state corresponding to Eq. [II-40] is given by $\pi = (RT/\sigma_0)\ \ell n\ [1 + (c/b)]$. The result is often called the Szyszkowski equation. Here $\sigma_0 = \Gamma_0^{-1}$.

II-12. A porous medium is sometimes modeled as a bundle of capillary tubes of circular cross section. Suppose that an air-filled porous medium is contacted with a liquid at atmospheric pressure. For what range of contact angles will liquid spontaneously enter the porous medium? When entry is not spontaneous, develop an expression for the liquid pressure required for entry in terms of the surface tension of the liquid, the contact angle, and tube diameter. In view of your answers, why do you think that fluorinated compounds are often used for waterproofing fabrics?

II-13. A porous medium with a distribution of pore sizes contains equal volumes of a wetting and a nonwetting fluid. With which fluid will most, if not all, of the small pores be filled? Why?

II-14. Contact angle hysteresis is one mechanism which hinders motion of fluid drops in a porous medium. Find an expression for the pressure drop required to initiate motion of a drop in a capillary tube of radius a if the advancing and receding contact angles of the drop material are λ_a and λ_r. Evaluate for a = 10 μm, $\lambda_a = 120^0$, $\lambda_r = 20^0$.

II-15. Consider a rough surface with a profile given by

$$\hat{z} = \hat{z}_0(1 + \cos \frac{2\pi\hat{x}}{\hat{x}_0}) \qquad \hat{x} = \frac{x}{V_0^{1/3}} \qquad \hat{z} = \frac{z}{V_0^{1/3}}$$

V_0 = volume of drop

It can be shown that when $(\hat{z}_0/\hat{x}_0) = 0.1$, Wenzel's factor r defined in Problem II-5 is 1.092. Suppose that the drop has an equilibrium contact angle λ_0 of 45° on a perfectly smooth surface.

(a) Find the apparent equilibrium contact angle $\hat{\lambda}$ on the rough surface. Also calculate the drop dimensions $\hat{h} = (h/V_0^{1/3})$ and $\hat{a} = (a/V_0^{1/3})$ at equilibrium from the following formulas for a spherical segment

$$\hat{a}^2 = \frac{\frac{6}{\pi} - \hat{h}^3}{3\,\hat{h}} \qquad \hat{r} = \frac{\hat{a}^2 + \hat{h}^2}{2\hat{h}} \qquad \sin\theta = \frac{\hat{a}}{\hat{r}}$$

where r = radius of curvature of the segment; h = maximum height of segment; a = radius on solid surface. Finally calculate the free energy $\hat{F} = (F/\gamma_{LG}V_0^{2/3})$ at equilibrium and find $(\hat{a}/\hat{x}_0)$, the ratio of drop radius to the wavelength of the roughness for the case $\hat{x}_0 = 3.6 \times 10^{-3}$.

(b) Show that the change in free energy when the contact line moves slightly along the wavy surface at constant drop volume is given by

$$d\hat{F} = d\hat{A}_{LG} - \cos\lambda_0\, d\hat{A}_{SL}$$

where $d\hat{A}_{LG} = (dA_{LG}/V_o^{2/3})$ and $d\hat{A}_{SL} = (dA_{SL}/V_o^{2/3})$

Show further that

$$d\hat{A}_{SL} = 2\pi \hat{x} \, d\hat{x} \, (1+\tan^2\alpha)^{1/2}$$

(where $\tan \alpha = (d\hat{z}/d\hat{x})$ along the rough surface)

$$d\hat{A}_{LG} = 2\pi \hat{x} \, d\hat{x} \, (1+\tan^2\alpha)^{1/2} \cos \lambda'$$

where λ' is the local angle between the solid-fluid and the liquid-gas interfaces.

(c) Combine the results of (b) to show that

$$\frac{d\hat{F}}{d\hat{x}} = 2\pi \hat{x} \left(1+4\pi^2 \left(\frac{z_o}{x_o}\right)^2 \sin^2 \frac{2\pi\hat{x}}{x_o}\right)^{1/2} (\cos(\theta+\alpha) - \cos \lambda_o)$$

where θ is the local angle of the fluid interface with the horizontal. Show that $\hat{F}$ has both a local maximum and a local minimum for $(\hat{a}/\hat{x}_o)$ between 328.5 and 329.0. Note that these extrema occur when $\lambda' = \lambda_o$, i.e., when the local contact angle has the equilibrium value λ_o.

II-16. Show that the BET isotherm, Eq. [II-36] may be rearranged to give

$$\frac{y}{v\,(1-y)} = \frac{1}{Kv_m} + \frac{(K-1)y}{Kv_m}$$

Thus a straight line can be expected if $[y/v(1-y)]$ is plotted as a function of y. The slope and intercept of this line can be used to calculate K and v_m.

III

Colloidal Dispersions

1. INTRODUCTION

Colloidal dispersions are those having particles or drops with at least one dimension greater than about 1 nm but less than about 1 μm. These systems are classified as emulsions, when a liquid phase is dispersed in a second liquid; suspensions, when a solid phase is dispersed in a liquid medium; foams, when a gas is dispersed in a liquid; or as aerosols when liquid droplets or solid particles are dispersed in a gas. Other combinations are less common.

Viscous, sticky or waxy materials are easier to dispense in the form of emulsions, as are solids in suspended form. Consequently, numerous consumer products are greatly influenced by the knowledge of how to make stable colloidal dispersions. Breaking such dispersions also has many interesting applications. In secondary oil recovery, for instance, petroleum is flushed out from the underground oil fields with water. The material that is extracted is frequently in the form of an emulsion: oil-in-water (o/w) or water-in-oil (w/o) depending on the relative amounts of the two liquids. As refinery feedstreams should be free of water, it is necessary to know how to break the emulsion into the two bulk phases.

Pollutants are also found in the form of colloidal dispersions: the haze in the atmosphere is the result of pollutants dispersed as aerosols. Oils or oily materials are emulsified in the waste water from refineries or chemical plants and have to be removed before discharging the effluent into rivers or seas.

These examples illustrate the diversity and widespread occurrence of colloidal dispersions in systems of practical interest. It is thus pertinent to consider their behavior and properties from a fundamental viewpoint.

Most colloidal dispersions are unstable from a thermodynamic point of view. That is, system free energy would be reduced if the dispersed material were collected into a single bulk phase. If the surface free energy is of the order of 20 ergs/cm^2, which corresponds to γ = 20 dynes/cm, one has a surface free energy of [20 (6/d) ϕ] ergs/cm^3. Here d is the particle diameter and of the order of 1 μm and ϕ is the volume fraction and of the order of 0.1. With these values the interfacial free energy is around 1.2 x 10^5 ergs/cm^3 or, in terms of the thermal energy per particle, 1.52 x 10^7 kT at 27°C. Such a value constitutes an overwhelming and positive contribution to the free energy of formation of the dispersion. Consequently the system attempts to go back into separate bulk phases. It is a matter of experience, however, that some colloidal dispersions are stable from a practical point of view in that they remain unchanged upon standing for time periods of weeks, months, or even years. Technically, they are metastable. It is possible, in principle, for them to attain a lower energy state, but a large energy barrier must be overcome before this state can be reached. If the energy barrier is sufficiently high, the dispersed state can be maintained almost indefinitely. Of course, not all colloidal dispersions are stable in this sense. Many are patently unstable and separate quickly upon standing.

For many reasons it is important to understand the causes of stability in colloidal systems. Manufacturers of cosmetics and paints, for example, want long-term stability so that their products do not separate into two or more bulk phases during the time period between manufacture and use. The operator of an oil field, on the other hand, needs to know how to rapidly destabilize the emulsion of oil and water produced by some of his wells.

Whether a colloidal dispersion is stable depends ultimately on interaction among the particles or drops. In the following sections we review the main types of interaction, present some equations for predicting their magnitude, and show how to use these equations to determine conditions for stability of colloidal dispersions. Other important aspects of colloid science, in particular experimental techniques such as light scattering, neutron scattering, sedimentation, and viscometry, are omitted here because they are not required as background for later chapters.

Information on these topics may be found in several of the general references listed at the end of this chapter.

First, a brief comment on terminology. When a dispersion is unstable, "flocculation" occurs. That is, the particles or drops aggregate to form clusters which move as a unit. If the dispersed phase is fluid, the drops in a cluster may coalesce to form a single drop. It is possible to have flocculation without coalescence in such systems.

2. ATTRACTIVE FORCES

Attractive forces between particles favor flocculation and oppose stability. Of prime interest in colloidal systems are attractive forces between particles due to attraction between the individual molecules comprising each particle. We consider here only London-van der Waals forces which are the chief source of attraction between molecules for nonpolar or slightly polar materials. These forces stem from induced dipole-induced dipole interactions. That is, a dipole which develops transiently in one molecule as a result of continual motion of its electrons induces a dipole in the second molecule. Although the time-averaged dipole moments are zero, the time-averaged interaction energy is not. The expression for the potential energy of interaction between a unit volume of species 1 and another of species 2 has the form which has been discussed in Chapter II:

$$\phi_{12} = - \frac{n_1 n_2 \beta_{12}}{r^6} \qquad \text{[III-1]}$$

where n_1 and n_2 are the moles per unit volume of species 1 and 2, β_{12} is a constant, and r is the distance between the two molecules. For large separation distances (a few tens of nanometers) "retardation effects" must be considered, i.e., the time required for electromagnetic radiation to travel between interacting molecules. Under these conditions it can be shown that ϕ_{12} varies inversely with r^7. A more extensive but nevertheless simplified discussion of the origin of London-van der Waals forces is given by Hiemenz (1).

The attractive interaction between colloidal particles is, to a first approximation, the sum of the attractive interaction between all pairs of molecules chosen such that one molecule of the pair is in each particle. The simplest situation is that shown in Figure III-1 in which two semi-infinite bodies of material are separated by a gap of thickness h. We

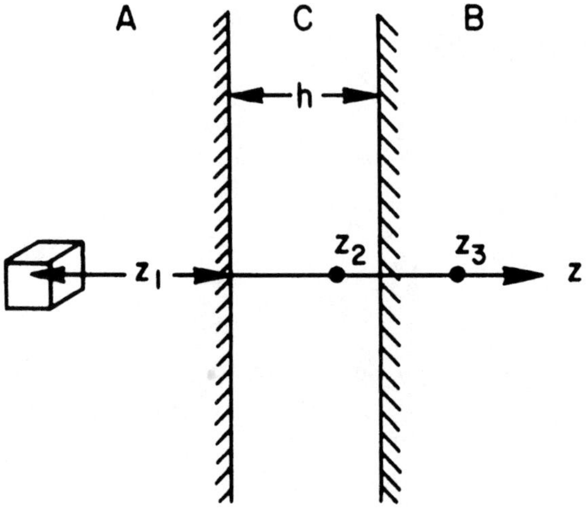

Figure III-1. Interacting semi-infinite slabs A and B with a separating medium C.

suppose in our initial derivation that the two bodies are made up of the same pure substance and that the gap is empty.

Let us first calculate the total energy of interaction ϕ_B between a unit volume in region A on the z-axis at coordinate $(-z_1)$, as shown in Figure III-1, and the entire region B. Using a cylindrical coordinate system, we find

$$\phi_B = -\int_h^\infty \int_o^\infty \frac{2\pi n^2 \beta \ r \ dr \ dz}{[r^2+(z+z_1)^2]^3} \qquad \text{[III-2]}$$

It is convenient to integrate first with respect to r. The result is

$$\phi_B = -\frac{\pi n^2 \beta}{2} \int_h^\infty \frac{dz}{(z+z_1)^4}$$

Integrating with respect to z, we obtain

$$\phi_B = -\frac{\pi n^2 \beta}{6(h+z_1)^3} \qquad \text{[III-3]}$$

Eq. [III-3] applies for a unit volume in region A. If we integrate this result over all of region A, i.e., between $z_1 = 0$ and $z_1 \to \infty$, we obtain the total interaction energy between the two semi-infinite regions. Actually it is convenient to calculate the energy ϕ_{AB} per unit area

$$\phi_{AB} = \lim_{R\to\infty} \frac{1}{\pi R^2} \int_0^R \int_0^\infty 2\pi r \phi_B \, dr \, dz_1 \qquad [III-4]$$

Carrying out the integration, we find that

$$\phi_{AB} = -\frac{\pi n^2 \beta}{12h^2} = -\frac{A_H}{12\pi h^2} \qquad [III-5]$$

Here the Hamaker constant A_H is given by $(\beta\pi^2 n^2)$. It is a key parameter characterizing the strength of the attractive forces.

It is noteworthy that the attractive interaction between two large bodies decreases much more slowly as they are separated than does the interaction between two molecules. According to Eqs. [III-1] and [III-5], the interaction is proportional to the inverse square of the separation distance in the former case and the inverse sixth power in the latter case. For colloidal particles London-van der Waals interactions are significant for separation distances h up to about 100 nm (1000 Å) .

Hamaker (2) used a similar procedure to derive an expression for the interaction energy between two spherical particles. He found that the attractive energy ϕ^S_{AB} for two identical spheres of radius R with centers a distance (2R+h) apart is given by

$$\phi^S_{AB} = -\frac{A_H}{12} \left[\frac{1}{\xi^2+2\xi} + \frac{1}{\xi^2+2\xi+1} + 2 \ln \frac{\xi^2+2\xi}{\xi^2+2\xi+1} \right] \qquad [III-6]$$

where $\xi = (h/2R)$. When the spheres are close together so that $\xi \ll 1$, Eq. [III-6] simplifies to

$$\phi^S_{AB} = -\frac{A_H R}{12h} \qquad [III-7]$$

In this last case note that the decrease in interaction energy with separation distance h is even slower than that given by Eq. [III-5] for semi-infinite bodies. In the opposite extreme where $(h/R) \gg 1$, Eq.[III-6] predicts that ϕ^S_{AB} is proportional to h^{-6}, as would be expected from the fundamental attraction law, Eq. [III-1].

Another useful result is that for two plates of finite thickness d separated by a distance h. The derivation leading to Eq. [III-5] is readily modified for this case and yields the following expression for the interaction energy ϕ^p_{AB}:

$$\phi^p_{AB} = -\frac{A_H}{12\pi} \left[\frac{1}{h^2} + \frac{1}{(h+2d)^2} - \frac{2}{(h+d)^2} \right] \qquad [III-8]$$

Clearly Eq. [III-8] simplifies to Eq. [III-5] in the limit of very thick plates, where $d \to \infty$. Finally Eqs. [III-6] and [III-7] can be generalized to the case of spheres of unequal radii R_1 and R_2. The results are

$$\phi_{AB}^{S} = -\frac{A_H}{12}\left[\frac{b}{\zeta^2+(b+1)\zeta} + \frac{b}{\zeta^2+(b+1)\zeta+b} + 2\ln\frac{\zeta^2+(b+1)\zeta}{\zeta^2+(b+1)\zeta+b}\right] \quad \text{[III-9]}$$

where $\zeta = (h/2R_1)$ and $b = (R_2/R_1)$, and

$$\phi_{AB}^{S} = -\frac{A_H R_1 R_2}{6h(R_1+R_2)} \quad \text{for } h \ll R_1, R_2 \quad \text{[III-10]}$$

Numerical evaluation of attractive forces using the above formulas requires that a value be assinged to the Hamaker constant A_H. Hiemenz (1) uses an order of magnitude argument to estimate β and concludes that A_H should be in the range of 10^{-13} to 10^{-12} erg (10^{-20} to 10^{-19} J) for most materials.

A complicating factor so far ignored is the effect of the medium separating the two particles. We return to the situation of Figure III-1. Let us find the excess energy of a unit volume in region A at a position $(-z_1)$ compared to a unit volume in an infinite bulk phase of A. We imagine starting with such an infinite phase. Now we replace molecules of A by those of B in the region $z > h$ and molecules of A by those of C in the region $0 < z < h$. Using the derivation leading to Eqs. [III-2] and [III-3] we find that the excess energy of our unit volume caused by these changes is

$$\phi_A' = -\frac{\pi n_A(n_C\beta_{AC} - n_A\beta_{AA})}{2}\int_0^h \frac{dz}{(z+z_1)^4} - \frac{\pi n_A(n_B\beta_{AB} - n_A\beta_{AA})}{2}\int_h^\infty \frac{dz}{(z+z_1)^4}$$

$$= -\frac{\pi n_A(n_C\beta_{AC} - n_A\beta_{AA})}{6\,z_1^3} - \frac{\pi n_A(n_B\beta_{AB} - n_A\beta_{AA})}{6\,(h+z_1)^3} \quad \text{[III-11]}$$

Similar expressions can be derived for a unit volume of C at position z_2 and a unit volume of B at position z_3 of Figure III-1. The results are

$$\phi_C' = -\frac{\pi n_C(n_A\beta_{AC} - n_C\beta_{CC})}{6\,z_2^3} - \frac{\pi}{6}\,\frac{n_C(n_B\beta_{BC} - n_C\beta_{CC})}{(h-z_2)^3} \quad \text{[III-12]}$$

$$\phi_B' = -\frac{\pi}{6}\,\frac{n_B(n_A\beta_{AB} - n_B\beta_{BB})}{z_3^3} - \frac{\pi}{6}\,\frac{n_B(n_C\beta_{BC} - n_B\beta_{BB})}{(z_3-h)^3} \quad \text{[III-13]}$$

We now wish to integrate Eqs. [III-11] - [III-13] for z_1, z_2, and z_3 having ranges corresponding to regions A, C, and B respectively of Figure III-1. We also want to calculate ϕ_{ACB}, the excess energy of the actual situation compared to that where the thickness h of region C becomes very large. Now ϕ_{ABC} is the sum of the three expressions obtained by integrating Eqs. [III-11] - [III-13] less the corresponding sum of these integrals in the limit $h \to \infty$ with the final result halved to correct for double counting of the interaction energy between each pair of molecules in separate regions. Using this procedure, we find that

$$\phi_{ACB} = -\frac{1}{12\pi h^2}\left[\pi^2 n_A n_B \beta_{AB} + \pi^2 n_C^2 \beta_{CC} - \pi^2 n_A n_C \beta_{AC} - \pi^2 n_B n_C \beta_{BC}\right] \qquad \text{[III-14]}$$

The expression in brackets in this equation is the effective Hamaker constant for two particles of A and B separated by a medium C.

If both particles are of the same material A, Eq. [III-14] simplifies to

$$\phi_{ACA} = -\frac{1}{12\pi h^2}\left[A_{HAA} + A_{HCC} - 2A_{HAC}\right] = -\frac{(A_H)_{eff}}{12\pi h^2} \qquad \text{[III-15]}$$

For London-van der Waals forces a frequently used approximation is

$$\beta_{AC} \cong (\beta_{AA}\beta_{CC})^{1/2} \qquad \text{[III-16]}$$

With this approximation [Eq. III-15] becomes

$$\phi_{ACA} \cong -\frac{1}{12\pi h^2}\left[A_{HAA}^{1/2} - A_{HCC}^{1/2}\right]^2 \qquad \text{[III-17]}$$

In this case the effective Hamaker constant is always positive, i.e., the interaction between like particles is an attractive one. But the strength of the attraction is less than if the dispersing medium C were absent, and effective Hamaker constants not far above 10^{-14} erg (10^{-21} J) have been reported. We note that effective Hamaker constants of the forms indicated in Eqs. [III-14] and [III-15] may be used in any of the expressions Eqs. [III-5] - [III-10] to account for the continuous phase C of a colloidal dispersion.

Since about 1970 considerable attention has been given to a different and more rigorous method for calculating attractive forces. Sometimes called Lifshitz theory, this method assumes that particle separation is sufficiently large that each appears as a continuous medium to the other.

Thus, concepts from the continuum theory of electrodynamics can be employed. We will not discuss here details of the theory. A readable review of the subject has been given by Parsegian (3). it is important to note, however, that with Lifshitz theory the attractive forces can be calculated from experimental spectral data that are obtainable although not yet available for all substances of interest. Thus, the greatest problem with using the simpler Hamaker procedure described above, viz., the uncertainty in what value to use for the Hamaker constant, is avoided.

Parsegian (3) has presented calculations for the polystyrene-water and polystyrene-vacuum systems for which the required spectral data are available. For simplicity he considered the situation of Figure III-1 with polystyrene in the two semi-infinite regions A and B. Results of his calculations are shown in Figures III-2 to III-4. As Figure III-2 shows, the attractive energy is about an order of magnitude less when the space between the polystyrene particles is occupied by water instead of a vacuum, at least for separation distances below 1000 Å, which is the range of greatest interest.

Figure III-3 shows that attraction is even less in salt water than in pure water. Since the Hamaker constant depends on separation distance, the form of Eq. [III-5] obtained by Hamaker's method is not quite correct. Figure III-4 makes this point more clearly by showing the power of h required in Eq. [III-5] to force it to agree with the results of Lifshitz theory. Note that in the vacuum case the exponent increases from about 2 to nearly 3 as the separation distance increases from 50 to 1000 Å. Indeed, Eq. [III-5] is often used with h^2 replaced by h^3 for large separation distances since the latter dependence is obtained when the integration is performed with the retardation effect mentioned above included. When pure water is the separating medium, we see from Figure III-4 that Eq. [III-5] should not be corrected in this way for retardation. And the curve for salt water indicates that, except for small separation distances, Eq. [III-5] is not a very good approximation with any power of h. We note that the salt water solution in these calculations was 0.1 M NaCl.

3. ELECTRICAL INTERACTION

An electrical interaction force exists between colloid particles which plays an important role in colloid stability. It is found that solid

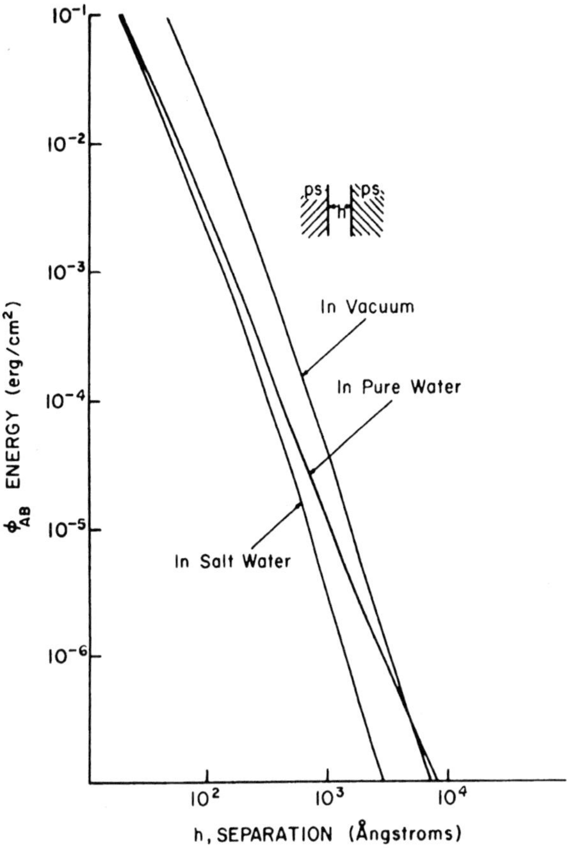

Figure III-2. Interaction potential of two semi-infinite polystyrene media, separated by vacuum, pure water and 0.1 M salt solution. Reprinted from Ref. (3) with permission.

surfaces electrify in presence of an electrolyte solution. This behavior is mainly due to unequal adsorption of the electrolyte ions on the surface, giving it a net positive charge or a net negative charge. The charge varies with electrolyte content of the solution. For instance, silver iodide particles are positively charged in the presence of a large excess of silver ions and negatively charged in the presence of a large excess of iodide ions. In more precise terms, it is the electrical potential at the particle surfaces which is determined by the electrolyte composition in this case, and silver and iodide ions are called "potential - determining ions." For metal oxide particles the surface potential varies with

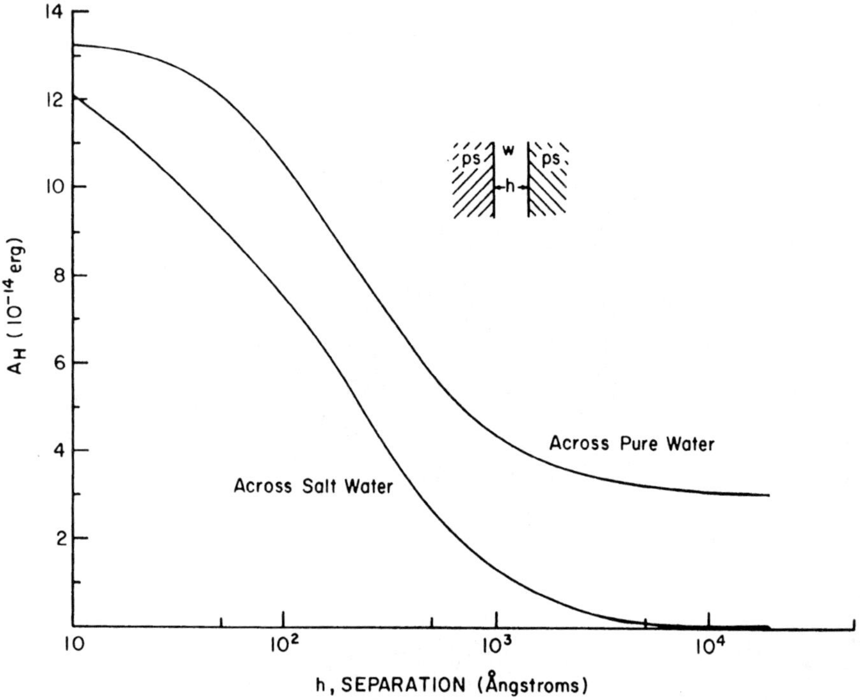

Figure III-3. Hamaker constant calculated from Eq. [III-5] and the results shown in Figure-III-2. Reprinted from Ref. (3) with permission.

solution pH, and H^+ is the potential determining ion. For oil-in-water emulsions the sign of the surface charge is that of adsorbed surface-active ions, e.g., negative for ionized carboxylic acids.

A charged surface attracts counterions from the solution while repelling ions of like sign. The result is a layer near the surface which has a net charge. Although this layer is diffuse, i.e., starts at the wall and extends outward into the fluid, its main portion is confined to the region close to the surface. The local charge separation gives rise to a capacitance, and the system can be modeled as a capacitor made up of the charged interface and an infinitely thin layer of counter-ions (4), hence the naming of this region the "electrical double layer." Since the charge separation cannot be measured, indirect means must be used to characterize the phenomenon as discussed below.

We begin with the simple situation shown in Figure III-5 with two infinite parallel plates having uniform surface potentials ψ_0 separated by

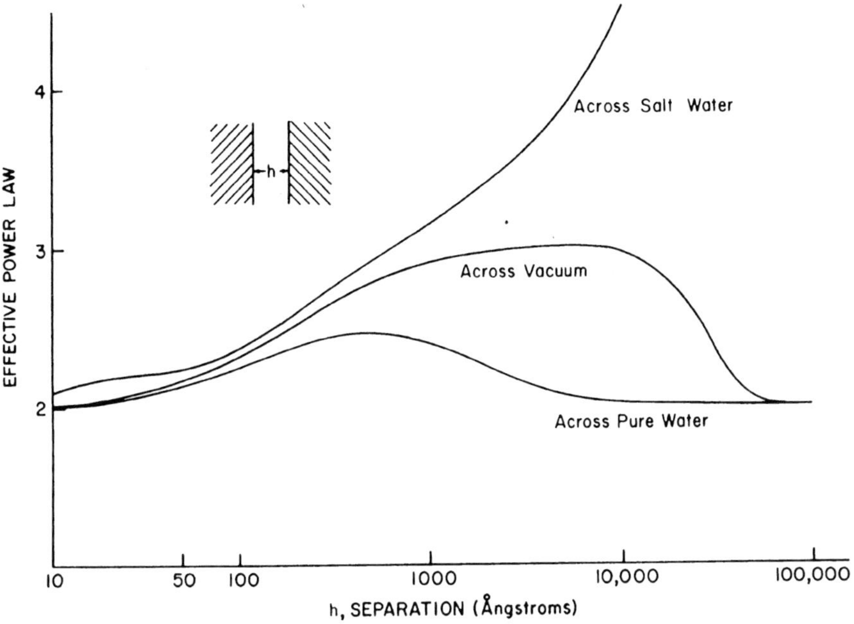

Figure III-4. Effective power (n) law calculated from a modified-version of Eq. [III-5], i.e., $\phi_{AB} = - A_H / 12\pi h^n$ and using the values from Figure III-2. The value for n as a function of separation distance h is chosen to keep A_H constant. Reprinted from Ref. (3) with permission.

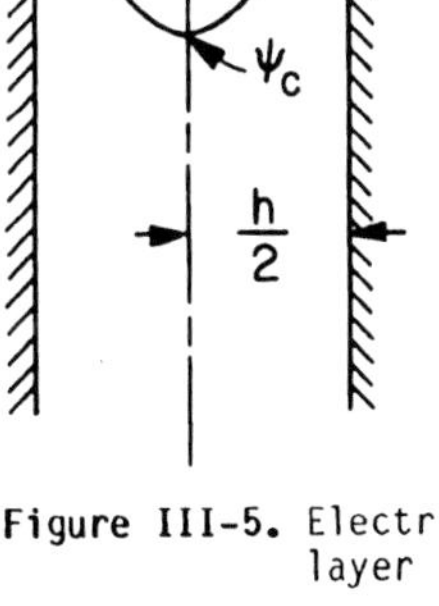

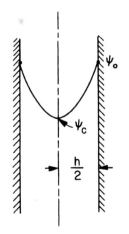

Figure III-5. Electrical potential distribution in an interacting double layer between two identical slabs.

a distance h. When h is small enough, the double layers of the two plates overlap and interaction occurs. At equilibrium the net force on each small element of fluid such as that shown in Figure III-6 must be zero. That is, pressure forces acting on the two surfaces of area A perpendicular the z-axis must balance the electrical body force, which is the product of the net electrical charge of the element and the local electrical field. This balance implies

$$pA|_z - pA|_{z+\Delta z} + (\rho_e A\Delta z)\left(-\frac{d\psi}{dz}\right) = 0 \qquad \text{[III-18]}$$

where ρ_e is the local free charge density, ψ is the electrical potential, and $(-d\psi/dz)$ is the electric field. Dividing by $(A\Delta z)$ and taking the limit as $z \to 0$, we find

$$\frac{dp}{dz} + \rho_e \frac{d\psi}{dz} = 0 \qquad \text{[III-19]}$$

Hence pressure varies with position in the electrical double-layer region where an electrical body force exists just as it varies with position in a static pool of liquid where the gravitational body force exists. In the latter case pressure increases with depth in accordance with the familiar rules of hydrostatics.

One of the governing equations in the electrical double-layer region is Poisson's equation, which can be derived from Coulomb's law of electrostatics (5):

$$\frac{d^2\psi}{dz^2} = -4\pi \frac{\rho_e}{\varepsilon} \qquad \text{[III-20]}$$

where ε is the dielectric constant and use of cgs units is implied. Substitution of Eq. [III-20] into Eq. [III-19] yields

$$\frac{dp}{dz} = \frac{\varepsilon}{4\pi}\frac{d\psi}{dz}\frac{d^2\psi}{dz^2} = \frac{\varepsilon}{8\pi}\frac{d}{dz}\left[\left(\frac{d\psi}{dz}\right)^2\right] \qquad \text{[III-21]}$$

In order to evaluate ψ, we find from Eq. [III-20] that knowledge of ρ_e is necessary. It can be obtained from the Boltzmann distribution for a symmetric electrolyte with bulk concentration c_0:

$$n_+ = N_A\, c_0 \exp(-e_0 \nu\psi/kT), \quad n_- = N_A c_0 \exp(e_0 \nu\psi/kT)$$

Here n_+ and n_- are the numbers of cations and anions per unit volume

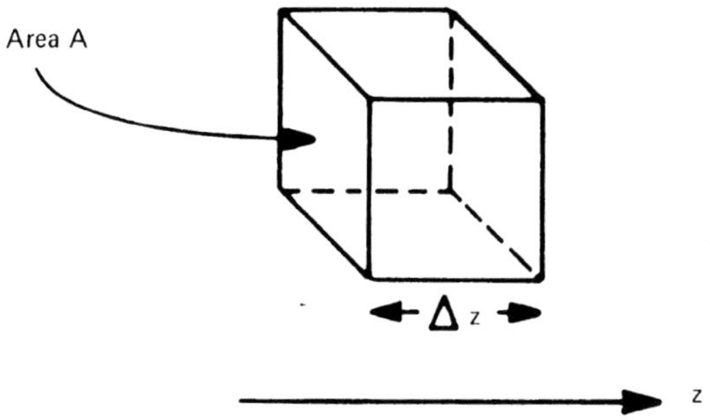

Figure III-6. A small rectangular element in a fluid is shown. The pressure forces in the z-direction act on the surface of area A. The electrical body forces are proportional to the volume ($A\Delta z$).

and ψ has been taken as zero in the bulk solution. Problem III-9 indicates how these expressions can be derived. Since $\rho_e = \nu e_0 n_+ - \nu e_0 n_-$, one has

$$\rho_e = -2\nu N_A c_0 e_0 \sinh(e_0 \nu\psi/kT) \qquad \text{[III-22]}$$

and the resulting Poisson-Boltzmann equation is

$$\frac{d^2u}{dz^2} = \left(\frac{8\pi\nu^2 e_0^2 N_A c_0}{\varepsilon kT}\right) \sinh u = \kappa^2 \sinh u \qquad \text{[III-23]}$$

$$u = \frac{\nu e_0 \psi}{kT} \qquad \text{[III-24]}$$

with ν the magnitude of ion valence, e_0 the electronic charge, κ^{-1} the Debye length, a measure of double layer thickness, N_A Avogadro's number, k Boltzmann's constant, and T the absolute temperature. Now if we multiply Eq. [III-23] by (du/dz) and rearrange, we find

$$\frac{d}{dz}\left[\left(\frac{du}{dz}\right)^2\right] = 2\kappa^2 \sinh u \frac{du}{dz} \qquad \text{[III-25]}$$

We note that this equation applies not only for the present case of two interacting plates but for individual plates as well. Its integration in the latter case to obtain the potential distribution in the electrical double layer region near a single plate is considered in Problem III-2. Some useful formulas applicable to double layers near single plane and spherical surfaces are given in Table I.

Table I. Formulas for Electrical Double Layers Near Single Particles

Plane Interface

(1) Gouy-Chapman solution for a symmetric electrolyte of valence ν

$$u = 2\,\ell n\left(\frac{1+\eta e^{-\kappa z}}{1-\eta e^{-\kappa z}}\right)$$

$$\sigma = \frac{\varepsilon\kappa kT}{2\pi\nu e_0}\sinh\left(\frac{\nu e_0\psi_0}{2kT}\right)$$

(2) Small potential approximation for general electrolyte mixture with κ given by Eq. [III-42]

$$\psi = \psi_0 e^{-\kappa z}$$

$$\sigma = \frac{\varepsilon\kappa\psi_0}{4\pi}$$

Spherical Interface (Double Layer Outside Spherical Particle)

(1) Small potential (Debye-Huckel) approximation for general electrolyte as above

$$\psi = \frac{\psi_0 R}{r}\, e^{-\kappa(r-R)}$$

$$\sigma = \frac{\varepsilon\psi_0}{4\pi R}(1+\kappa R)$$

(2) Result for $\kappa R > 1$ including first correction to Gouy-Chapman solution to account for curvature (7). Note that no exact solution is available for the general Poisson-Boltzmann equation ($\nabla^2 u = \kappa^2 \sinh u$) in spherical coordinates

$$\sigma = \frac{\varepsilon\kappa kT}{2\pi\nu e_0}\left[\sinh\left(\frac{\nu e_0\psi_0}{2\kappa T}\right) + \frac{2\eta}{\kappa R}\right]$$

Combining Eqs. [III-21] and [III-25] and making use of Eq. [III-24], we obtain the pressure distribution in the region between interacting plates:

$$\frac{dp}{dz} = \frac{\varepsilon\kappa^2}{4\pi}\left(\frac{kT}{\nu e_0}\right)^2 \sinh u \frac{du}{dz} \qquad \text{[III-26]}$$

Integrating between the centerline position z = 0 in Figure III-5 where $u = u_c$ and $p = p_c$ and the bulk phase outside the double layer where u = 0 and $p = p_b$, we obtain the following result:

$$p_c - p_b = \frac{\varepsilon\kappa^2}{4\pi}\left(\frac{kT}{\nu e_0}\right)^2 (\cosh u_c - 1) \qquad \text{[III-27]}$$

If the first limit of integration is the particle surfaces where $u = u_o$ and $p = p_o$ instead of the centerline position, the result is

$$p_o - p_b = \frac{\varepsilon\kappa^2}{4\pi}\left(\frac{kT}{\nu e_0}\right)^2 (\cosh u_o - 1) \qquad \text{[III-28]}$$

Thus, fluid pressure at both the centerline and the particle surfaces exceeds the bulk phase pressure, an indication that the interaction due to overlap of double layers having the same charge is a repulsive one, the expected result.

In the next section we shall look at the combined effect of attractive and electrical interaction. For this purpose it is desirable to know the potential energy E_e of electrical interaction. Let F_e be the force per unit area in excess of bulk fluid pressure p_b which the fluid exerts on the particle surfaces of Figure III-5 with a positive F_e denoting a repulsive force. It is known from electrostatics that

$$F_e = (p_o - p_b) - \frac{\varepsilon}{8\pi}\left(\frac{d\psi}{dz}\right)^2 \Big|_{z = \frac{h}{2}} \qquad \text{[III-29]}$$

The last term in this equation is an electrical stress (negative pressure) which can be viewed as resulting from the attraction between dipoles induced in the solvent by the electric field and lying adjacent to one another in the direction of the field, here the z-direction. The electrical stress term can be evaluated by integrating Eq. [III-21] between the centerline z = 0 and the particle surface z = (h/2), noting that $(d\psi/dz)$ vanishes at the centerline. When the result is substituted into Eq. [III-29], we find

$$F_e = p_c - p_b \qquad [III-30]$$

The potential energy E_e per unit area is related to F_e by

$$\frac{dE_e}{dh} = - F_e \qquad [III-31]$$

If the energy is taken as zero when the separation distance between the particles is large, integration of Eq. [III-31] yields

$$E_e = - \int_{\infty}^{h} F_e \, dh = \int_{h}^{\infty} F_e \, dh \qquad [III-32]$$

We see from Eqs. [III-31] and [III-27] that knowledge of centerline potential u_c is needed for evaluation of F_e and E_e. The potential distribution between parallel plates can be determined exactly in terms of elliptic integrals (4,6). It is more convenient, however, to use simpler expressions, and we present two approximate solutions.

The first approximation applies when surface potential ψ_o is small in magnitude, i.e., $u \ll 1$. In this case the Poisson-Boltzmann equation [III-23] may be linearized by expanding the hyperbolic function and retaining only the first term of the series:

$$\frac{d^2\psi}{dz^2} = \kappa^2\psi \qquad [III-33]$$

This differential equation is readily solved. Since the potential distribution is symmetric about the centerline, the solution may be written as

$$\psi = \psi_o \frac{\cosh \kappa z}{\cosh (\kappa h/2)} \qquad [III-34]$$

In this case Eq. [III-27] simplifies to

$$p_c - p_b \cong \frac{\varepsilon\kappa^2\psi_c^2}{8\pi} \quad . \qquad [III-35]$$

From Eqs. [III-30], [III-34], and [III-35] we have

$$F_e = p_c - p_b = \frac{\varepsilon\kappa^2\psi_o^2}{8\pi \cosh^2(\kappa h/2)} \qquad [III-36]$$

If ψ_o is independent of separation distance h, we invoke Eq. [III-32] to find

$$E_e = \frac{\varepsilon\kappa\psi_0^2}{4\pi} [1 - \tanh(\kappa h/2)] \qquad \text{[III-37]}$$

It is clear from this equation that electrical interaction is significant only when the separation distance h does not greatly exceed the Debye length κ^{-1}. The latter is about 10nm (100 Å) in a 0.001 M aqueous solution of a univalent electrolyte at room temperature and about 1nm (10 Å) in a 0.1 M solution. Thus, double layer thickness decreases with increasing electrolyte concentration.

The surface charge density σ can be obtained from the potential distribution by invoking a basic boundary condition of electrostatics which can be obtained by integrating Poisson's equation, Eq. [III-20], over a thin "pillbox" control volume such as that of Figure I-3(5). In the absence of an electric field within the colloidal particles themselves it takes the following form for the present situation

$$\sigma = \frac{\varepsilon}{4\pi} \left(\frac{d\psi}{dz}\right)\Big|_{z = \frac{h}{2}} \qquad \text{[III-38]}$$

We note that by integrating Poisson's equation between the centerline at z = 0 and the particle surface at z = (h/2), we can demonstrate that the overall free charge built up in the solution is equal in magnitude but opposite in sign to the total surface charge. That is, the condition of global electroneutrality is satisfied.

Substitution of Eq. [III-34] into Eq. [III-38] yields

$$\sigma = \frac{\varepsilon\kappa\psi_0}{4\pi} \tanh(\kappa h/2) \qquad \text{[III-39]}$$

Thus, we may rewrite Eq. [III-36] as

$$F_e = p_c - p_b = \frac{2\pi\sigma^2}{\varepsilon \sinh^2(\kappa h/2)} \qquad \text{[III-40]}$$

If σ is independent of h, integration in accordance with Eq. [III-32] yields

$$E_e = \frac{4\pi\sigma^2}{\varepsilon\kappa} [(\coth(\kappa h/2) - 1)] \qquad \text{[III-41]}$$

The constant potential and constant charge density results given by Eqs. [III-37] and [III-41] are limiting cases of behavior which may occur as colloidal particles approach one another. In the constant potential case the approach is slow enough that equilibrium of the potential-determining ion is maintained between surface and bulk solution. Adsorption or desorption occurs as necessary to maintain the equilibrium potential ψ_0. The opposite extreme is the constant charge density case where the particles approach so rapidly that no adsorption or desorption has time to occur. Clearly, intermediate situations are possible as well when the time constant for adsorption or desorption and double-layer relaxation is comparable to the approach time of the particles.

It is readily shown that Eqs. [III-34] - [III-41] apply to two identical particles of small surface potential even when the bulk solution contains anions and cations of different valence. In this case the inverse Debye length κ is given by

$$\kappa^2 = \frac{4\pi e_0^2 N_A}{\varepsilon kT} \sum_i \nu_i^2 c_{io} \qquad \text{[III-42]}$$

where the sum is over all ionic species in the bulk solution whose valences are ν_i and bulk concentrations c_{io}.

The other widely used approximation for calculation of u_c in Eq. [III-27] is not restricted to small surface potentials, but it is restricted to small degrees of overlap of the double layers between the two particles. In this case the centerline potential is assumed to be small ($u_c \ll 1$) and to be approximately equal to the sum of the potentials due to the two double layers individually, neglecting interaction. The solution of the Poisson-Boltzmann equation for a single double layer can be used to develop the following expression applicable for small u_c:

$$u_c = 8\eta\, e^{-\kappa h/2} \qquad \text{[III-43]}$$

where $\eta = \tanh(\nu e_0 \psi_0 / 4kT)$. Combining this equation with Eqs. [III-27] and [III-30], we find

$$F_e \cong \frac{8\varepsilon\kappa^2}{\pi} \left(\frac{kT}{\nu e_0}\right)^2 \eta^2 e^{-\kappa h} \qquad \text{[III-44]}$$

Using Eq. [III-32], we can calculate the interaction energy E_e when the surface potential ψ_0 is independent of separation distance:

$$E_e = \frac{8\varepsilon\kappa}{\pi} \left(\frac{kT}{\nu e_0}\right)^2 \eta^2 e^{-\kappa h} = \frac{64 N_A c_0 kT}{\kappa} \eta^2 e^{-\kappa h} \qquad \text{[III-45]}$$

No exact general solution exists for the interaction between electrical double layers of spherical particles. Deryagin has developed an approximation useful when particle radius R is much greater than the Debye length κ^{-1}, i.e., $\kappa R \gg 1$. In this case the radius of curvature of each surface is much greater than the separation distance between surfaces when the particles are close enough to interact appreciably. Deryagin's idea was to model each spherical surface as a central flat disk and a series of surrounding flat rings as indicated in Figure III-7. The equations developed above for flat surfaces could then be used to calculate forces between the central disks of the two surfaces and between the various pairs of corresponding rings. When the small overlap approximation given by Eq. [III-44] is used with Deryagin's basic scheme to calculate these forces, the total repulsive force F_e^S between the spheres is found to be

$$F_e^S = \frac{64\pi R c_o N_A k T}{\kappa} \eta^2 e^{-\kappa h} \qquad \text{[III-46]}$$

If the surface potential of the two spheres remains constant during their approach, the corresponding energy of interaction is

$$E_e^S = \frac{64\pi R c_o N_A k T}{\kappa^2} \eta^2 e^{-\kappa h} = 8R\varepsilon \left(\frac{kT}{\nu e_o}\right)^2 \eta^2 e^{-\kappa h} \qquad \text{[III-47]}$$

We note that F_e^S and E_e^S have units of force and energy since they represent

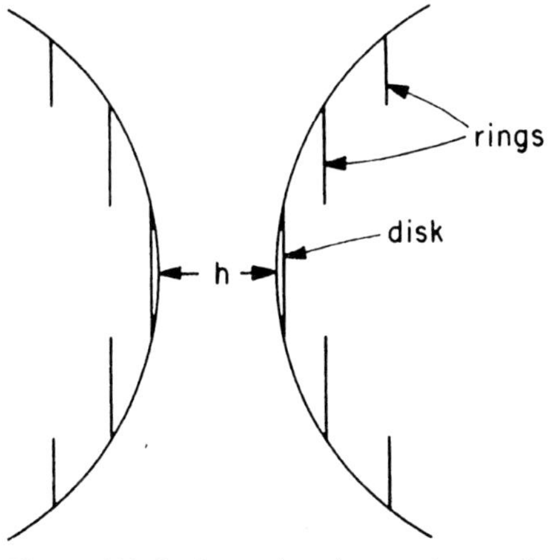

Figure III-7. Approximation used to calculate the electrical energy of interaction between two spheres by a central disk and surrounding rings for $(h/R) \ll 1$.

the entire interaction between spheres. In contrast, F_e and E_e for the parallel plate case are the force and energy of interaction per unit area.

Deryagin's approach and the small overlap approximation have also been used for the case of two spheres having different radii and surface potentials. The resulting interaction energy is found to be

$$E_e^S = 16\varepsilon \left(\frac{kT}{\nu e_o}\right)^2 \left(\frac{R_1R_2}{R_1+R_2}\right) \eta_1 \eta_2 \, e^{-\kappa h} \qquad \text{[III-48]}$$

As before, the surface potentials have been taken as independent of separation distance h. Note that E_e^S is positive and the interaction is repulsive when η_1 and η_2 have the same sign. In this case both surfaces have charges of the same sign. But, as might be expected, the interaction energy is negative indicating an attractive interaction when η_1 and η_2 have opposite signs, so that one surface has a positive charge and the other a negative charge.

Finally, for completeness, we give the corresponding electrical interaction energy E_e per unit area between flat plates with different surface potentials. The small overlap approximation is employed to obtain:

$$E_e^p = 64 \frac{c_0 N_A kT}{\kappa} \eta_1 \eta_2 \, e^{-\kappa h} \qquad \text{[III-49]}$$

We emphasize that Eqs. [III-43] - [III-49] apply only for the case of a symmetric electrolyte. But they are useful for practical purposes when the valence ν is that of the counterion, i.e., the ion whose charge is opposite to that of the surface.

So far it has been implicitly assumed that the Poisson-Boltzmann equation [III-23] applies throughout the liquid phase near the solid-liquid interface and that ψ_0 is the potential at the interface itself. A major limitation of this approach is that Eq. [III-23] neglects the finite size of ions and thus ignores the fact that ion centers can approach no closer to the interface than the ionic radius. A first approximation to account for this effect involves dividing the double layer region into two parts: a "compact" or "Stern" layer adjacent to the interface which is devoid of ions and which has a thickness of the order of molecular dimensions, and a "diffuse" layer in which Eq. [III-23] applies. With this model the potential ψ_0 of the above equations is that at the (imaginary) surface separating the compact and diffuse layers (see Problem III-3).

The approximate experimental determination of ψ_0 is based on measurement of the velocity of a charged particle in a solvent subjected to an applied voltage. Such a particle experiences an electrical force which initiates motion. Since a hydrodynamic frictional force acts on the particle as it moves, a steady state is reached with the particle moving with a constant velocity U. To calculate this electrophoretic velocity U theoretically, it is, in general, necessary to solve Poisson's equation, Eq. [III-20] and the governing equations for ion transport subject to the condition that the electric field is constant far away from the particle. The appropriate viscous drag on the particle can be calculated from the velocity field and the electrical force on the particle from the electrical potential distribution. The fact that the sum of the two is zero provides the electrophoretic velocity U. Actual solutions are complex, and the electrical properties of the particle, e.g., polarizability, conductivity, surface conductivity, etc., come into play. Details are given by Levich (8).

The electrophoretic velocity is given by

$$U = \frac{f \zeta \varepsilon E}{\pi \mu}$$

where f is a function of κR and R is the particle radius. For small values of κR, $f \rightarrow 1/6$, which is the Huckel limit and for large κR, $f \rightarrow 1/4$, which is von Smoluchowski's result (see Problem VII-7). For conducting spheres, f drops from 1/6 to zero with increasing κR. Values of f for various κR have been calculated (6,9).

An important feature of the detailed analysis is that the potential ζ in the above equation is the value which exists at the surface where the no-slip boundary condition applies. Since the first layer of solvent molecules and any adsorbed counterions present are normally rather strongly bound to the surface, it is plausible to assume that they move with the particle. Hence, the surface separating the particle and these bound molecules from the remaining liquid is likely close to that separating the compact and diffuse portions of the electrical double layer. If these surfaces coincide, $\zeta = \psi_0$, and ψ_0 can be obtained from the above equation for an experiment where f is known and where U, E, and μ are measured. If they do not coincide the zeta potential ζ determined in this way is not equal to ψ_0.

It is clear from this discussion that the assumption involved in equating the measured potential ζ with the potential ψ_0 at the inner boundary of the diffuse layer is open to question. Nevertheless, it is widely made because it is the only available method of estimating ψ_o from a relatively simple experiment.

4. COMBINED ATTRACTIVE AND ELECTRICAL INTERACTION - DLVO THEORY

In the absence of adsorbed polymeric molecules which are discussed below, colloid stability is in many cases governed by the combination of London-van der Waals attractive forces and the repulsive forces produced by double layer overlap. This concept is the basis of the famous DLVO theory, developed independently in the late 1930's and early 1940's by Deryagin and Landau in the Soviet Union and by Verwey and Overbeek in the Netherlands.

To find the total interaction energy between two particles, we simply add the contributions ϕ and E_e from attractive and electrical effects. For example, the total interaction energy per unit area E_T between identical semi-infinite blocks can be found from Eqs. [III-5] and [III-45]:

$$E_T = -\frac{A_H}{12\pi h^2} + 64N_A c_o kT\,\kappa^{-1}\,\eta^2\,e^{-\kappa h} \qquad \text{[III-50]}$$

Figure III-8 shows variation of E_T with separation distance for several values of surface potential ψ_o with all other parameters fixed. Several general features of these curves are of interest. In the first place, E_T becomes large and negative for small values of h. Attractive forces dominate under these conditions as a brief inspection of Eq. [III-50] indicates. The curve must eventually, of course, reach a minimum since short range repulsive forces which come into play when the particles are virtually in contact have been ignored. The existence of this "primary" minimum shows that the particles will adhere to one another if they can ever be brought sufficiently close together.

Most of the curves in Figure III-8 exhibit a maximum at somewhat larger values of h. The increase in the height of this maximum with increasing surface potential demonstrates that it is produced by electrical repulsion. It amounts to an energy barrier which must be surmounted as paricles approach if they are to adhere. In the limit $\psi_o = 0$, Eq. [III-50] predicts that E_T is inversely proportional to $(-h^2)$.

Another feature of most of the curves in Figure III-8 is a shallow "secondary" minimum at a separation distance somewhat larger than that of the maximum. As Eq. [III-50] suggests, the electrical forces diminish more rapidly than the attractive forces with increasing h. Thus, the interac-

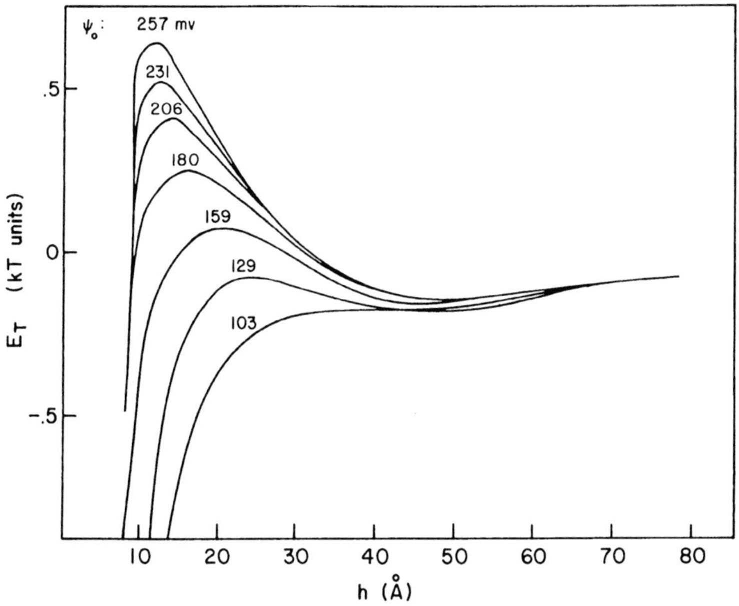

Figure III-8. Total interaction energy for two semi-infinite flat plates for various surface potentials and $\kappa = 10^7 cm^{-1}$ and $A_H = 2 \times 10^{-12}$ ergs. E_T is expressed in terms of the thermal energy kT for a 400 Å^2 area. Eq. [III-50] has been used. Reprinted from Ref. (1), Fig. 10.11 by courtesy of Marcel Dekker, Inc.

tion energy is small but negative at large separation distances where attractive forces dominate. As h decreases and electrical forces become important, E_T reaches a minimum and begins to increase.

Figure III-9 illustrates the effect of varying the Hamaker constant A_H. As would be expected, increasing A_H lowers the height of the maximum in the curve. Figure III-10 show the dramatic changes caused by varying electrolyte concentration and hence the inverse Debye length κ . These curves are for spherical particles, so that E_T^S is the sum of the expressions given by Eqs. [III-6] and [III-47]. Adding electrolyte decreases the Debye length and hence double-layer thickness with the result that electrical forces do not become significant until the particles are closer together. When they finally do become important, attractive interaction is greater and the height of the maximum in the E_T^S curve is less. Adding sufficient electrolyte removes the maximum altogether, as Figure III-10 shows.

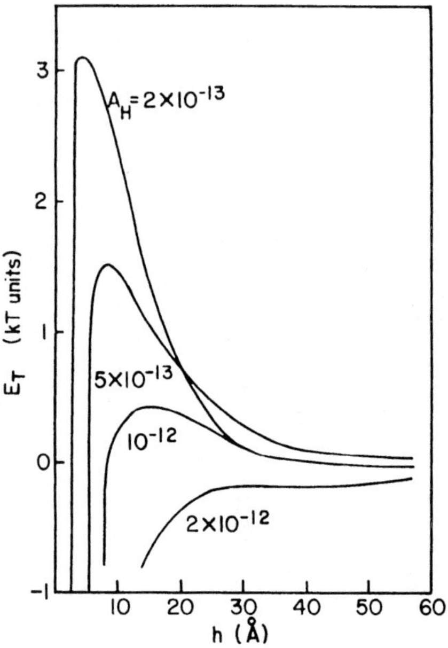

Figure III-9. E_T is shown here with varying A_H. The values of ψ_0 = 103 mV and $\kappa = 10^7$ cm^{-1} have been used. Reprinted from Ref. (1), Fig. 10.12 by courtesy of Marcel Dekker, Inc.

Curves of this type provide a basic understanding of colloid stability. Attractive forces favor flocculation and oppose stability while electrical repulsion has the opposite effect. Flocculation occurs if particles can reach the separation corresponding to the primary minimum. The single most important factor influencing flocculation is the height of the maximum in the E_T curve. If it is nonexistent or small, particles have little trouble reaching the primary minimum as a result of ordinary thermal motion and the dispersion is unstable. But if the maximum is sufficiently high -- several times the effective energy of thermal motion kT -- particles almost never have sufficient energy to surmount this energy barrier and the dispersion is usually stable. The qualifying word "usually" is needed because flocculation can, in principle, occur at separations corresponding to the secondary minimum. Such flocs cannot persist, however, unless the depth of the secondary minimum is several times kT. This condition is usually not satisfied with the result that no flocculation takes place even though the E_T curve has a secondary minimum.

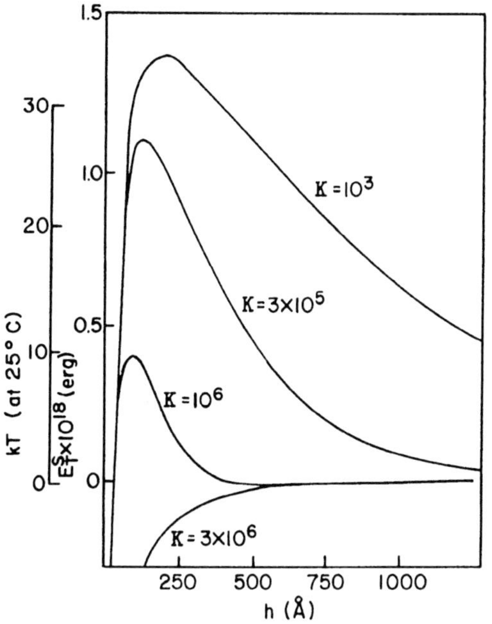

Figure III-10. E_T is shown here with varying electrolyte strengths, κ in cm^{-1}. The sphere radii are 1000 Å, $A_H = 10^{-12}$ ergs and $\psi_0 = 25$ mV. Reprinted from Ref. (4) with permission.

Several general conclusions can be drawn concerning colloid stability. Larger particles are more likely to flocculate since attractive forces are greater (see Figure III-11). This assertion is most easily justified from Eq. [III-7]. At a separation distance h comparable to the Debye length κ^{-1}, the dimensionless parameter ξ decreases with increasing particle radius R. According to Eq. [III-7], the attractive energy is greater for small values of ξ. Reducing the magnitude of the surface potential by adjusting the concentration of the potential-determining ion in solution also promotes flocculation. For instance the pH can be adjusted for metal oxide particles where H^+ is the potential-determining ion. Finally, adding electrolyte strongly promotes flocculation, as indicated previously. Changes in the opposite direction, e.g., a decrease in electrolyte content, can naturally be made if a stable dispersion is desired.

A major success of the DLVO theory has been its ability to predict the very large and striking effect of counterion valence on colloid

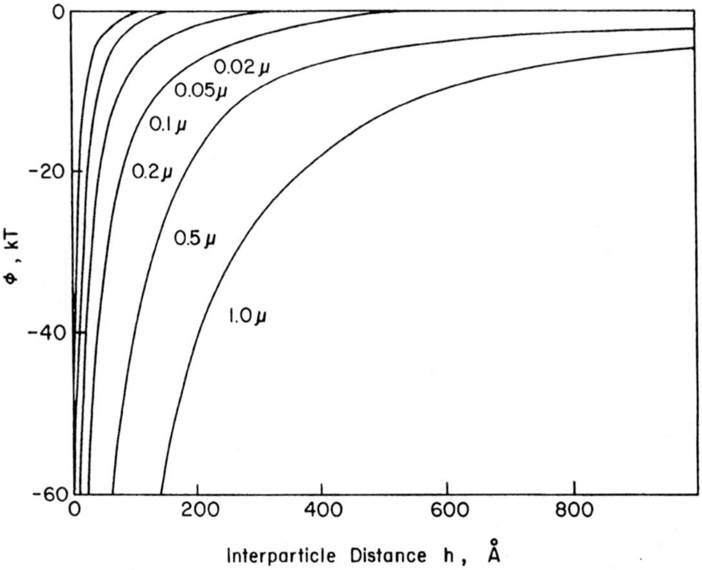

Figure III-11. Hamaker interaction potential between two particles, as a function of the particle size. Reprinted from Ref. (14), p. 66, by courtesy of Marcel Dekker, Inc.

stability. To see this, we focus on the maximum in the E_T curve. We first use Eq. [III-50] to find the separation distance h_m at which (dE_T/dh) vanishes. The result is

$$\frac{A_H}{6\pi h_m^3} = 64\ N_A\ c_o kT\ \eta^2\ e^{-\kappa h_m} \qquad \text{[III-51]}$$

As a reasonable initial estimate of the condition separating stable and unstable dispersions we set $E_T = 0$ at $h = h_m$. Accordingly, we have

$$\frac{A_H}{12\pi h_m^2} = 64\ N_A\ c_o kT\ \kappa^{-1}\ \eta^2\ e^{-\kappa h_m} \qquad \text{[III-52]}$$

If we take the ratio of these equations, we find

$$\kappa h_m = 2 \qquad \text{[III-53]}$$

Substitution of Eq. [III-53] into Eq. [III-51] yields

$$c_o = \frac{A_H}{3072\pi N_A kT\eta^2\ e^{-2}}\ \kappa^3 = B'\ \kappa^3 \qquad \text{[III-54]}$$

Using the definition of κ given at Eq. [III-23], this equation can be written as

$$c_0 = \frac{B}{\nu^6} \qquad \text{[III-55]}$$

where B is a constant. The electrolyte concentration c_0 for which E_T vanishes at h_m is referred to as the critical flocculation concentration concentration (CFC).

According to Eq. [III-55], the concentration required to produce flocculation in a colloidal dispersion with a (nearly) constant surface potential is inversely proportional to the sixth power of the valence. That is, the concentration of a symmetric electrolye containing divalent ions required for flocculation should be only (1/64) of that required with monovalent ions. We note that since the double layer of a negatively charged colloidal particle contains mostly cations and that of a positively charged particle contains mostly anions, it suffices, for practical purposes, to consider only the valence of the counterion in applying Eq. [III-55]. Hence, the requirement of a symmetric electrolyte may be relaxed. In Figure III-12, Overbeek's (6) compliation of the CFC data for a wide range of flocculating electrolytes has been plotted for negatively charged colloid particles (a) as well as for positively charged colloid particles (b). The CFC data used are the averages over a number of flocculating electrolytes with the same value of ν. Eq. [III-53] is known as the Schulze-Hardy rule.

It should be noted that an assumption of the derivation leading to Eq. [III-55] is that η does not change greatly when electrolyte valence ν is varied at constant surface potential ψ_0. Inspection of the definition of η given at Eq. [III-43] reveals that this condition is satisfied only for sufficiently large ψ_0. For small values of ψ_0 the dependence of the CFC on valence is weaker than predicted by the Schulze-Hardy rule.

In the last decade or so techniques have been developed for actually measuring forces between molecularly smooth mica sheets separated by distances ranging from a few Angstroms to a few thousand Angstroms (10). These measurements show that, in the absence of adsorbed polymeric molecules, DLVO theory is valid for separation distances h exceeding about 30 Å (3 nm).

Figure III-13 shows measured forces in KNO_3 solutions of various concentrations. At large separation distances attractive forces are small

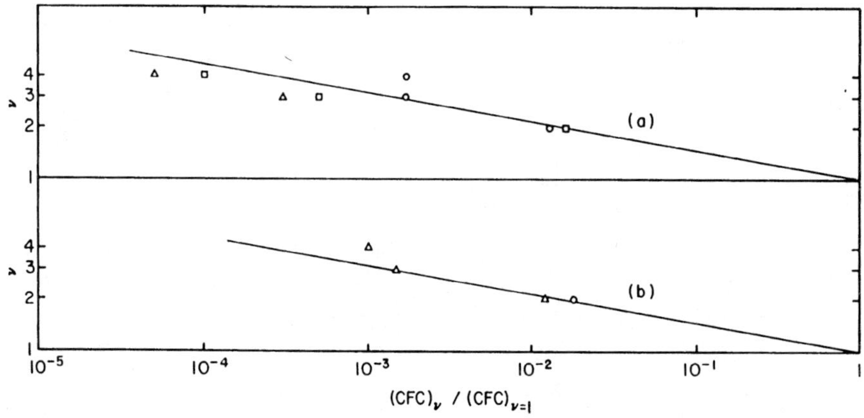

Figure III-12. Schulze-Hardy rule for colloids is shown in bold lines with slopes of 1/6, (a) negatively charged and (b) positively charged particles. From data compiled in (6).

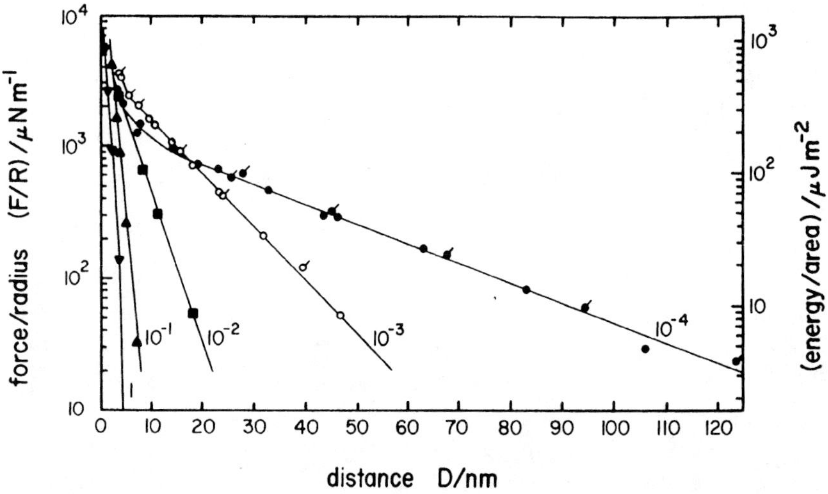

Figure III-13. Experimental results of direct measurements of repulsive forces F as a function of separation D between two crossed mica cylinders of radius R in aqueous KNO_3 solutions (concentrations marked in mol dm^{-3}). The right hand ordinate gives the interaction energy per unit area for two parallel plates, calculated according to the Deryagin approximation. The results in 10^{-4} - 10^{-1} mol dm^{-3} solutions are for the same pair of mica sheets. The points with tails in 10^{-4} and 10^{-3} mol dm^{-3} solutions are for a different pair of sheets cut from the same sheet as the first pair. In 1 mol dm^{-3} KNO_3 the force was attractive above 4 nm. Reproduced from Ref. (10) with permission.

and the force decreases exponentially with increasing separation as predicted by electrical double layer theory Eq. [III-44]. Moreover, at the two lowest KNO_3 concentrations, the measured slopes are within 10% of the predicted values. At the two higher KNO_3 concentrations measured and predicted slopes agree within 25%. In all cases the surface potential ψ_0 is found from the data to be about 75mv.

At 0.1 M KNO_3 the attractive force can be calculated as the difference between the measured force and the electrical force at small separation distances, in the range of 40 Å to 70 Å, where attractive forces are appreciable. The results are consistent with the formulas given above with a value of 2.2×10^{-20} J for the Hamaker constant A_H .

At separation distances below about 30 Å "hydration forces" and other effects of molecular structure sometimes cause large deviations from the predictions of DLVO theory. At very small separation distances, for instance, oscillatory forces are sometimes observed which are evidently associated with the ease of packing molecules between the two surfaces (11,12). At certain definite separation distances molecular dimensions are such that an integral number of molecular layers fits precisely into the avialable space, and the energy has a local minimum. At intermediate separations the solvent structure must be distorted to occupy the available space and energy increases. Such additional forces and the hydration forces mentioned above are a current subject of active research.

5. EFFECT OF ADSORBED POLYMER MOLECULES ON STABILITY OF COLLOIDAL DISPERSIONS

Polymers have been used both to stabilize and to destablize colloidal dispersions. Both charged and uncharged molecules have been employed. Their stabilizing effect is particularly important in nonaqueous dispersions where the very low concentration of ionic species normally renders electrical effects incapable of providing stability.

In aqueous dispersions the effect of polymer addition can sometimes be explained without resort to special properties of polymer solutions. For example, cationic polyelectrolytes are often used in wastewater applications to flocculate negatively charged particles or drops. Their primary effect upon being adsorbed is to reduce surface potential and charge density and hence electrical repulsion between particles. Care is required to avoid adding too much polyelectrolyte as the particles then become positively charged and stability is restored. Such polyelectrolytes are often used instead of inorganic salts of trivalent iron and aluminum, which cause flocculation by decreasing double-layer thickness, because much smaller quantities of additive are required.

Even in such cases, however, indications exist that surface charge neutralization is not the whole story. Gregory (13) found that flocculation occurred over a wider range of conditions than would be expected based on charge neutralization. He suggested that particle surface charge was non-uniform, being positive in the vicinity of adsorbed polymer molecules but remaining negative elsewhere. Attraction between positive and negative regions of adjacent particles would promote flocculation in this case.

Another mechanism by which polymers can cause flocculation and which does not depend on electrical effects is bridging, i.e., adsorption by two particles of portions of a single polymer molecule. Bridging occurs only at very low polymer concentrations where the fraction of the particle surface covered by polymer is low. At higher polymer concentrations each particle is covered by an adsorbed film which is not easily penetrated by polymer chains attached to nearby particles.

Still another mechanism of flocculation is polymer-induced phase separation. Phase separation occurs upon addition of sufficient polymer to a very poor solvent, i.e., one in which attraction between polymer segments is much greater than that between polymer and solvent, even in the absence of colloidal particles. When particles are present, they presumably concentrate in the polymer-rich phase if they interact strongly with the polymer and in the polymer-lean phase if they do not.

A closely related situation is phase separation brought about by the presence of both colloidal particles and polymers. When the polymer is not strongly adsorbed by the particles, for instance because of like electrical charge, and when the diameter of the polymer molecules in solution is much greater than the mean separation distance between particle surfaces, phase separaton can be expected. For if both polymer and particles were to reside in a single phase, the polymer chains would have to undergo considerable distortion to conform to the spaces between the particles with a resulting large decrease in entropy and hence increase in free energy. This distortion is avoided if separation occurs into particle-rich and polymer-rich phases (14). Recent work indicates that this mechanism, which is sometimes called "depletion flocculation", is responsible for phase separation observed when high molecular weight, water-soluble polymers are added to aqueous surfactant solutions containing liquid crystalline particles or to water-continuous microemulsions in systems being considered for enhanced oil recovery processes (15).

When, in contrast, the mean separation distance between particle surfaces is greater than the effective diameter d_p of non-adsorbing polymer molecules, "depletion stabilization" is possible in good solvents. Basically, an energy barrier must be overcome to bring the particles much closer together than d_p because polymer molecules must be removed from the interparticle regions. This process is energetically unfavorable because it decreases the number of polymer segment-solvent molecule contacts--at least when the polymer concentration in the solvent is sufficiently large (16-18).

We now turn to the stabilization of colloidal dispersions by adsorbed polymer molecules, the most common method of polymeric stabilization. Two qualitative effects are important. One is usually referred to as the "volume restriction effect" or as "entropic stabilization". In its simplest form the adsorbed polymer film is compressed as another particle approaches but there is no interpenetration of polymer molecules from the two particles (Figure III-14a). We note that the adsorbed film is a mixture of polymer and solvent. The polymer is present as "loops", i.e., portions of a polymer chain between adsorbed segments, and as "tails", i.e., portions of a chain between either end and the first adsorbed segment. The effect of compression is to distort the loops and tails and thus to decrease their entropy in a manner similar to that described above.

The second important effect is often called a "mixing" or an "osmotic" effect. Its simplest form is shown in Figure III-14b where, in contrast to the volume restriction case, interpenetration occurs without compression. The density of polymer segments obviously increases in the region where the adsorbed layers overlap. The result is a local decrease in the entropy of the polymer chains and hence an increase in free energy. The total attractive interaction among the polymer segments and solvent molecules in the overlap region is also affected by the increase in the number of polymer segments there. The total free energy increase in the overlap region is, accordingly, greater for a "good" solvent than for a "poor" solvent.

In actuality both compression and interpenetration occur so that both volume restriction and osmotic effect should be considered. Various theoretical models have been presented to describe compression alone, interpenetration alone, or some combination of the two. Their main features have been reviewed (16-19).

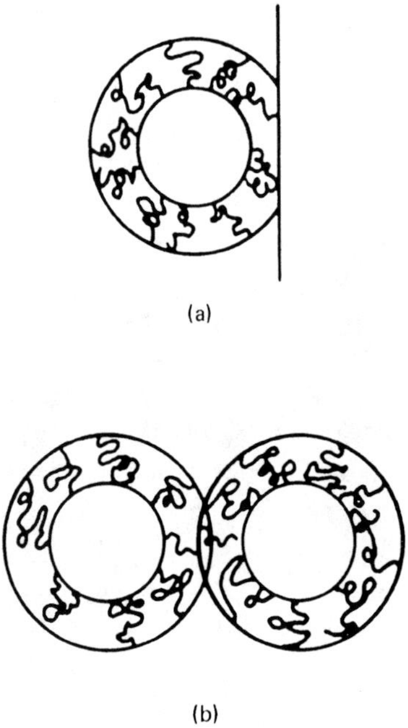

Figure III-14. Schematic depiction of (a) compression and (b) interpenetration of adsorbed polymer layers. Reprinted from Ref. (18), p. 49, by courtesy of Marcel Dekker, Inc.

We shall omit theoretical details here but present a few results which provide some insight on polymer stabilization effects. Figure III-15 taken from the work of Hesselink *et al* (20), shows how a "tail" is compressed as a plane, impenetrable surface approaches another surface on which polymer is adsorbed. In this graph the local density of polymer segments is plotted as a function of position. The calculations are based on a random walk scheme. Figure III-16 from another paper by Hesselink (21), gives the free energy increase due to volume restriction for a loop, a tail, and a bridge as a function of the distance between parallel plates. As might be expected, tail compression begins at greater separation distances than loop compression. Note the minimum in free energy for a bridge at a position which would represent the equilibrium separation distance if London-van der Waals forces were neglected. When the latter are taken into account, the equilibrium separation distance decreases.

Figure III-17 shows the total energy, including attractive forces, for a particular case of particles stabilized by loop interactions (curve a) or

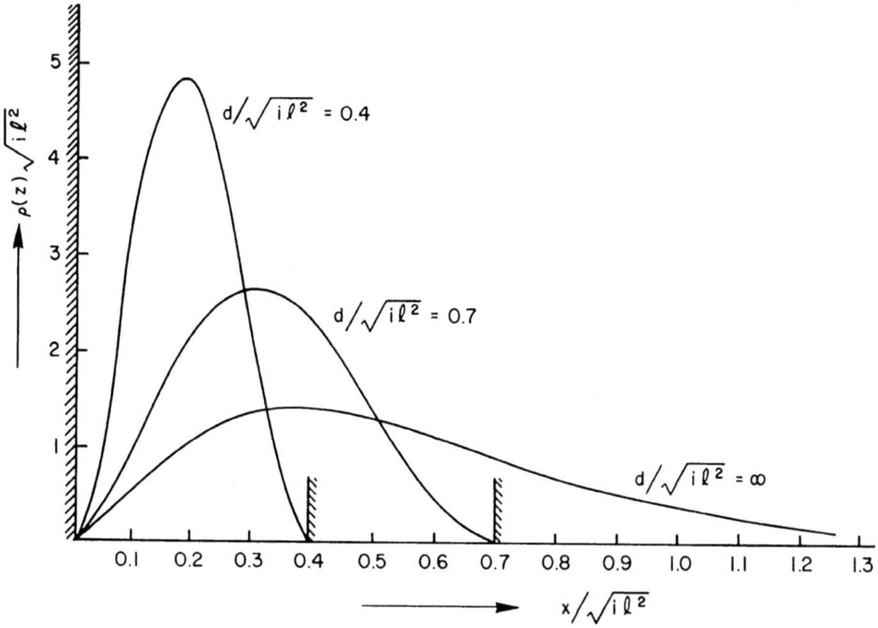

Figure III-15. Normalized density distribution of the segments of an adsorbed polymer tail with a second impenetrable interface at $d/\sqrt{i\ell^2}$ = 0.4, 0.7, ∞. Here d is the distance between interfaces, i the number of polymer segments in a molecule, and ℓ the length of each segment. Reprinted with permission from Ref. (20). Copyright 1971 American Chemical Society.

by tail interactions (curve f). In the latter case the volume restriction and osmotic effects are shown separately (curves b and c). Note that, in contrast to the case of electrical stabilization considered earlier, there is no maximum in the free-energy curve and hence no primary minimum. Flocculation is possible if the single minimum shown is of sufficient depth, just as was true for the secondary minimum found using the DLVO theory.

Various experiments and calculations have shown that in most nonpolar solvents the primary stabilization effect is the decrease in entropy of the polymer-solvent mixture in the region between the particle surfaces. In many aqueous systems, however, the situation is quite different. The local increase in polymer segment density between the particles causes some breakdown in water structure with a consequent increase in system entropy. In this case stability is a consequence of an increase in

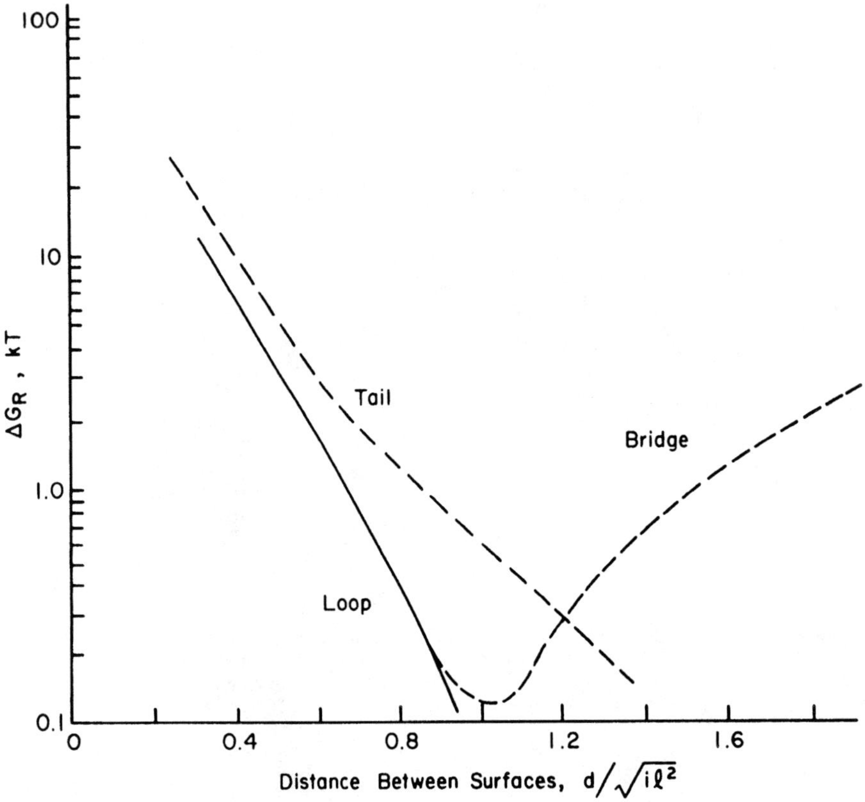

Figure III-16. Free energy of a tail, a loop, and a bridge, of adsorbed polymers between two slabs. Reprinted with permission from Ref. (21). Copyright 1971 American Chemical Society.

potential energy produced by the increase in the number of polymer-polymer and the decrease in the number of polymer-solvent interactions. Such an increase can occur only if polymer-solvent interactions are strong, i.e., the solvent is a very good one. We note that entropically stabilized dispersions tend to flocculate on cooling since entropy effects decrease with decreasing temperature. But dispersions stabilized by enthalpic effects tend to flocculate on heating because the stabilizing interaction effects are eventually outweighed by the destabilizing entropic effects.

According to the theory of polymer solutions, a temperature usually denoted by θ exists for which the free energy of mixing of polymer with a given solvent vanishes in the limit of high polymer molecular weight. This

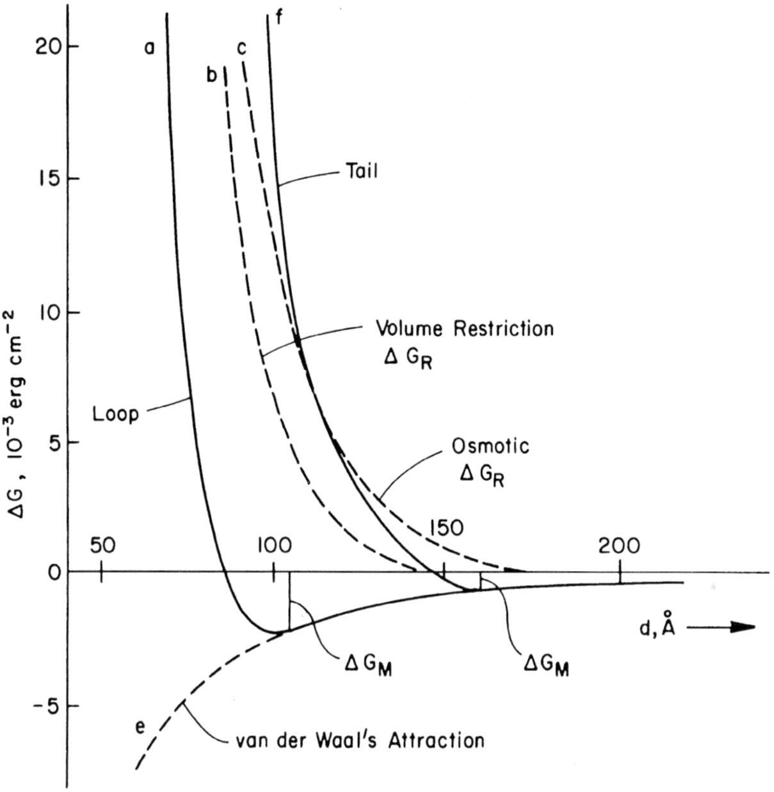

Figure III-17. The free energy of interaction due to polymer adsorption between particles covered by tails (f) and equal loops (a). For particles covered by equal tails, (b) gives the volume restriction effect and (c) the osmotic repulsion; (f) is the resultant of adding (b), (c), and (e). A value of $A_H = 10^{-12}$ erg has been used. Reprinted with permission from Ref. (21). Copyright 1971 American Chemical Society.

temperature θ is a function of solvent composition for a given polymer. If the above concept of how polymers stabilize colloidal dispersions is correct, we might expect such dispersions to flocculate when the temperature approaches θ as a result of varying system temperature or composition. Such behavior has been found as reported by Napper and co-workers (19, 22).

We note that, from the above discussion, a polymer for which the dispersion medium is a "good" solvent is desired for purposes of stabilization in both nonpolar and aqueous systems. On the other hand, adsorption on particle surfaces is favored if the dispersion medium is a "poor" solvent. One method of resolving these conflicting requirements is to use a copolymer having some groups which adsorb strongly on the colloidal particles and "anchor" the polymer and other groups which interact strongly with the solvent and make up the loops and tails. Discussion of polymer adsorption may be found in references (16-18).

In summary, it is fair to say that the effect of polymers on colloidal dispersions is generally understood from a qualitative point of view. But further improvements in mathematical models are required before polymeric stabilization effects can be predicted quantitatively with the same confidence as attractive and elecrical repulsion interactions can with the DLVO theory. Measurements with the technique described above of the forces between surfaces with adsorbed polymers under various conditions are also needed.

6. KINETICS OF COAGULATION

Colloidal particles exhibit random Brownian motion as a result of which a net diffusive flux can be generated. If the thermodynamic force on each particle which causes such a flux is the gradient of chemical potential $-\nabla\mu_c$, it must, under steady conditions, be balanced by frictional forces. These forces are taken to be hydrodynamic drag forces and are obtained from Stokes' law.

Making the balance, one obtains

$$6\pi\mu R \mathbf{U} = -\nabla\mu_c$$

Substituting $(\mu_c^0 + kT \ln c)$ for μ_c, where μ_c^0 is the standard state chemical potential, one has

$$c\mathbf{U} = -\frac{kT}{6\pi\mu R} \nabla c$$

This equation may be rewritten as

$$j = -D \nabla c \qquad \text{[III-56]}$$

where $j = c\mathbf{U}$ represents the diffusive flux, and the diffusion coefficient D is given by

$$D = \frac{kT}{6\pi\mu R} \qquad [III\text{-}57]$$

Eq. [III-55] is known as the Stokes-Einstein equation.

If external forces act on the particles, as electrical fields will for charged colloidal particles, Eq. [III-56] is modified to

$$\mathbf{j} = - D\nabla c + cM\mathbf{F} \qquad [III\text{-}58]$$

where F is the force on a single particle and M is its mobility. M is usually taken as (D/kT), and the force F is usually written as the gradient of a scalar potential ϕ. Eq. [III-58] thus becomes

$$\mathbf{j} = -D[\nabla c + \frac{c}{kT}\nabla\phi] \qquad [III\text{-}59]$$

For equilibrium situations the flux is zero, and it follows from Eq. [III-59] that the concentration distribution is of the Boltzmann type:

$$c = c_\infty \exp[-\frac{\Phi}{kT}] \qquad [III\text{-}60]$$

A simple case where flocculation takes place among single colloidal spheres of initial concentration c_∞ is considered below. The concentration decreases with time because the collisions between two spheres lead to flocculation. Since collisions involving two-particle flocs are not considered, the analysis is restricted to collisions between two individual particles and their consequences. Hence the solution is valid only at short times.

We consider diffusion in the vicinity of a single fixed colloidal particle. If the quasi-steady state approximation is satisfactory, the conservation equation simplifies to

$$0 = -\nabla\cdot\mathbf{j} = -\frac{1}{r^2}\frac{\partial}{\partial r}(r^2 j_r) \qquad [III\text{-}61]$$

so that $r^2 j_r$ is a constant. Here j_r is the flux in the radial direction using a spherical coordinate system with the center of the fixed particle as the origin. We write, on integrating the above equation

$$z = -4\pi r^2 j_r \qquad [III\text{-}62]$$

where z is a constant and represents the total number of particles moving towards the fixed single colloidal particle. Substituting Eq. [III-59] into [III-62] and rearranging exp $(\phi/kT)[dc/dr + (c/kT)(d\Phi/dr)]$ as $d\,(ce^{\Phi/kT})/dr$, we have

$$\int_0^{c_\infty} d(ce^{\frac{\Phi}{kT}}) = c_\infty = \frac{3\mu Rz}{2kT} \int_{2R}^{\infty} \frac{1}{r^2} \exp\left(\frac{\Phi}{kT}\right) dr \qquad \text{[III-63]}$$

In Eq. [III-60] we have assumed that as $r \to \infty$, Φ tends to some constant value assumed to be its datum, and that c goes to some uniform value c_∞. These assumptions imply that as $r \to \infty$, $j_r \to 0$, as may be verified from Eq. [III-62] for a constant z.

It has also been assumed that the distance of closest approach between two single colloidal spheres is 2R, where 2R is the hard sphere diameter of the particles. At the distance of closest approach the particles are assumed to have flocculated and the concentration of the colloid particles that have approached the *fixed* colloid is zero.

The quantity z can be evaluated from Eq. [III-63]. It represents the collision frequency of any *one* colloid particle. In the given system the total number of collisions per second is $\frac{1}{2} z c_\infty$, the factor $\frac{1}{2}$ being included to avoid double counting. If *every* collision gives rise to flocculation, then the initial depletion rate in a monodisperse suspension is twice the collision rate since each collision removes two particles:

$$-\frac{dc_\infty}{dt} = zc_\infty = k_2 c_\infty^2 \qquad \text{[III-64]}$$

Here k_2 is a second order reaction rate constant given by

$$k_2 = \frac{4kT}{3\mu} \left[\int_1^{\infty} \frac{1}{r^{*2}} \exp\left(\frac{\Phi}{kT}\right) dr^* \right]^{-1} \qquad \text{[III-65]}$$

where $r^* = r/2R$ and r is the center-to-center distance between spheres. As discussed below, the integral in this equation is, under many conditions, proportional to exp $[\Phi_{max}/kT]$ where Φ_{max} is the potential energy at the maximum of an interaction curve such as those in Figures III-8 through III-10. Since k_2 is a rate constant, Φ_{max} is analogous to the activation energy in chemical kinetics.

Eq. [III-65] is often modified to account for the fact that we have assumed in the derivation that the central particle remains fixed. Obviously, it in fact moves due to diffusion and each particle of a pair may be considered to diffuse toward the other. It turns out that to account for the relative motion, 2D should be used instead of D alone, and the value of k_2 becomes twice that given by Eq. [III-65] (see Eq. [III-66]).

This method of formulation by von Smoluchowski and Fuchs is limited to small concentrations of the particles. Then the fixed particle can at most feel the presence of one other particle, and Φ is equal to the sum of the van der Waals attraction and the electrical double layer repulsion potential or E_T^S as discussed in previous sections. In this limit, it is also legitimate to model the reaction as a second order reaction, i.e., only two-particle collisions can occur and the higher body collisions are virtually non-existent. In aerosols, which are colloidal dispersions in air, there is no significant electrical repulsion between particles. Hence, the effect of interparticle forces on the initial coagulation rate is negligible, and we find

$$k_2 = \frac{8kT}{3\mu} \qquad \text{[III-66]}$$

Eq. [III-66] is the "fast coagulation" limit of the general expression for "slow coagulation" represented by Eq. [III-65], in which it has been implicitly assumed that Φ provides a potential energy barrier hindering coagulation. For this case the appropriate value of k_2 is twice that given by Eq. [III-65] as discussed earlier.

The obvious restriction on the theory that flocs are made out of only two colloidal spheres can be relaxed (6,23) but the mechanism of flocculation is still based on bi-particle (floc) collision. Consequently, the improved solutions (see Problems III-4, 5) are still restricted to small dilutions but valid over longer times. The particle concentration may have a reasonably large range, but if the comparison with chemical reaction kinetics holds, then a different theory would be needed to obtain the kinetics of flocculation at large colloid concentrations where tri-particle (floc) and higher order collisions become more important (24).

The mechanism also fails when the distance over which two flocs (or single particles) first feel the presence of one another due to their interaction potential is much greater than the average center-to-center distance among flocs. For then the term Φ , previously taken as the interaction potential between two particles, has to be modified to account for the presence of other particles even for a bi-particle collision. Such effects are observed in nonaqueous media (25) and could occur in concentrated systems as well.

Comparing Eqs. [III-65] and [III-66], we find that the term

$$W = \int_1^\infty \frac{1}{r^{*2}} \exp\left(\frac{\Phi}{kT}\right) dr^* \qquad \text{[III-67]}$$

in Eq. [III-65] decreases the flocculation rate. W is called the stability ratio. As mentioned previously, Φ is approximated as the interaction potential between two spheres. For systems which are difficult to flocculate, the function Φ has a maximum as discussed in Section 4 above. Under these conditions, the integrand of Eq. [III-67] has a sharp maximum and the integral W is mainly composed of the area under this peak. This allows an asymptotic expansion for W in terms of Φ_{max}. Either the approximate expression for W given in Problem III-6 or Overbeek's approximate form (6) $W \sim (2\kappa R)^{-1} \exp[\Phi_{max}/kT]$ can also be used together with Φ as E_T^S to obtain

$$\log W = K_1 + K_2 \log c \qquad \text{[III-68]}$$

This linear form is verified by experiment (6, see Figure III-18). Further, log W = 0 at c = CFC as required. It is also seen from Figure III-18 that existence of a highly stable colloidal dispersion requires that W be at least as great as 10^4. In Figure III-18, W has been plotted against the concentrations c of the counterions, where these have valences of 1, 2, 3. Although Eq. [III-68] cannot be used to predict the Schulze-Hardy rule, CFC versus ν values obtained from the figure are in resonable agreement with the rule.

It was anticipated by von Smoluchowski that Brownian diffusion is not the only mechanism by which one particle can come into contact with another. There can be convection in the system as well. Convection aids flocculation when it enhances the relative velocity between nearby particles, making it easier for them to surmount the maximum in the potential energy curve and reach the primary minimum corresponding to flocculation. Application of a velocity gradient (shear) is especially effective in flocculation of relatively large particles (1,26). It can also break up flocs associated with the secondary minimum of the particle interaction curve (26).

Another important problem which has attracted the attention of a host of investigators (27-30) is particle collection. In the deep bed

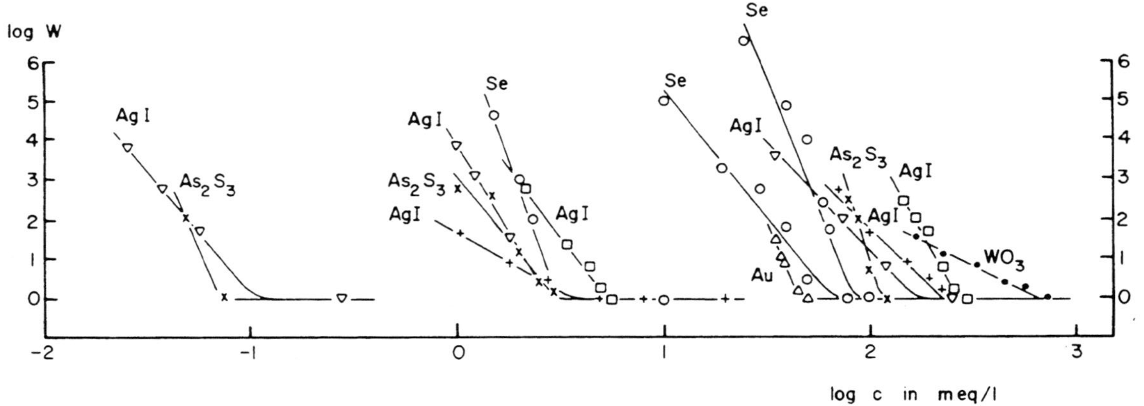

Figure III-18. Relation between rate of flocculation and concentration of electrolyte as determined experimentally. The three groups of curves relate to monovalent, divalent and trivalent electrolytes (from the right). Reproduced from Ref. (6) with permission.

filtration of colloidal particles, one seeks to describe the interaction and the collision between one colloidal particle and one grain of the packing material which forms the bed. The latter, called the collector, is immobile. The liquid containing the suspended colloidal particles flows past the collectors and flocculation of the colloid particles with the grains of the packing material is called particle capture. The particles are brought to the collector surface both by convection and diffusion.

From fluid mechanical calculations, if $\mathbf{v}^{(0)}$, the velocity field around the collector in the absence of the colloidal particles is known, then $\mathbf{v}_p$, the velocity field of the particles, can also be found, at least to a good approximation. Moreover, the particle diffusion coefficient becomes position dependent since the Stokes drag force used to derive Eq. [III-57] is modified by the presence of the nearby collector surface. The conservation equations [III-59] and [III-61] must therefore be replaced with the appropriate one:

$$\nabla \cdot (c\mathbf{v}_p) = \nabla \cdot [D(r) \{\nabla c + \frac{c}{kT} \nabla \Phi\}] \qquad \text{[III-69]}$$

Eq. [III-69] is solved for a single collector with the same boundary conditions as before. The flux to a single collector can thus be calculated. In some cases it is possible to solve Eq. [III-69] using the bulk value of D given by Eq. [III-57] and formulate a suitable boundary condition which includes particle-collector interactions and hydrodynamic effects in the immediate vicinity of the wall (27).

Finally, we note that experimental investigation of flocculation, particle capture, and indeed most aspects of behavior of colloidal dispersions has been greatly aided in recent years by the availability of spherical latex particles having extremely narrow size distributions. The ability to carry out experiments with well-characterized colloidal dispersions which are virtually monodisperse has been of great value. Moreover, latex particles having different diameters can be mixed to form dispersions with size distributions that are specially tailored and accurately known.

7. REFERENCES

GENERAL REFERENCES ON COLLOIDAL DISPERSIONS AND THEIR INTERACTION FORCES

Goodwin, J.W., ed. (1982) *Colloidal Dispersions*, Royal Society of Chemistry, London.

Hiemenz, P.C. (1977) *Principles of Colloid and Surface Chemistry*, Marcel Dekker, N.Y.

Kruyt, H.R., ed. (1952) *Colloid Science*, **V. 1-2**, Elsevier, Amsterdam.

Levich, V.G. (1962) *Physico-Chemical Hydrodynamics*, Prentice-Hall, Inc., N.J.

Sato, T. and Ruch, R. (1980) *Stabilization of Colloidal Dispersions by Polymer Adsorption*, Marcel Dekker, N.Y.

Saville, D.A. (1977) *Ann. Rev. Fluid Mech.* **9**, 321.

Sheludko, A. (1966) *Colloid Chemistry*, Elsevier, Amsterdam.

Van Olphen, H. and Mysels, K., eds. (1975) *Physical Chemistry: Enriching Topics from Colloid and Surface Science*, Theorex, La Jolla, California.

Verwey, E.J.W. and Overbeek, J.Th.G. (1948) *Theory of Stability of Lyophobic Colloids*, Elsevier, Amsterdam.

Textual References

1. Hiemenz, P.C. (1977) *Principles of Colloid and Surface Chemistry*, Marcel Dekker, N.Y.

2. Hamaker, H.C. (1937) *Physica* **4**, 1058.

3. Parsegian, V.A. (1975) in *Physical Chemistry: Enriching Topics from Colloid and Surface Science*, H. van Olphen and K. Mysels, eds., Theorex, La Jolla, California, pp 27 ff.

4. Verwey, E.J.W. and Overbeek, J.Th.G. (1948) *Theory of Stability of Lyophobic Colloids*, Elsevier, Amsterdam.

5. Jackson, J.D. (1975), *Classical Electrodynamics*, 2nd ed., Wiley, N.Y.

6. Overbeek, J.Th.G. (1952) in *Colloid Science* **V. 1**, H.R. Kruyt, ed., Elsevier, Amsterdam, pp. 245 ff.

7. Mitchell, D.J. and Ninham, B.W. (1983) *J. Phys. Chem.* **87**, 1996.

8. Levich, V.G. (1962) *Physico-Chemical Hydrodynamics*, Prentice-Hall, Inc. N.J., pp 472 ff.

9. O'Brien, R.W. and White, L.R. (1978) *J. Chem. Soc. Faraday II*, **74**, 1607.

10. Israelachvili, J.N. and Adams, G.E. (1978) *J. Chem. Soc. Faraday I* **74**, 975.

11. Ninham, B.W. (1980) *J. Phys. Chem.* **84** 1423.

12. Christenson, H.K., Horn, R.G., and Israelachvili, J.N. (1982) *J. Colloid Interface Sci.* **88**, 79.

13. Gregory, J. (1973) *J. Colloid Interface Sci.* **42**, 448.

14. Vrij, A. (1976) *Pure Appl. Chem.* **48**, 471.

15. Qutubuddin, S., Benton, W.J., Miller, C.A., and Fort, Jr., T. (1981) "A Proposed Mechanism for Polymer-Surfactant Interactions in Enhanced Oil Recovery", Preprint for ACS National Mtg., Polymer Chemistry Division, N.Y., Aug., 1981. To appear in ACS Symposium Series volume on Emulsions and Microemulsions, 1984.

16. Vincent, B. (1974) *Adv. Colloid Interface Sci.* **4**, 193.

17. Parfitt, G.D. and Peacock, J. (1978) *Surface and Colloid Science*, **V. 10**, E. Matijevic, ed., Plenum, N.Y., pp. 163 ff.

18. Sato, T. and Ruch, R. (1980) *Stabilization of Colloidal Dispersions by Polymer Adsorption*, Marcel Dekker, N.Y.

19. Napper, D.H. (1982) "Polymeric Stabilization", pp 99 ff in *Colloidal Dispersions*, J.W. Goodwin, ed., Royal Soc. of Chem., London.

20. Hesselink, F.Th., Vrij, A., and Overbeek, J.Th.G. (1971) *J. Phys. Chem.* **75**, 2094.

21. Hesselink, F.Th. (1971) *J. Phys. Chem* . **75**, 65.

22. Napper, D.H. and Hunter, R.J. (1972) *MTI International Reviews of Science, Physical Chemistry, Surface Chemistry and Colloids*, **Ser. 1, V. 7**, M. Kerker, ed., Butterworths, London.

23. Sheludko, A. (1966) *Colloid Chemistry*, Elsevier, Amsterdam, pp 208 ff.

24. Gardiner, Jr., W.C. (1972) *Rates and Mechanisms of Chemical Reaction*, 2nd ed., Benjamin Press, Menlo Park California.

25. Albers, W. and Overbeek, J.Th.G. (1959) *J. Colloid Sci.* **14**, 501, 510.

26. Zeichner, G.R. and Schowalter, W.R. (1977) *AIChE J.* **23**, 243, *J. Colloid Interface Sci.* **71**, 237 (1979).

27. Ruckenstein, E. and Prieve, D.C. (1973) *J. Chem. Soc. Faraday II* **69**, 1522.

28. Spielman, L.A. (1977) *Ann. Rev. Fluid Mech.* **9**, 297.

29. Saville, D.A. (1977) *Ann. Rev. Fluid Mech.* **9**, 321.

30. Tien, C. and Payatakes, A.C. (1979) *AIChE J.* **25**, 737.

PROBLEMS

III-1. Use Eqs. [III-6] and [III-47] to calculate the number of extrema in the interaction potential E_T^S between two identical spheres in an aqueous solution. A_H is 10^{-14} ergs or higher and ψ_0 ranges from 5 ~ 200 millivolts. Further,

N_A = 6.023×10^{23} molecules/mole
e_0 = 4.803×10^{-10} esu or $erg^{1/2}\ cm^{1/2}$
k = 1.380×10^{-16} erg/°K
e_0V = (charge on an electron times 1 volt) = 1.602×10^{-12} erg
ε = 80

For this problem use the values of R = 80 nm and c = 0.001 mole/ℓ and 0.1 mole/ℓ.

III-2. Integrate Eq. [III-25] to calculate u as a function of z, for a double layer emanating from a single plate. For $\nu = 1$ plot u as a function of κz. If ψ_0 = 300 millivolts and c_0 = 0.0001 mole/ℓ, what is the value of n- at z = 0? Can an explanation be offered for this result?

Derive Eq. [III-43] from the results.

Integrate Poisson's equation Eq. [III-20] over a thin pillbox control volume as in Figure I-3 and obtain the following relation between the surface charge and the potential gradients at an interface:

$$\sigma = \frac{\varepsilon_A}{4\pi}\,\mathbf{n}\cdot\nabla\psi_A\Big|_S - \frac{\varepsilon_B}{4\pi}\,\mathbf{n}\cdot\nabla\psi_B\Big|_S$$

where **n** is the unit outward normal at the interface from phase A. When phase B is a good conductor (or insulator), ψ_B is a constant. Under this assumption, σ reduces to

$$\sigma = -\frac{\varepsilon}{4\pi}\frac{d\psi}{dz}\Big|_{z=0}$$

where the subscript A has been dropped. Evaluate σ when ψ_0 and c_0 are as given above.

III-3. Consider an electrical double layer made up of a compact layer with a dielectric constant ε' and a thickness δ and a diffuse layer with dielectric constant ε. As explained in the text δ is approximately the radius of a hydrated ion. Calculate ψ_0, the potential at the junction of the two layers, in a 0.001 M aqueous solution of a univalent electrolyte at 25°C where the potential ψ_s on the solid surface is 137 mv. Further, ε' = 10, and δ = 2.5 Å.

III-4. If c_k is the number density of flocs with k-particles in them, it can be shown that for rapid coagulation,

$$c_k = \frac{c_\infty (t/T^*)^{k-1}}{(1+\frac{t}{T^*})^{k-1}}$$

where $T^* = (3\mu/4kTc_\infty)$ and c_∞ is the number density in the original monodisperse system. Furthermore, the total number density is given by

$$\sum_{k=1}^{\infty} c_k = \frac{c_\infty}{(1+\frac{t}{T^*})}$$

Calculate T* for each of the two sets of the data reported by Tuorila [*Kolloidchem. Beihefte* 22, 191 (1926)].

	t(sec)	number of flocs x 10^{-8}
Gold sol	0	20.20
	30	14.70
	60	10.80
	120	8.25
	240	4.89
	480	3.03
Kaolin	0	5.0
	105	3.90
	180	3.18
	255	2.92
	335	2.52
	420	2.00
	510	1.92
	600	1.75
	600	1.75
	1020	1.54
	2340	1.15

How would one plot the data to get a straight line?

III-5. Overbeek (6) argues that

$$W = \frac{1}{2\kappa R} \exp \; (\Phi_{max}/kT)$$

Calculate Φ_{max} from Eqs. [III-7] and [III-47] and obtain Eq. [III-68].

III-6. (a) The stability ratio W is given by Eq. [III-67]. Let us assume that the main contribution to W is for values of r* where (Φ/kT) is near its maximum value (Φ_m/kT). In particular, use a Taylor series to write

$$\frac{\Phi}{kT} \;\hat{=}\; \frac{\Phi_m}{kT} - p^2 \, (r - r_m)^2$$

$$\text{where } p^2 = -\frac{1}{2kT} \left. \frac{d^2\Phi}{dr^2} \right|_{r_m}$$

Recalling that for $\kappa R \gg 1$, the separation distance h_m for $\Phi = \Phi_m$ is much less than particle radius R and assuming that ph_m is relatively large, show that

$$W \;\hat{=}\; \frac{2R\,\pi^{1/2}}{p r_m^2} \exp \left(\frac{\Phi_m}{kT}\right) \qquad \text{(i)}$$

Hint: manipulate the integral to get the usual form used in defining the error function.

(b) Taking $\kappa = 10^6$ cm^{-1} and other values as in Figure III-10, calculate W from Eq. (i) above and from Overbeek's approximate expression given in Problem III-5.

III-7. Eq. [III-64] can be written as

$$-\frac{dc_1}{dt} = \frac{1}{WT^*c_\infty} \, c_1^2$$

using the notation of Problem III-4. For slow coagulation one may also assume that the total number of particles is the same as in Problem III-4 with WT* replacing T*. Analyze the data of Kruyt and van Arkel [*Rec. Trav. Chim.* **39**, 656 (1920), *Kolloid-Z.* **32**, 29 (1923)], on the coagulation of selenium sol of 52 μm diameter particles with 50 millimols/ℓ KCℓ.

t(hrs)	number of particles per $cm^3 \times 10^{-8}$
0	33.5
0.25	32.3
22.5	28.6
42.5	19.1
67.5	14.6
187.5	7.5
239	7.5
335	4.7
1008	1.46

Obtain the value of W if T* is calculated to be 20 sec.

III-8. Why are river waters so muddy while the sea water is virtually devoid of particulate material? Suggest a reason for the delta formation at the confluence.

III-9. In an electrochemical system the chemical potential μ_i of the ith ion is replaced by the electrochemical potential η_i, where

$$\eta_i = \mu_i + z_i e_0 \psi \tag{i}$$

and z_i is the algebraic charge of the ith ion.

(a) using the method described in Section 6, show that the flux of the ith ion is

$$j_i = -D[\nabla c_i + \frac{c_i e_0 z_i}{kT} \nabla\psi] \tag{ii}$$

This is known as the Nernst-Planck equation. Compare with Eq. [III-59] and interpret the factor $e_0 z_i \nabla\psi$. Show also that the Boltzmann distributions given following Eq. [III-21] are obtained at equilibrium when the fluxes are zero.

(b) Using the definition of γ from Eq. [I-13] and Gibbs' adsorption equation [I-24] at constant temperature, and Eq. (i), show that

$$\Delta F^s = - \int_0^{\psi_0} \sigma(\psi'_0) d\psi'_0 \tag{iii}$$

where ΔF^s is the change in the surface excess free energy per unit area if the interface acquires charge, all other things remaining the same. Using the results of Problem III-2 show that the change is negative, i.e., that the electrical double layer forms spontaneously.

IV

Surfactants

1. INTRODUCTION

Surfactants have long been known to us in the form of soaps, which are sodium salts of naturally occurring fatty acids. Some of their properties have also long been recognized, e.g., the ability of alkali to convert fatty acids to soaps and the precipitation of insoluble compounds in hard water (bathtub ring) with an accompanying loss in cleaning power. The advent of petrochemicals has brought synthetic detergents, which are typically sodium sulfates and sulfonates of long-chain hydrocarbons.

Although some basic study of surfactants has been carried out for many years, such work has recently been boosted by the possibility of using surfactants to recover crude oil from underground petroleum fields (1-2). It is found that after conventional methods for recovering oil have been used, more than half of the original oil is left behind in the pores of the underground oil fields, the latter being made of porous limestone or sandstone. It is also found that immiscible fluids, such as water or brine, are unable to displace the residual oil because of the high interfacial tension between the fluid and the oil. One way to overcome this problem is to reduce interfacial tensions to "ultralow" values, i.e., from a typical crude oil-brine tension of 30 dynes/cm to 10^{-3} to 10^{-4} dyne/cm by the use of surfactants. Surfactant-aided oil recovery as a process is still far from being technically and economically feasible. However, if and when it arrives, the need for surfactants there will swamp the current needs and production levels. The surfactants employed in first-generation processes are called petroleum sulfonates and are obtained by sulfonating a refinery stream or crude oil and then neutralizing the resulting sulfonic acids with caustic.

Surfactants also play an important role in biological systems. They occur as a primary component of cell walls, as lining material in lungs, and as bile salts which play a major role in digestive processes. Consequently, diseases associated with malfunctioning of these tissues or processes often involve surfactant behavior.

A brief introduction to surfactants has been given in Chapter I. Because water has a high surface tension, many materials, both soluble and insoluble, display some degree of surface activity, i.e., they reduce the surface tension of water. (The strong electrolytes at high concentrations form notable exceptions.) Surfactants of practical interest reduce the surface tension dramatically at low concentrations and, in addition, possess aggregation or structure-forming capabilities. As noted previously, a surfactant molecule has a long chain hydrocarbon tail and a short ionizable or polar group. The hydrophobic tail makes the molecule virtually insoluble, with some slight solubility imparted by the ionizable or polar group. Frequently, a surfactant has only one hydrocarbon tail, but molecules with two tails joined to the same polar group also exist and indeed are common in biological membranes.

However, the more interesting part of the amphiphile (so called because it contains a hydrophobic part and a hydrophilic part), from the point of view of classification, is the polar hydrophilic group. The surfactant is characterized as anionic if on ionization in water the surface-active portion containing the hydrophobic chain has a negative charge. An example is the alkyl sodium sulfonate, $R\text{-}SO_3N_a$ which splits into $R\text{-}SO_3^-$ and Na^+. Anionic surfactants are by far the most commonly found surfactants among the various types. Cationic surfactants are usually soluble in acid solutions. The primary amine $R\text{-}NH_2$, for instance, gives rise to the $R\text{-}NH_3^+$ ion in such solutions. In some applications quaternary ammonium compounds, which are less sensitive to pH changes, are used, e.g., $R\text{-}N(CH_3)_3^+Br^-$ (or Cl^-). As their name implies, nonionic surfactants contain only electrically neutral polar groups, notably the ethoxy group $\text{-}OCH_2CH_2\text{-}$. An example is the ethoxylated fatty alcohol $R(OCH_2CH_2)_nOH$. Other classes also exist. Detailed description of surfactant chemistry has been given by Rosen (3), Osipow (4), Shinoda *et al* (5), and McBain and Hutchinson (6). In the subsequent discussion cationic surfactants will not be considered separately because of their similarity in behavior to anionic surfactants.

2. ANIONIC SURFACTANTS

Figure IV-1 shows the surface tension of potassium laurate as a function of its concentration (7). The fact that beyond a certain concentration the surface tension of the aqueous solution does not change any more is the distinctive feature of the plot. The sudden change in the slope of the plot should not be construed as the point where the interface becomes saturated with surfactant as the discussion on adsorption in Chapter II might suggest. It is found that at or in the vicinity of this concentration a host of properties of the bulk solution also change, e.g., density, solubility, osmotic pressure, electrical resistance, light scattering properties, detergency, etc. The light scattering experiments show that aggregates form beyond this concentration. The former are called micelles and the latter is called the critical micelle concentration (CMC). Addition of surfactant to the solution beyond the CMC gives rise, for the most part, only to further micelles. Consequently the surface properties remain almost constant.

In Figure IV-2, the concentration - temperature diagram is shown for sodium dodecyl sulfate in water. The bold line denotes the solubility limit of the sulfate in water. The dashed line is the behavior expected on

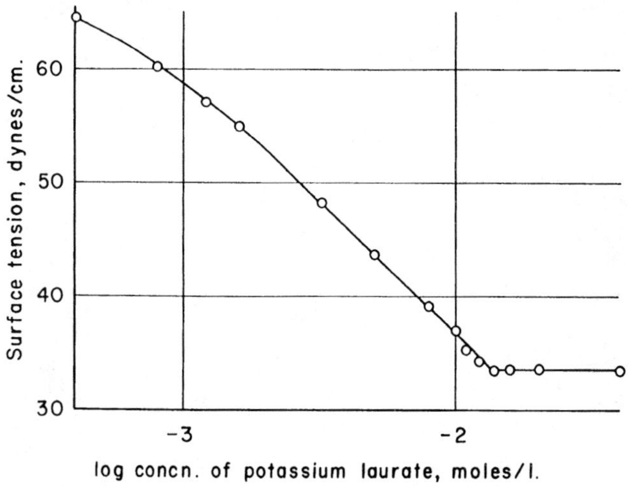

Figure IV-1. Surface tension as a function of surfactant concentration. The break in the slope is seen distinctly. Reprinted with permission from Ref. (7). Copyright 1954 American Chemical Society.

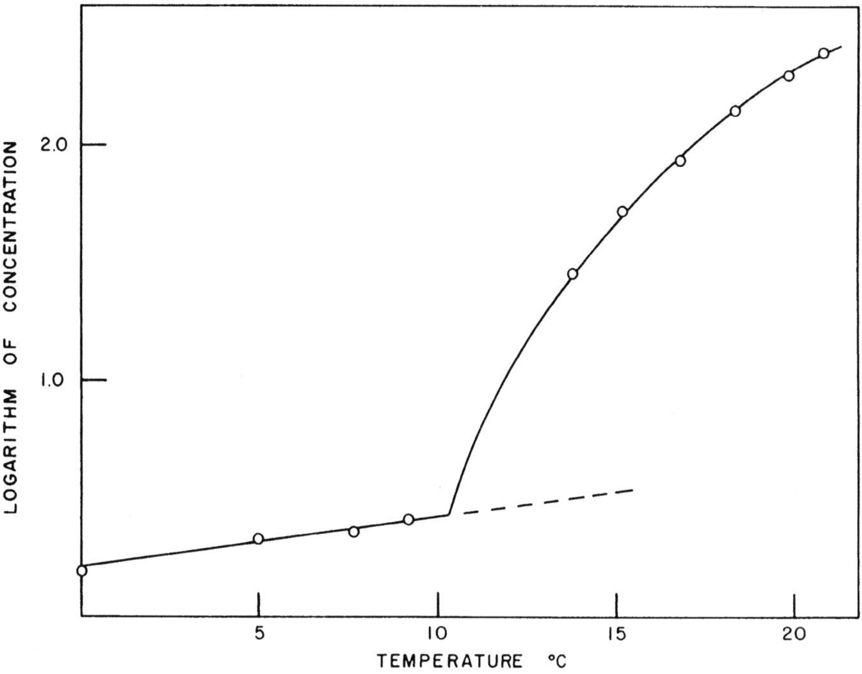

Figure IV-2. Solubility of sodium dodecyl sulfate in water. The break in the slope represents the Kraft point. Reprinted from Ref. (6) with permission.

extrapolation of the low concentration behavior. Thus one expects that the jump in the solubility at about 10°C is due to the onset of micellization. The temperature where this jump takes place -- which is the minimum temperature for micelles to exist -- is called the Kraft point.

It is noteworthy that it is easier to dissolve micelles than single amphiphiles in water. The solubility of individual molecules is given by the curve below the Kraft point and by its extrapolation above that temperature (approximately). It may be recalled now that the difficulty in dissolving surfactants in water is due to the hydrocarbon tails. Together with the fact that micelles are aggregates, one postulates their *structure* (that surfactants are structure-forming species has been alluded to earlier). A section of a spherical micelle is shown schematically in Figure IV-3. The conventional explanation is that in this form the hydrocarbon chains are shielded from water and the entire structure, as seen by water, is hydrophilic and compatible with water. While basically

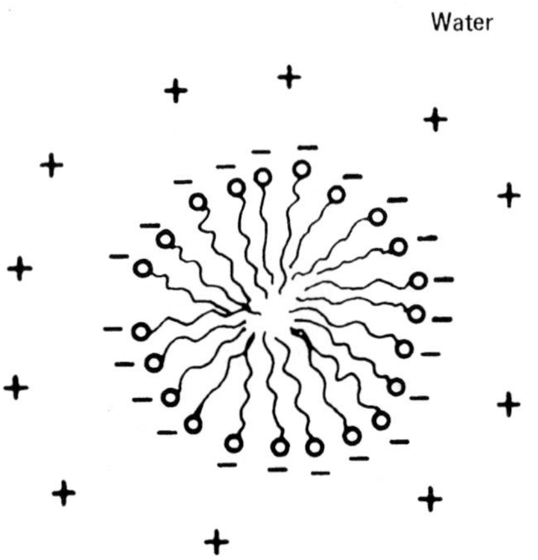

Figure IV-3. A schematic view of micellar aggregation of anionic surfactants is shown.

correct, this picture is an oversimplification. Detailed consideration of micelle geometry indicates that the chains are more randomly arranged throughout the micelle interior than shown in Figure IV-3 and that they are not totally shielded by the head groups in typical systems.

Figure IV-2 has been plotted on a linear scale in Figure IV-4. The region marked A is the region of single amphiphiles. Region B is the micellar region separated from A by the line of CMCs, the dashed line. In region C a swollen, solid ionic phase or a liquid crystalline phase coexists with the aqueous phase. More discussion on this phase "map" appears in Section 4.

To review the conventional reasons cited to account for micelle formation, one finds that single amphiphiles have a low solubility because of the hydrophobic nature of the hydrocarbon tails. In a micelle the hydrocarbon tails are largely shielded from water, and hence the overall solubility of the amphiphiles is enhanced. Thermodynamic aspects of surfactants have been treated very elegantly by Anacker (8) and by others (5, 9-11).

It is evident that for micelles to exist the Gibbs free energy of their formation has to be negative. That micelles form only above the CMC indicates that there are both positive and negative contributions to the Gibbs energy, and that only above the CMC do the latter prevail. One may write $\Delta G = \Delta H - T\Delta S$ at constant temperature to separate an energy

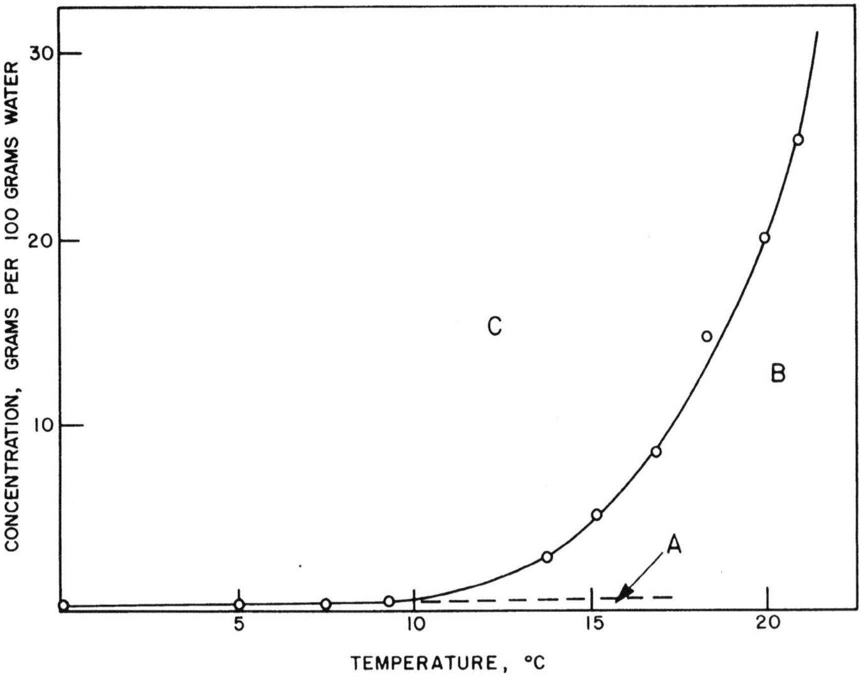

Figure IV-4. The various "phases" of sodium dodecyl sulfate in water are shown. Reprinted from Ref. (6) with permission.

contribution ΔH from an entropic contribution ΔS. Clearly, negative values of ΔH and positive values of ΔS favor negative values of ΔG. One expects that if the hydrophobic parts are shielded, H will decrease, as the hydrocarbon and water are incompatible whereas hydrocarbon tails amongst themselves are not. This decrease is partly offset by the fact that like ionic head groups are brought close to one another and they repel, i.e., give rise to a positive interaction energy. Noting that S is a measure of randomness, one might expect ΔS to be negative on micellization, at least as far as the amphiphiles are concerned. Thus one supposes that only when ΔH is sufficiently negative can micellization be realized.

The thermodynamics of formation of micelles can be investigated on the basis of the model

$$N\, M_1^- + N\alpha\, C^+ \rightleftarrows M_N^{-N(1-\alpha)} \qquad [IV\text{-}1]$$

where N monomers M_1^- aggregate to form a micelle M_N, the charge of which is determined by the number $N\alpha$ of the counter-ions C^+ it binds with. From thermodynamics

$$K' = \frac{a_N}{a_C^{N\alpha} a_1^N} \qquad [IV-2]$$

where K' is the equilibrium constant, a is the activity, and the subscripts N, C, and 1 denote the micelles, counterions, and single amphiphiles. If the activity coefficients are taken to be 1, then after some rearrangement one has

$$K = \frac{c_N}{c_C^{N\alpha} c_1^N}$$

where c represents concentration, and finally

$$\frac{c_N^*}{c_C^{*N\alpha} c_1^{*N}} = 1 \qquad [IV-3]$$

where $c^* = c/K^{\frac{1}{1-N(1+\alpha)}}$

Plots of c_N^* and c_1^* versus total surfactant concentration show that c_1^* increases up to some value of the total surfactant concentration beyond which the increase is negligible. Further, c_N^* is almost zero up to point but increases continuously above that concentration of the surfactant. Thus the concentration of the surfactant where any further addition mainly gives rise to micelles is the CMC. However, the CMC is seen not be a point but a narrow region beyond which micelle formation occurs rapidly. Some micelles can exist below the CMC, and some single amphiphiles can exist above the CMC. The transition region is sharper for larger values of N.

If it is assumed that the activity coefficients are 1, that the concentration of micelles is constant at the CMC, and that the standard state for the micelles is chosen to make $a_N \cong 1$ under these conditions,

$$K' \simeq \frac{1}{x_C^{N\alpha} x_1^N} \qquad [IV-4]$$

where x denotes a mole fraction. Immediately below the CMC, $x_1 \sim x_{CMC}$, the mole fraction of surfactant ions at the CMC. When there are no added electrolytes, $x_C = x_1 = x_{CMC}$, hence

$$-\ln K' = N(1+\alpha) \ln (x_{CMC}) \qquad [IV-5]$$

If there is complete charge neutralization, $\alpha = 1$, and

$$- \ln K' = 2N \ln (x_{CMC}) \qquad [IV-6]$$

If there is no binding with the counterions, or for the case of the nonionic surfactants, $\alpha = 0$ and

$$- \ln K' = N \ln (x_{CMC}) \qquad [IV-7]$$

Now the change in the standard state Gibbs free energy per mole of surfactant is

$$\Delta\bar{G}^0 = - (RT/N) \ln K' \qquad [IV-8]$$

which is required to be negative for the formation of a significant number of micelles. Eqs. [IV-5] - [IV-8] provide the means for obtaining $\Delta\bar{G}^0$ as a function of temperature, from which $\Delta\bar{H}^0$ can be calculated (see Problem IV-1). In view of the previous discussion, $\Delta\bar{H}^0$ is expected to be negative. From the experimental results this is found not to be the case at low temperatures in many systems. Indeed, the data show that a negative $\Delta\bar{G}^0$ is obtained from large and positive changes in entropy.

The source of the positive entropy changes was clarified by Ben Naim (12,13). Using a combination of thermodynamic reasoning and experimental data, he obtained the contribution δA^{HI} of "hydrophobic interactions" to the free energy of dimerization of simple hydrocarbons in various solvents (see Figure IV-5). This contribution represents the effect of the solvent on the dimerization process, i.e., it excludes the effect of the dispersion forces between the solute molecules which would be present even if dimerization occurred in a vacuum. Figure IV-5 shows that hydrophobic interactions favor dimerization and that the effect is greater in water than in less polar solvents. It should be noted that the term "hydrophobic interactions" survives in the literature even through the physical effect it describes is now known to be chiefly an entropic one.

Ben Naim (12,13) and O'Connell and Brugman (11) showed by simple statistical mechanical calculations using a hard sphere model that one source of the solvent effect on dimerization (and, by extension, on micellization) is that the nature of the volume available to the solvent molecules changes in such a way as to increase the entropy of the

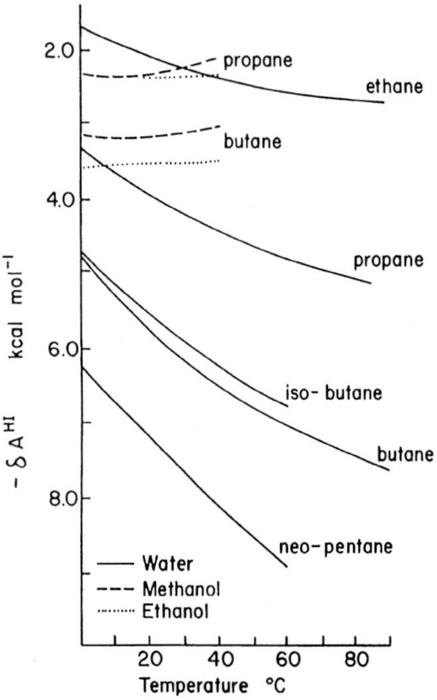

Figure IV-5. The free energy of dimerization of simple hydrocarbons in various solvents. The behavior in water is remarkably different. The data are from Ref. (13) as plotted in Ref. (14). Reprinted from Ref. (14) with permission.

solvent. In particular, inability of the relatively small solvent molecules to penetrate the relatively large solute domains occupied either by individual solute molecules or by micelles reduces the entropy of solvent molecules near the surfaces of these domains. The total entropy decrease of the solvent is approximately proportional to the overall area of the solute domains. Since formation of micelles decreases this overall area, entropic effects of this type favor micellization. They are also greater for solvents having smaller molecules, a prediction consistent with the observed large magnitude of hydrophobic interactions in water. Thus, entropy changes for both the surfactant and water must be considered in calculating $\Delta\bar{G}^0$ for micellization.

There is one other effect which gives rise to an increase in entropy on formation of micelles. It is known that water molecules form clusters around hydrophobic molecules (14). Since these form a structure, i.e., have an order, their entropies are lower than in the bulk. On the formation of micelles the hydrocarbon tails are no longer in contact with water, i.e., the molecules from the cluster are released into the water phase with an increase in entropy. Of course, there can be no structure formation at high temperatures and this effect disappears (8). Current research on the formation of micelles (15) indicates that at high temperatures, energy effects do become dominant and $\Delta \bar{H}^0$ becomes negative.

A short discussion of the aggregation number is now in order. In principle the aggregation number can vary from one to infinity with the concentration of each aggregate dictated by K_N, the equilibrium constant for an N-amphiphile aggregate, exp $[-\Delta G_N^0 /NRT]$. The results of semi-statistical mechanical and semi-phenomenological calculations for sodium octyl sulfate are shown in Figure IV-6, (16), where ln c_N has been plotted against N at 25°C for various values of c_1. Here c_N is the concentration in moles/ℓ. The question arises as to how an average $\bar{N}$ may be defined. Obviously different experimental methods give different values for the CMC and $\bar{N}$, since there are mass averages, averages over chemical potentials, detergency effects, sizes, electrical activities, etc. However, they all fall within a narrow band (4 - 6). With plots such as that of Figure IV-6, such averages can be calculated (16). At low surfactant concentrations the average shifts towards $\bar{N} = 1$, i.e., no significant micellization. At the CMC, $\bar{N}$ increases sharply to about 50 in Figure IV-6. Typical values of $\bar{N}$ range from 30 or 40 to a few hundred.

The CMC decreases and the aggregation number increases with increasing surfactant chain length and with increasing salinity since both these effects cause the surfactant to become less hydrophilic. On the other hand, the presence of branched chains or double bonds hinders micelle formation and thus increases the CMC. The same is true for surfactants with two polar groups. Addition of propanol or longer chain alcohols promotes micelle formation and lowers the CMC. Data supporting all these statements may be found in reference (5).

3. NONIONIC SURFACTANTS

As mentioned previously, the most frequently used nonionic surfactants are prepared by adding ethylene oxide to long chain hydrocarbons with terminal polar groups, e.g., -OH, -COOH, etc. This procedure introduces ethoxy groups, which are polar in nature (17) and form hydrogen bonds with

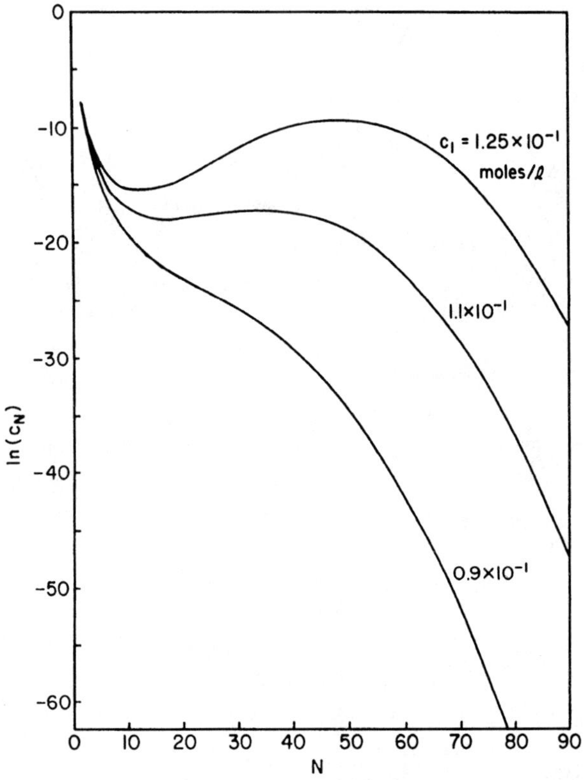

Figure IV-6. The calculated distribution of micelles of different aggregation numbers N of sodium octyl sulfate in water. All concentrations are in moles/ℓ. Reprinted from Ref. (16) with permission.

water. It increases the solubility of nonionic surfactants in water. However, the resulting molecules still have amphiphilic character and micelle formation takes place. As the reaction of the addition of the ethoxy groups is a polymerization reaction, these surfactants as produced commercially do not exist in pure form. Consequently the CMC is less well marked as shown in Figure IV-7. Of course, the commercial products can be separated if desired into fractions having narrower distributions of molecular weights (5). And small quantities of isomerically pure surfactants can be synthesized for laboratory studies.

It is worth pointing out that with comparable hydrocarbon chain lengths, CMC's are much lower for nonionics, e.g. 6.8×10^{-5} moles/ℓ for

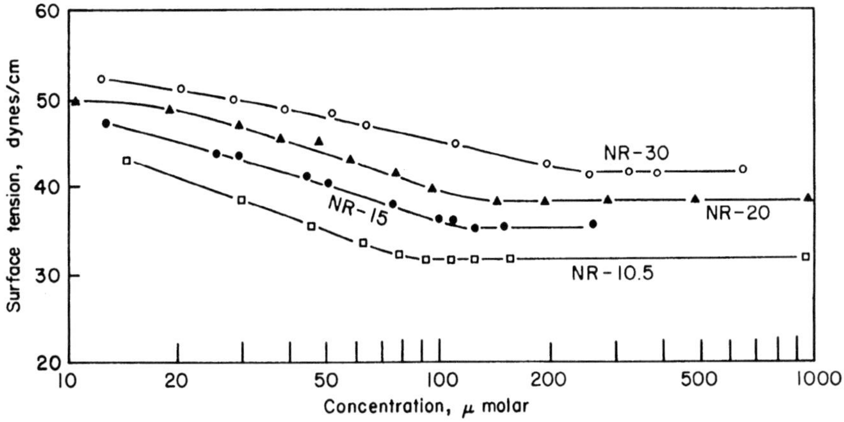

Figure IV-7. Surface tensions shown as functions of nonylphenol concentrations and the *average* number of moles of ethylene oxide per mole of alkyphenol (NR). The CMC's are less well marked (compare with Figure IV-1) because of the dispersity of the surfactants. Reprinted from Ref. (18) with permission. Copyright 1956 American Chemical Society.

$C_{12}E_5$ (an ethoxylated alcohol with a twelve-carbon tail and five ethoxy groups) at 23°C (2) but 8 x 10^{-3} moles/ℓ for sodium dodecyl sulfate at room temperature. The reason is that electrical repulsion strongly opposes micelle formation for anionic surfactants.

Sensitivity to temperature changes is a distinctive feature of the behavior of nonionic surfactants. A micellar solution turns turbid as the temperature is raised beyond a point called the cloud point temperature. The cloud point appears to be less well-defined than the Kraft point, but it marks the condition where a surfactant-rich liquid begins to form in equilibrium with the micellar solution. The structure of the surfactant-rich liquid is not presently known. It may be assumed that in micelles the ethoxy groups are hydrogen bonded with water. As the temperature is raised these relatively weak bonds begin to break and water is gradually driven out of the region occupied by the ethylene oxide chains. This picture is partly borne out by measurements which show that the volume per amphiphile in a micelle is decreased with increased temperatures (19).

Due to the length of the ethoxy groups, their conformations are of interest (20). In fact, calculations show that the various quantities, e.g., CMC, $\bar{N}$, etc., are affected by it (21). The main effect of the length of the polar groups is on the cloud point. For nearly pure surfactants, these are, for example, about 10°C for $C_{12}E_4$, about 30°C for $C_{12}E_5$, and about 50°C for $C_{12}E_6$ (3). The CMC is by far more sensitive to carbon number than to the number of ethoxy groups (9).

4. OTHER PHASES INVOLVING SURFACTANT AGGREGATES

So far we have dealt mainly with formation of small micelles at relatively low surfactant concentrations in aqueous solutions. For a typical anionic surfactant the micelles remain more or less spherical over a substantial concentration range above the CMC but ultimately become rodlike as the concentration continues to increase. At surfactant concentrations of perhaps 20% to 30% by weight a new phase appears which is birefringent and quite viscous. X-ray diffraction experiments demonstrate that this phase consists of many long, parallel, rodlike micelles arranged in a hexagonal array (Figure IV-8). The micelle interiors are apparently rather fluid, resembling liquid hydrocarbons in many respects. This phase is a "liquid crystal", i.e., it possesses substantial order but is not truly crystalline. It is usually called the normal hexagonal or simply the hexagonal phase. Occasionally it is referred to as the middle phase, a term which comes from the soapmaking industry.

At even higher surfactant concentrations the arrangement of surfactant molecules into bilayers becomes favorable, and another liquid crystalline phase known as the lamellar phase forms. Figure IV-9 shows the structure of this phase, which is called the neat phase by soapmakers. Because the basic structure of biological membranes is a bilayer of phospholipid molecules, which are surfactants, the lamellar phase has sometimes been used as a "model system" to gain insights regarding membrane behavior.

Figure IV-10 shows the effects on phase behavior of both surfactant concentration and temperature for an ionic surfactant. The liquid crystalline phases melt at sufficiently high temperatures. A similar diagram for a particular nonionic surfactant is given in Figure IV-11. A hexagonal phase occurs at moderately high surfactant concentrations at low temperatures, while the lamellar phase occurs at high surfactant concentrations and at temperatures where the ethylene oxide chains are somewhat dehydrated, as indicated above, and thus can pack more readily into bilayers. We note that the cloud point temperature mentioned in Section 3 lies, for a given surfactant concentration, at the boundary between the I_1 and $W+I_1$ regions in Figure IV-11. For this surfactant the cloud point is about 30°C in dilute solutions.

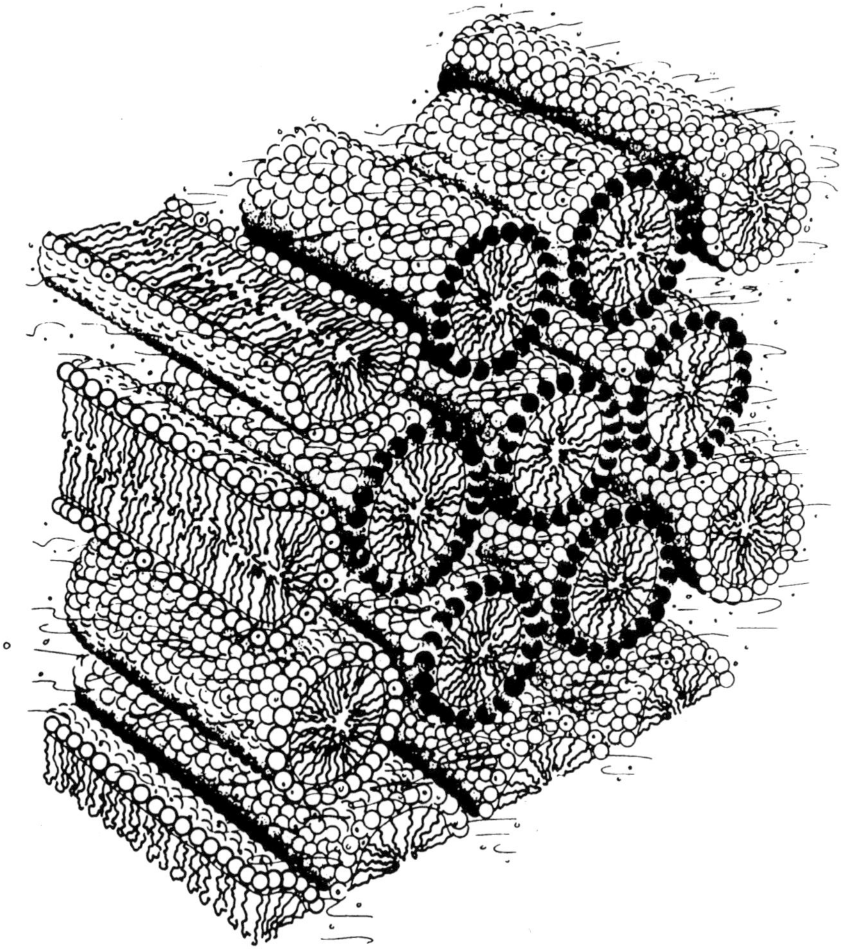

Figure IV-8. Schematic representation of normal hexagonal phase.

Another interesting feature of Figure IV-11 is the existence of an isotropic phase labelled I_3 at temperatures above those where water and the lamellar (L) phase coexist. The structure of this phase has not been established but it likely contains platelike micelles produced by breakup of the large bilayer sheets of the L phase. Similar behavior occurs in anionic surfactant systems, an increase in salinity or in the concentration of an oil-soluble alcohol effecting transformation from a lamellar phase to an isotropic phase which exhibits streaming birefringence (22). Note that

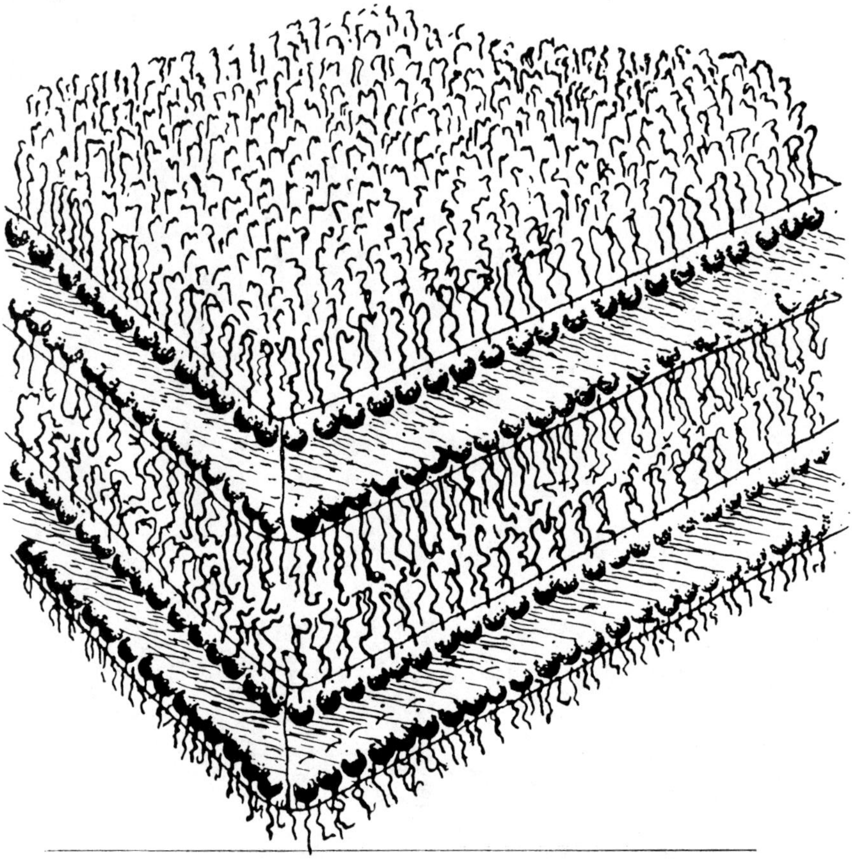

Figure IV-9. Schematic representation of the structure of lamellar phase.

for both nonionic and anionic systems the transformation accompanies changes which make the surfactant films more hydrophobic.

Other phase changes between micellar solutions and liquid crystalline phases or among different liquid crystalline phases can also be explained, to a large extent, in terms of changes in head group repulsion and the resulting changes in aggregate area-to-volume ratio (10,23). For example, electrical repulsion between adjacent charged polar groups decreases with increasing salinity owing to charge screening effects. The result is a shift toward aggregates with smaller areas per head group, i.e., away from spheres and toward cylinders and bilayers. Similarly, addition of a long-

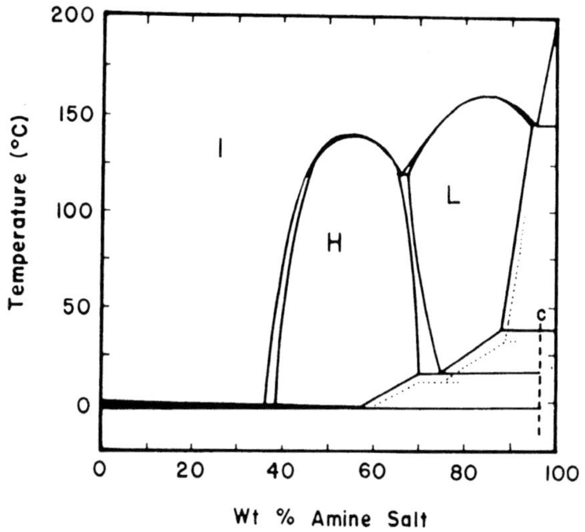

Figure IV-10. Binary phase diagram for dodecyl dimethylammonium chloride - water. I : micellar solution, H : normal hexagonal phase, L : lamellar phase. From Ref. (70) with permission.

chain alcohol or other molecule with a small, neutral head group favors aggregates with denser packing of head groups and hence smaller area-to-volume ratios. Figure IV-12 shows, for instance, that addition of decanol to sodium octanoate-water mixtures causes transformation from a hexagonal to a lamellar phase and even to a reverse hexagonal phase. In this last phase water is inside the cylindrical micelles and the surfactant tails are directed outward, just the opposite of the arrangement in the normal hexagonal phase of Figure IV-8.

"Cubic" phases, probably with differing structures, constitute yet another type of liquid crystal known to exist in many systems containing surfactants (24,25). Indeed, it is plain that, even in the absence of oil, the phase behavior of systems containing water and one or more surface-active compounds can be quite complex.

5. SURFACE FILMS OF INSOLUBLE SURFACTANTS

Let us turn away briefly from the solution properties and bulk phase behavior of surfactants. Much has been learned about the properties of

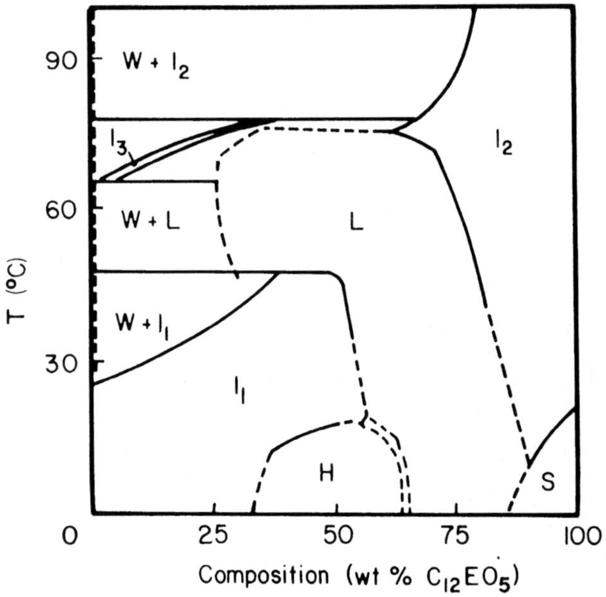

Figure IV-11. Phase diagram of the $C_{12}E_5$-water system over the temperature range 0-100°C (H: normal hexagonal phase, L: lamellar phase, W: water, I_1, I_2, I_3: isotropic liquid phases, S: solid phase). Reprinted from (23) with permission.

surfactant molecules at interfaces by study of molecules with hydrocarbon chains so long that they are virtually insoluble in water.

The main experimental tool for these studies has been the film balance (see Figure IV-13). Various workers including Pockels, Langmuir, and Adam made major contributions to its development (see (26)). A small, known quantity of the surfactant to be studied is dissolved in a volatile solvent and deposited carefully by pipette on the surface of a pool of water. The solvent is chosen so that it spreads rapidly over the water and then evaporates, leaving the surfactant uniformly distributed as a monomolecular layer or monolayer in the region between the two barriers. One of the barriers is movable, so that the area occupied by the surfactant film can be varied. A torsion balance is provided to measure the surface pressure, i.e., the difference between the surface tension of pure water and that of the film-covered interface. Alternately, a Wilhelmy plate may be used to measure surface tension in the film region.

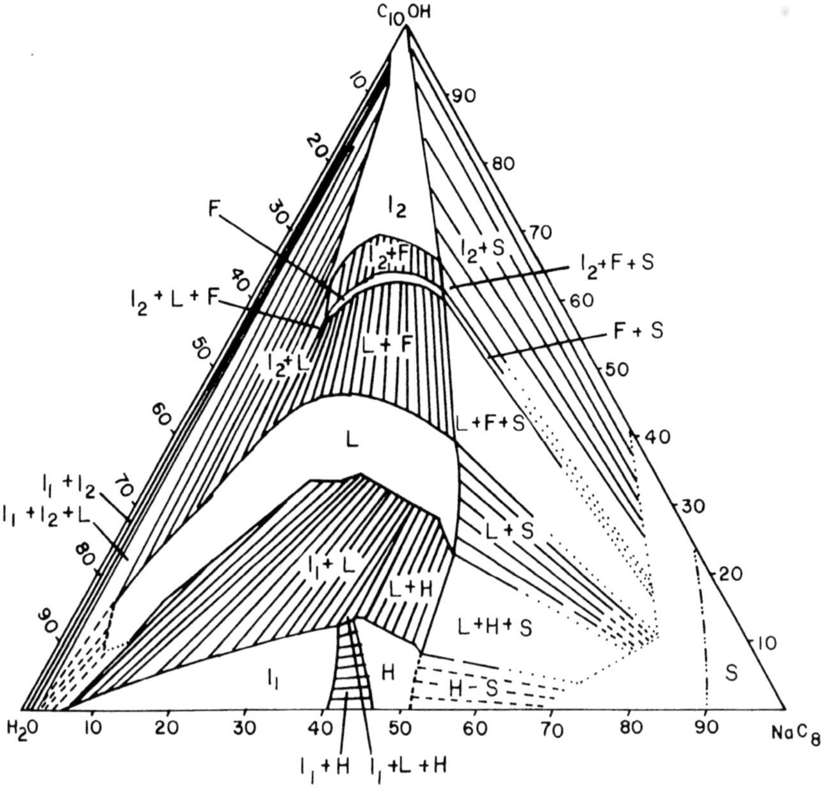

Figure IV-12. Phase diagram for the three-component system sodium octanoate-water-decanol at 293°K. Symbols as in Figure IV-11 but with F : reverse hexagonal phase. After (71) with permission.

Film balance data are normally presented as plots of the surface pressure π as a function of the area σ per molecule. The latter is obtained by dividing the total area of the region between the barriers by the known number of surfactant molecules deposited.

Figure IV-14 is illustrative of key aspects of results frequently obtained. For very large values of σ, e.g., several nm^2 per molecule, the surfactant molecules move independently along the surface owing to random thermal motion, and the two-dimensional ideal gas equation applies

$$\pi \sigma = kT \qquad [IV-9]$$

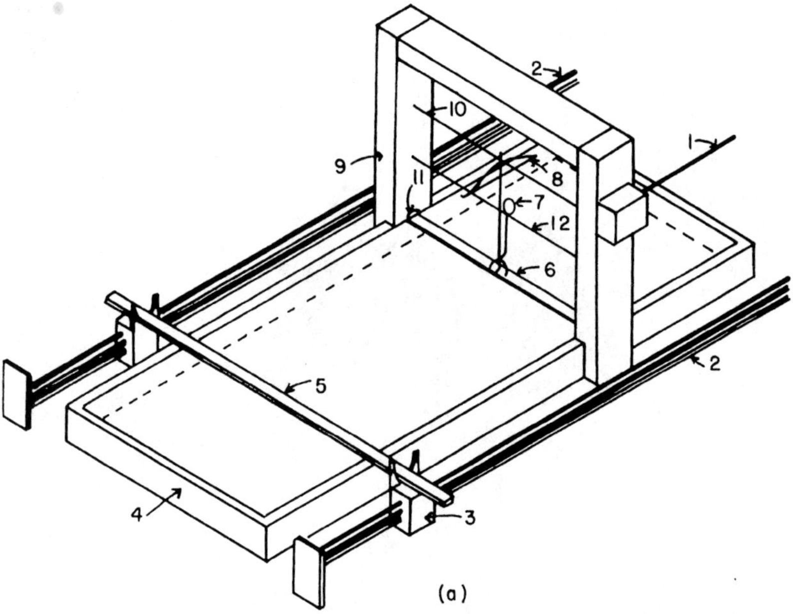

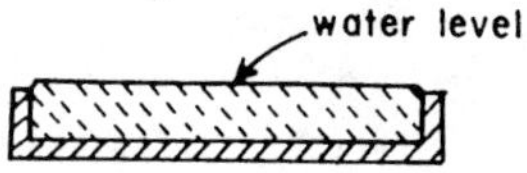

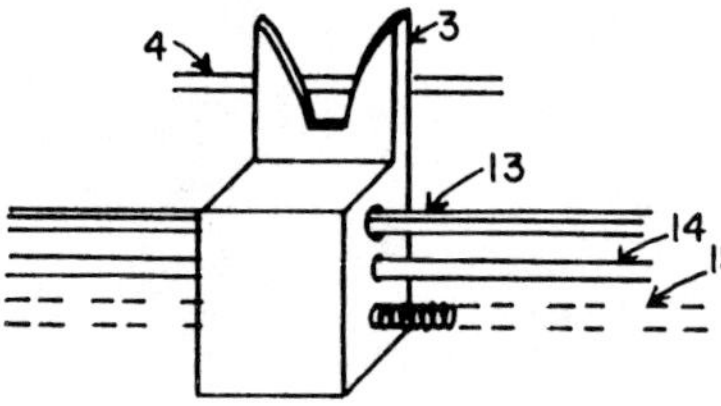

Figure IV-13. A modern film balance; 1. torsion wire control; 2. sweep control; 3. sweep holder; 4. trough; 5. sweep; 6 float; 7. mirror; 8. calibration arm; 9. head; 10. main torsion wire; 11, gold foil barriers; 12. wire for mirror; 13. elevation control; 14. guide; 15. traverse. Reprinted with permission from Ref. (72).

As indicated in Chapter II, this equation can be modified to account for nonideal effects at smaller values of σ by methods similar to those used for three-dimensional equations of state for nonideal gases.

For many insoluble surfactants it is found that π remains constant when σ is varied in a particular range (line L_1-G in Figure IV-14). When

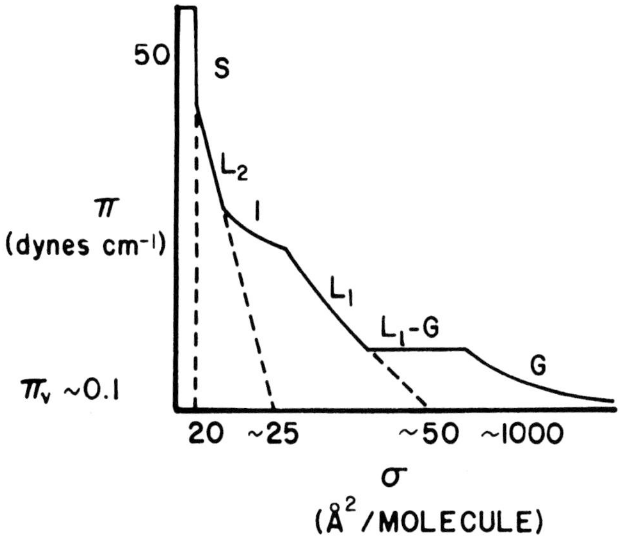

Figure IV-14. Composite π-σ isotherm which includes a wide assortment of monolayer phenomena. Note that the scale of the figure is not uniform so that various features may be included on one set of coordinates. Reprinted for Ref. (73) with permission.

the film is of tetradecanol, for instance, such behavior occurs for π = 0.11 mN/m at 15°C. Such behavior has been interpreted as a transition from a film which behaves as a gas to one which behaves as a liquid. There is, as in ordinary three-dimensional fluids, a critical temperature above which this condensation behavior is not seen.

Highly compact films are only slightly compressible and behave as two-dimensional liquids or solids. The surfactant molecules are oriented nearly perpendicular to the surface, in contrast to the situation for two-dimensional gases where the molecules are far apart and the hydrocarbon chains lie along the surface. One use of film balance data for compact films is to estimate the cross sectional area of a surfactant molecule. As shown in Figure IV-15, the π - σ relationship for compact films is nearly linear. Its intersection with the horizontal axis upon extrapolation yields the area per molecule of a compact film subject to no lateral pressure. The value obtained for straight-chain uncharged compounds such as alcohols and fatty acids is about 0.20 nm^2 per molecule.

Continued compression of a compact film produces film "collapse", in which some molecules are ejected from the monolayer. From a thermodynamic

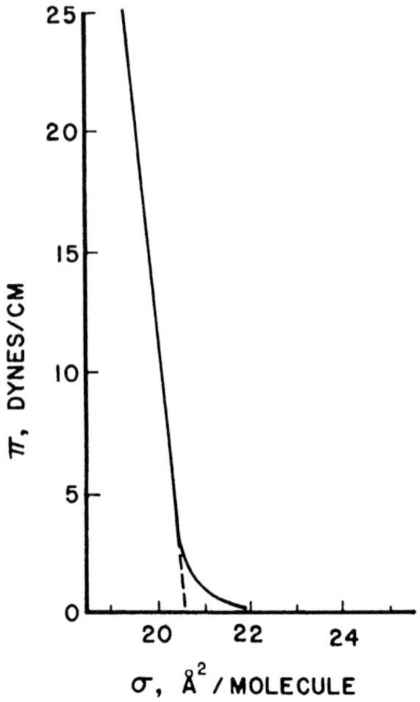

Figure IV-15. π-σ curve of n-octadecylamine on 10^{-3}N NaOH at 20°C (compression at 3 Å²/molecule/min.). After Ref. (26).

point of view, a bulk phase of liquid or solid surfactant should begin to form at the "equilibrium spreading pressure" where it is in equilibrium with the compressed monolayer. While such behavior is often found when the bulk phase is a liquid, the more typical behavior when the bulk phase is a solid is that the monolayer retains its integrity until pressures well beyond the equilibrium spreading pressure are reached. Ries and co-workers (27) have studied film collapse using electron microscopy. Harris (28) developed a model of collapse based on the theory of buckling of an elastic plate. Smith and Berg (29) have treated collapse in terms of nucleation and growth of the bulk surfactant phase.

6. SOLUBILIZATION AND MICROEMULSIONS

Surfactant solutions with concentrations above the CMC can dissolve considerably larger quantities of organic materials than can pure water or surfactant solutions at concentrations below the CMC. This enhanced solubility is of great importance to the pharmaceutical and cosmetics industries and plays a role in detergency processes. The implication of this behavior is clearly that the added materials are "solubilized" in the micelles. Hydrocarbons and other nonpolar compounds are thought to be incorporated in the micelle interiors (30; see Figure IV-16a). Molecules having some surface activity distribute themselves among the surfactant molecules (Figure IV-16b). Rather polar substances may even occupy positions at the micelle surfaces (Figure IV-16c).

As discussed previously, a decrease in repulsion between adjacent head groups favors surfactant aggregates of smaller curvature and leads to the formation of liquid crystalline phases in the absence of oil. When oil is present, the curvature reduction is achieved instead by an increase in drop size. Oil molecules occupy the interior of the drops, the surfactant molecules remaining at the surfaces where their polar groups can contact water. When the swollen micelles reach diameters exceeding 4 or 5 nm, the resulting phase is frequently called a "microemulsion" (see Figure IV-17). Often solubilization of oil increases greatly for small changes in system composition or temperature, so that a definite region where microemulsions exist can be identified (31). Use of the separate terms micellar solution and microemulsion to describe systems containing small and large drops respectively should not be allowed to obscure the fact that both can lie in different parts of the region of existence of a single thermodynamically stable phase.

Surfactant aggregation in oleic phases is more gradual than in water, formation of various small oligomers typically preceding that of larger aggregates. The presence of appreciable quantities of water generally promotes formation of "reverse micelles" with the polar groups directed

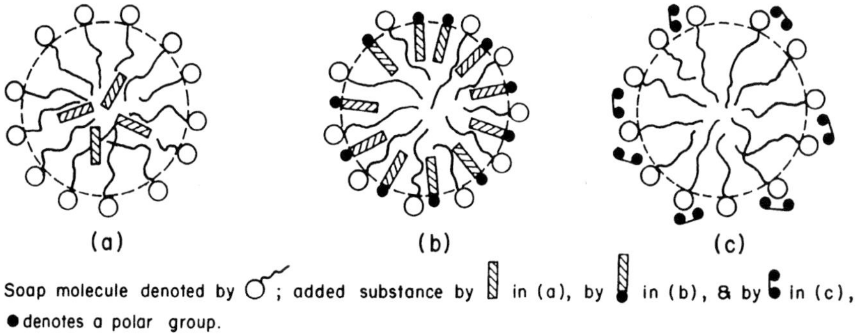
(a)
(b)
(c)
Soap molecule denoted by ; added substance by in (a), by in (b), & by in (c),
● denotes a polar group.

Figure IV-16. Schematic view of solubilization. Reprinted from Ref. (30) with permission.

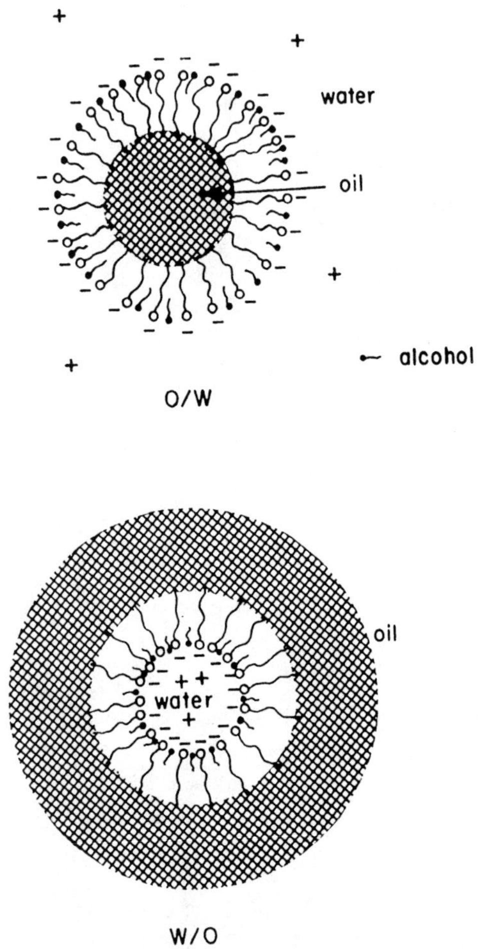

Figure IV-17. Schematic diagrams of oil-in-water (o/w) and water-in-oil (w/o) microemulsions. The small molecules shown are the co-surfactant.

inward since water solubilized within the micelles is effectively removed from contact with the oil. As would be expected, solutions containing reverse micelles are capable of solubilizing various polar compounds. We note that a decrease in micelle curvature is favored by an increase in repulsion between head groups in this case, and water-in-oil microemulsions form under appropriate conditions (Figure IV-17). Indeed, most of the pioneering studies of microemulsions by Schulman and co-workers (32) dealt with oil-continuous phases.

A wide range of experimental techniques has been employed in recent years to gain a better understanding of microemulsions and to investigate the possibility of using micellar solutions and microemulsions to dramatically alter the rates of certain chemical reactions (see, for example, the collections of papers edited by Mittal and by Robb under general references at the end of this chapter). A detailed review of such work is beyond the scope of this book. Important aspects of microemulsion thermodynamics and phase behavior are considered, however, in the next three sections.

7. THERMODYNAMICS OF MICROEMULSIONS

As pointed out in Chapter III, most colloidal dispersions, including emulsions, are unstable from a thermodynamic point of view owing to their large interfacial area. How then can microemulsions be thermodynamically stable? Some years ago Shchukin and Rehbinder (33) suggested that a colloidal dispersion could be thermodynamically stable provided that interfacial tension was low enough that the increase in interfacial energy accompanying dispersion of one phase in the other could be outweighed by the free energy decrease associated with the entropy of dispersion. Ruckenstein and Chi (34) recognized the importance of this effect for microemulsions and developed a suitable analysis to describe it quantitatively. Subsequently, Ruckenstein (35) also pointed out that the free energy decrease accompanying adsorption of surfactant molecules from a bulk phase favors the existence of a large interfacial area and hence plays a major role in stabilizing microemulsions. In his actual calculations attractive and electrical repulsive forces between nearby drops were also included using the methods described in Chapter III.

Figure IV-18 shows the general form of the calculated free energy of microemulsion formation as a function of drop radius (34). Interfacial tension is lowest for curve A and highest for curve D. For curve B the predicted radius is R* where the free energy is minimized. The negative value of ΔG_M^* implies that the microemulsion is thermodynamically stable. The same is true for curve A, but the minimum for curve C occurs for a positive ΔG_M indicating that any microemulsion which forms is metastable. Not even a metastable microemulsion is possible for curve D. For given amounts of oil, water, and surfactant such curves can be calculated for both oil-in-water and water-in-oil microemulsions. The arrangement having the lower value of ΔG_M^* would be expected to occur.

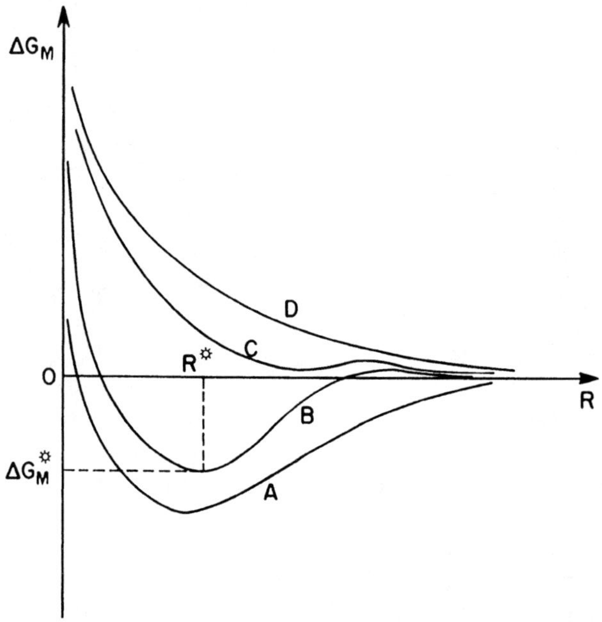

Figure IV-18. The Gibbs free energy of formation of microemulsions ΔG_M has been shown as a function of the droplet radii R, under various conditions. The curves A and B show that negative values ΔG_M can be obtained, i.e., that microemulsions can form spontaneously. The actual radius is R* corresponding to a minimum ΔG_M^*. The curve C shows kinetic stability and D is unstable. Reprinted for Ref. (34) with permission.

An additional factor which frequently is important is the work required to bend a surfactant film at constant area (36). As discussed in Problem I-4, an additional term should be added to Eq. [I-9] for interfaces which are of low tension or high curvature--conditions which are clearly fulfilled for microemulsions. Especially when a microemulsion is in equilibrium with an excess phase, e.g., an oil-in-water microemulsion with excess oil, bending effects play a major role in determining drop size. That is, drop size is typically not far from the "natural" radius the film assumes at an interface between oil and water. When head group repulsion is large, for example, the natural radius is small with oil the interior phase.

Miller and Neogi and Mukherjee *et al* (37) developed the appropriate thermodynamic relations for this case using concepts based on Hill's thermodynamics of small systems (38) and and explicitly including bending effects. They found that with a given amount of surfactant present, more drops of smaller size are predicted than would be expected based on the pertinent natural radius. In this case the energy required to bend the film beyond its natural radius is more than offset by the decrease in free energy associated with the increased entropy of the more numerous drops. A simple analysis of the thermodynamics of microemulsions including bending effects is outlined in Problem IV-6. Other analyses of a microemulsion with an excess phase have also been given (39-42).

8. PHASE BEHAVIOR OF OIL-WATER-SURFACTANT SYSTEMS

Let us consider mixtures containing a few percent of an anionic surfactant, approximately equal volumes of oil and NaCl brine, and usually some short-chain alcohol used as a "cosurfactant". If only the salinity is varied a general pattern of phase behavior has been observed (43, Figure IV-19).

At low salinities an oil-in-water microemulsion coexists with nearly pure oil, as shown in Figure IV-19. Solubilization of oil and hence microemulsion drop size increase with increasing salinity, i.e., as repulsion between the charged head groups decreases. The interfacial tension between phases is low by ordinary standards but is normally above 0.01 mN/m. Because the microemulsion is below the oil in the test tubes of Figure IV-19, it is sometimes called a "lower phase" microemulsion.

At high salinities the situation is reversed and an oil-continuous "upper phase" microemulsion coexists with excess brine, as the figure shows. Drop size increases with decreasing salinity, i.e., with increasing repulsion between head groups. Interfacial tensions in this case as well usually exceed 0.01 mN/m.

At intermediate salinities three phases are present, as Figure IV-19 indicates. Nearly pure oil and brine are in equilibrium with a microemulsion phase which contains almost all the surfactant in the system (except very near the transitions to the above mentioned two-phase regions) as well as significant quantities of oil and brine. This interesting and important phase is called the "surfactant" or "middle" phase. In the low salinity portion of its region of existence its interfacial tension with brine is extremely low (see Figure IV-20). Conversely, in the high salinity portion of its region of existence its interfacial tension with oil is ultralow. Between these extremes in suitably chosen systems is a

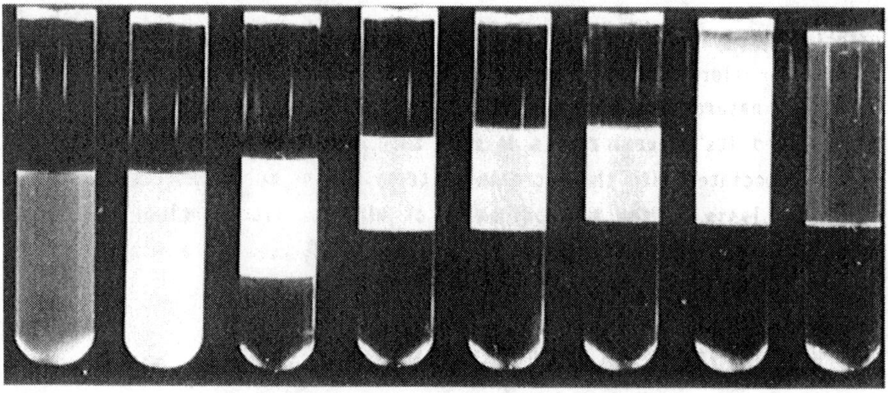

Figure IV-19. Phase behavior as a function of salinity for a system containing a petroleum sulfonate surfactant, a short-chain alcohol, and approximately equal volumes of oil and brine.

region where the surfactant phase has the remarkable property of exhibiting ultralow tension with oil and brine simultaneously (Figure IV-20). The salinity where the two tensions are equal is called the "optimum" salinity and is of great importance in designing surfactant processes for enhanced oil recovery.

At the optimum salinity the amounts of oil and brine solubilized in the surfactant phase are approximately equal. These quantities can be obtained to a good approximation from measurements of phase volumes in samples such as those shown in Figure IV-19 provided that all surfactant is assumed to reside in the microemulsion phases. Oil solubilization increases with increasing salinity (Figure IV-20) while brine solubilization increases with decreasing salinity. Indeed, correlations of wide (but not universal) applicability have been developed for relating microemulsion-oil interfacial tension to oil solubilization and microemulsion-brine interfacial tension to brine solubilization (43). With these correlations interfacial tensions can be estimated from information on phase volumes as a function of salinity.

The structure of the surfactant phase has not been determined definitively. Based on their ultracentrifuge results Hwan *et al* (44) proposed that it was water-continuous near the transition to the water-continuous lower phase and oil-continuous near the transition to the oil-continuous upper phase. A theoretical model was also presented which showed, using simple statistical mechanics and the principles of colloid

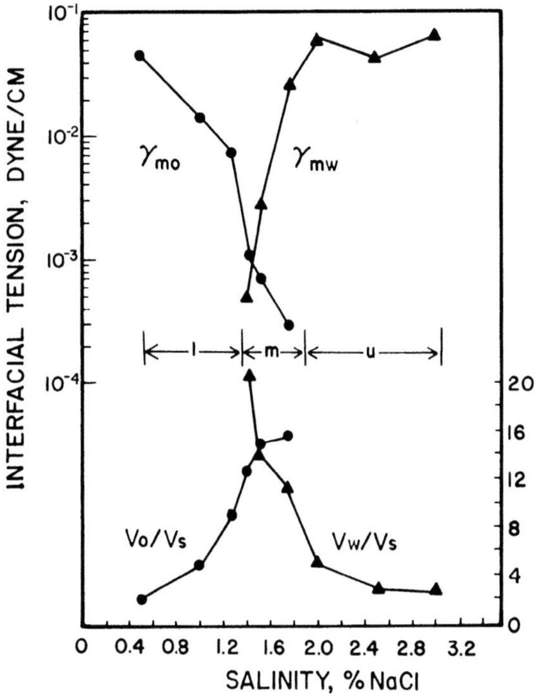

Figure IV-20. Interfacial tensions and solubilization parameters for a system containing 2% of a petroleum sulfonate surfactant and 1% of a short-chain alcohol. γ_{mo} is the interfacial tension between the microemulsion and an excess oil phase. (V_o/V_s) is the volumetric ratio of solubilized oil to surfactant in the microemulsion phase. γ_{mw} and (V_w/V_s) are defined in a similar manner. Reprinted from Ref. (43) with permission.

stability discussed in Chapter III, how an oil-in-water microemulsion can separate into a more concentrated water-continuous microemulsion and excess water or brine (45). Ultralow interfacial tensions were predicted between the two phases formed in agreement with the observed low tensions between the middle phase and brine near the lower-to-middle phase transition.

Scriven (46) has proposed that the surfactant phase is bicontinuous, perhaps resembling consolidated porous media where both solid and pore space are continuous or certain mathematical models of bicontinuous structures. Theoretical models of bicontinuous microemulsions have been developed which are capable of predicting ultralow tensions (47). It seems almost certain that the surfactant phase is bicontinuous near the optimum salinity though the dimensions and approximate geometry of the oil-continuous and water-continuous regions are not known. Probably their shapes change continually as a result of thermal fluctuations.

Yet another proposal is that since the natural curvature of the surfactant films is quite low near optimum salinity, the surfactant phase is basically lamellar although with considerable breakage and distortion of the oil and water layers due to thermal motion (48,49). It is to be hoped that further studies with various techniques including neutron scattering will resolve the question of surfactant phase structure.

One additional feature of the phase behavior described above merits comment. Variation of phase volumes and interfacial tension in the surfactant phase region suggests that both the lower-to-middle and middle-to-upper phase transitions are associated with critical phenomena (50-52). Light scattering results near the transition also support such an interpretation (52-53). Thus, to the extent that the systems involved may be represented by ternary phase diagrams, the variation of phase behavior with salinity is as illustrated schematically in Figure IV-21. Note that critical end points where a microemulsion separates into two phases are associated with the appearance and disappearance of the three-phase region. For simplicity other aspects of the phase behavior including the existence of liquid crystalline phases have been ignored in the phase diagrams of Figure IV-20.

9. EFFECT OF COMPOSITION CHANGES

So far we have considered only the effect of salinity on microemulsion phase behavior. Effects of other compositional variables and of temperature can be understood in terms of a single unifying principle. First we summarize the results of numerous experiments by several workers in systems containing oil, brine, an anionic surfactant, and a short-chain alcohol.

a. If the oil phase is a straight chain hydrocarbon, increasing oil chain length causes optimum salinity to increase (with temperature and other compositional variables fixed).
b. Increasing the hydrocarbon chain length of the surfactant causes optimum salinity to decrease.
c. Increasing the concentration of a relatively oil-soluble alcohol such as n-pentanol or n-hexanol causes optimum salinity to decrease.

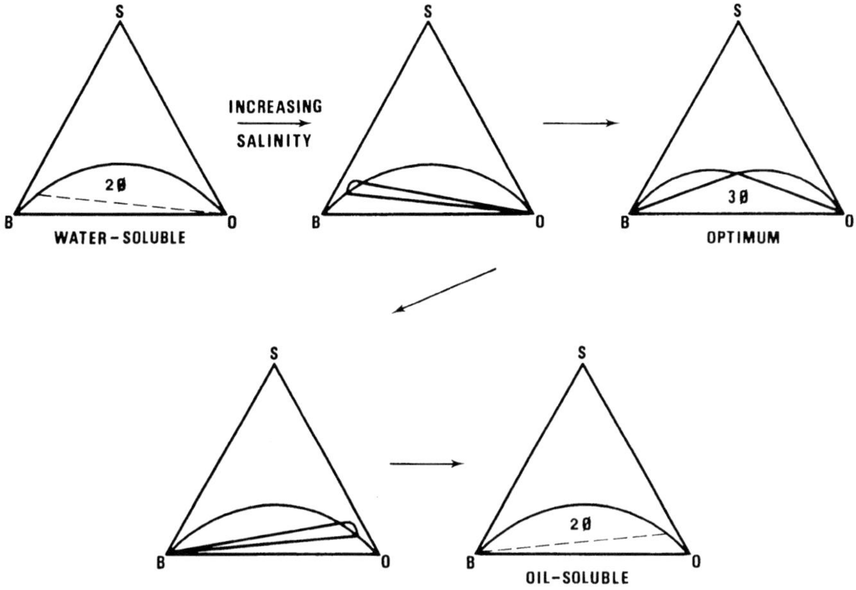

Figure IV-21. Schematic pseudoternary phase behavior. B, O, and S designate brine, oil, and surfactant respectively.

d. Increasing the chain length of a relatively oil-soluble alcohol, e.g., from n-pentanol to n-hexanol, causes optimum salinity to decrease.

e. Increasing the concentration of a relatively water-soluble alcohol such as isopropanol causes optimum salinity to increase.

f. Increasing temperature causes optimum salinity to increase.

The key to understanding these results, as Robbins (54) and others have noted, is recognition that optimum salinity represents a balance between oil-soluble (hydrophobic) and water-soluble (hydrophilic) tendencies of the surfactant-alcohol films present at the surfaces of the microemulsion drops. If optimum salinity is to be maintained, any change (in a variable other than salinity) which makes the surfactant-alcohol film more water-soluble must be balanced by an increase in salinity, which make it more oil-soluble. An example is statement (a) above, where an increase in oil chain length makes the surfactant more water-soluble. In a similar manner, any change which makes the surfactant-alcohol film more oil-soluble must be balanced by a decrease in salinity, which makes it more water-soluble. This situation is represented, for example, by statement (b) since increasing the chain length of a surfactant clearly makes it more oil-soluble.

The statements involving alcohol deserve some comment. Salter (55) showed that, with a given surfactant, adding one particular alcohol might produce no change in optimum salinity. Adding longer chain alcohols caused decreases in optimum salinity, however, and adding shorter-chain alcohols caused increases in optimum salinity. It is noteworthy that while the surfactant molecules are probably almost all at the surfaces of the microemulsion drops, the alcohols partition between the drop surfaces and the oil and/or brine regions (56). Partitioning behavior depends, of course, on such variables as salinity and temperature. In any case, the alcohol in the surfactant films at the drop surfaces is an important factor influencing phase behavior, as indicated above.

A quantitative correlation of how various compositional variables influence optimum salinity for petroleum sulfonate surfactants has been developed by Salager *et al* (57).

$$\ell n\ S_{opt} = 0.16\ (ACN) + f(A) - \sigma + 0.01\ (T-25) \qquad [IV-10]$$

where S_{opt} is optimum salinity in g/dℓ, ACN is the equivalent alkane carbon number of the oil, e.g., the chain length for straight chain hydrocarbons, f(A) is a function depending only on alcohol composition and concentration, σ is a parameter depending only on surfactant structure, and T is temperature in °C. It is interesting that the oil, alcohol, surfactant, and temperature can be considered, for practical purposes, as making independent contributions to optimum salinity.

Wade, Schechter, and co-workers (58) have proposed that mixtures of oils and of surfactants can be described by Eq. [IV-10] provided that average values of ACN and σ are used:

$$(ACN)_{mix} = \sum_i x_i\ (ACN)_i \qquad [IV-11]$$

$$\sigma_{mix} = \sum_i x_i\ \sigma_i \qquad [IV-12]$$

Here x_i is the mole fraction of each component. Further discussion of oil mixtures and of handling oils with naphthenic and aromatic groups may be found in references (59) and (60).

For certain classes of anionic surfactants other effects are also important. For example, changes in pH affect the degree of ionization and hence the phase behavior when the surfactants include organic acids, amines, or other pH-sensitive compounds. An increase in the degree of ionization produces a more hydrophilic surfactant film and hence, other things being equal, increases optimum salinity (61).

To avoid complicating the discussion, we have thus far dealt only with anionic surfactants. But Shinoda and co-workers have found that nonionic surfactants such as ethoxylated alcohols exhibit the lower-to-middle-to-upper phase transition with increasing temperature (62). Variation of interfacial tension and solubilization with temperature is similar to that described above for anionics with varying salinity. The temperature where the surfactant phase has equal interfacial tensions with oil and water is called the phase inversion temperature (PIT).

The effects of surfactant and oil chain length on phase behavior of nonionic surfactants are the same as given above for anionics, i.e., rules a and b above apply with PIT replacing optimum salinity. As might be expected, nonionics are much less sensitive to salt than anionics, but the direction of shift in phase behavior is the same. Finally, we remark that an increase in the length of the ethylene oxide chain makes the surfactant more hydrophilic, i.e., it increases the PIT.

Example IV-1. Effect of Composition and Temperature on Optimum Salinity.

A given surfactant-alcohol formulation has an optimum salinity of 1.7 wt % NaCl at 25°C with n-decane as the oil. Estimate the optimum salinity if (a) the oil is changed to n-tetradecane or (b) the temperature is raised to 40°C.

Solution Equation [IV-10] applies.

(a) If ACN changes from 10 to 14, $\ell n\ S_{opt}$ increases by 0.16 x 4 = 0.64 from 0.53 to 1.17. S_{opt} = 3.2 wt % NaCl
(b) If T increases from 25° to 40°, $\ell n\ S_{opt}$ increases by 0.15. S_{opt} = 2.0% NaCl

Note that the temperature effect is relatively weak for anionic surfactants.

10. APPLICATIONS OF SURFACTANTS - EMULSIONS

In contrast to microemulsions, ordinary emulsions are thermodynamically unstable. But they can be stable in a practical sense if the energy barrier to flocculation is sufficiently high. As with other colloidal dispersions, this energy barrier may be electrical in nature if the drops are charged, if water is the continuous phase, and if the ionic strength is not too high (cf. the discussion of DLVO theory in Chapter III). Adsorbed polymers and even very small solid particles which are not completely wet by either phase and thus accumulate at the drop surfaces can provide stability as well.

The phase behavior of oil-water-surfactant systems also has an important influence on emulsion stability, however. Friberg and co-workers have shown that the presence of a liquid crystalline phase instead of a simple surfactant monolayer at the drop surfaces can greatly enhance stability (63). Thus, shifting overall system composition from a two-phase region of the equilibrium phase diagram with aqueous and oleic phases present to a three-phase region where liquid crystal is also found increases emulsion stability dramatically. In a similar manner foams containing some liquid crystalline material are frequently quite stable. Manev *et al* (64) have shown that even in some concentrated aqueous surfactant solutions where no bulk liquid crystal exists at equilibrium, the surfactant in a thin liquid film can form a liquid crystal and thus impart stability to a foam.

Even in portions of the phase diagram where microemulsions are found but not liquid crystals, emulsion stability is strongly correlated with phase behavior. As several groups have observed, emulsions break more rapidly in the region near optimum salinity where the surfactant phase coexists with excess oil and brine than in the regions at lower and higher salinities where a microemulsion and a single excess phase are present (65). It is likely that far from the optimum salinity, local compositions during mixing are such that a liquid crystalline phase forms. Although not an equilibrium phase, it apparently dissolves very slowly once formed and so can provide enhanced stability. Or perhaps it exists only in thin films as in the situation mentioned above for foams (64).

Generally speaking, surfactants preferentially soluble in water are found to be best for making oil-in-water (o/w) emulsions and surfactants preferentially soluble in oil for water-in-oil (w/o) emulsions. Davies (67) has proposed an explanation for this behavior. Basically, both types of emulsions are formed during mixing, but only the type which flocculates and coalesces more slowly persists after agitation ceases. One might expect, for instance, o/w emulsions to flocculate and coalesce slowly for water soluble surfactants because such surfactants are strongly hydrated and/or electrically charged, providing a substantial energy barrier for flocculation and for rupture of the thin aqueous films between nearby oil drops. Since any w/o emulsion present coalesces rapidly when mixing stops in this case, the final dispersion is of the o/w type.

One method of characterizing the relative oil and water solubility of surfactants is the "HLB" value (for hydrophilic-lipophilic balance) originally proposed by Griffin (66). The smaller the HLB value, the more oil soluble the surfactant with a HLB value of about 7 corresponding to a surfactant having equal oil and water solubilities. Methods of calculating HLB values from knowledge of surfactant structure are described in reference (67). Two surfactants with different HLB values may be blended to obtain any desired intermediate value.

The relative amounts of oil and water employed also influence phase continuity. Other factors being equal, o/w emulsions are formed when only small amounts of oil are present and w/o emulsions when only small amounts of water are present. Indeed, stable, dilute emulsions can be formed without any surfactant at all if a procedure such as ultrasonic vibration is used which generates very small drops. On the other hand, stable emulsions with high concentrations of the dispersed phase - upwards of 90% by volume - can be produced with suitable choice of surfactant and mixing procedure, e.g., gradual addition of the dispersed phase. We note that for such high volume fractions to be reached the drops must be distorted from spheres into polyhedra, the arrangement becoming much the same as in a foam of high gas content (see Chapter V).

11. APPLICATIONS OF SURFACTANTS - DETERGENCY

The detergent action of surfactants proceeds by different mechanisms under different conditions. Liquid soils of an oily nature are often removed from the solid being cleaned by a "rollback" mechanism. Typically, the soiled material is poorly wet by ordinary water. But if a surfactant is present, it adsorbs at the water-solid interface, lowering the interfacial tension there. According to Young's Equation [II-1], the contact angle measured through the water is thereby reduced. With sufficient reduction in contact angle the aqueous solution spreads along the solid, causing the initial film of oil to form a drop. Ultimately, the drop is completely displaced from the solid or at least made large enough that it is readily removed by the agitation of the detergent solution.

In some cases where soils are highly viscous under laundering conditions, the rollback mechanism is ineffective. Soils which have relatively low viscosities and thus are readily removed by rollback at high temperatures may thus require a different detergency mechanism if low temperatures are desired for the cleaning process. One such mechanism involves formation of a liquid crystalline or microemulsion phase near the original interface between soil and surfactant solution. This phase contains water, surfactant, and the soil. If it forms during "spontaneous emulsification" and streams away from the interface continuously (68), or if it can be removed from the solid by the mechanical action of the cleaning bath, it provides an efficient means of soil removal. Intelligent application of this mechanism clearly requires knowledge of the phase behavior of surfactant systems discussed above.

Another possibility is that the soil slowly dissolves in the surfactant solution. Since the solubility of rather nonpolar compounds in water is low, a mechanism by which micelles are transported to the solid surface, acquire some solubilized soil molecules while temporarily adsorbed, and then return to the bulk solution after desorption has been proposed (69).

From this brief discussion it is clear that much remains to be learned about detergency. Since it is a nonequilibrium process, information on spreading rates and on spontaneous emulsification and related diffusional phenomena are pertinent. These and other nonequilibrium phenomena are considered in the following chapters.

REFERENCES

General References on Surfactants and Their Behavior

Becher, P. (1966) *Emulsions: Theory and Practice*, Robert E. Krieger, Huntington, New York.

Gaines, G.L. (1966) *Insoluble Monolayers at Liquid-Gas Interfaces*, Wiley, New York.

Mittal, K.L. (ed) (1976) *Micellization, Solubilization, and Microemulsions*, Plemum, New York.

Mittal, K.L. and Lindman, B. (ed) (1984) *Surfactants in Solution*, Plenum, New York.

Osipow, L.I. (1977) *Surface Chemistry*, Robert E. Krieger, Huntington, New York.

Prince, L.M. (1977) *Microemulsions*, Academic Press, New York.

Robb, I.D. (ed) (1982) *Microemulsions*, Plenum, New York.

Schick, M.J. and Fowkes, F.M., consulting editors. Surfactant Science Series, **vols. 1-13**, Marcel Dekker, New York.

Tanford, C. (1973) *The Hydrophobic Effect: Formation of Micelles and Biological Membranes*, Wiley, New York.

Textual References

1. Shah, D.O. and Schechter, R.S., ed. (1977) *Improved Oil Recovery by Surfactant and Polymer Flooding*, Academic Press, Inc., New York.

2. Johansen, R.T. and Berg, R.L., ed., (1979), *Chemistry of Oil Recovery*, ACS Symposium Series **91**, ACS, Washington, D.C.

3. Rosen, M.J. (1978) *Surfactants and Interfacial Phenomena*, Wiley-Interscience, New York.

4. Osipow, L.I. (1977) *Surface Chemistry*, Robert E. Kieger Pub. Co., Huntington, New York.

5. Shinoda, K., Nakagawa, T., Tamamushi, B.I., and Isemura T. (1963) *Colloid Surfactants*, Academic Press, New York.

6. McBain, M.E.L. and Hutchinson, E. (1955) *Solubilization*, Academic Press, New York.

7. Roe, C.P. and Brass, P.D. (1954) *J. Am. Chem. Soc.* **76**, 4703.

8. Anacker, E.W. (1970) in *Cationic Surfactants*, Marcel Dekker, Inc., New York, p. 203.

9. Hall, D.G. and Pethica, B.A. (1967) "Thermodynamics of Micelle Formation", in *Nonionic Surfactants*, M.J. Schick, ed., Marcel Dekker, Inc., New York, p. 516.

10. Tanford, C. (1973) *The Hydrophobic Effect: Formation of Micelles and Biological Membranes*, Wiley, New York.

11. O'Connell, J.P. and Brugman, R.J. (1977) "Some Thermodynamic Aspects and Models of Micelles, Microemulsions, and Liquid Crystals", Ref. (1),pp 339 ff.

12. Ben-Naim, A. (1971) *J. Chem. Phys.* **54**, 1387.

13. Ben-Nain, A. (1971) *J. Chem. Phys.* **54**, 3696.

14. Franks, F. (1975) in *Water* **V. 4**, Franks, F., ed., Plenum Press, New York, p. 1.

15. Evans, D.F. and Wightman, P.J. (1982) *J. Colloid Interface Sci.* **86**, 515.

16. Nagarajan, R. and Ruckenstein, E. (1977) *J. Colloid Interface Sci.* **60**, 221.

17. Schick, M.J. (1967) in *Nonionic Surfactants*, M.J. Schick ed., Marcel Dekker, Inc. New York, p.l.

18. Hsiano, L., Dunning, H.N., and Lorenz, P.B. (1956) *J. Phys. Chem.* **60**, 657.

19. Nakagawa, T., and Shinoda, K. (1963) in Ref. (5).

20. Rosch, M. (1967) "Configuration of the Polyoxy-ethylene Chain in Bulk," in *Nonionic Surfactants*, M.J. Schick, ed., Marcel Dekker, Inc. New York, p. 753.

21. Nagarajan, R. and Ruckenstein, E. (1979) *J. Colloid Interface Sci* **71**, 580.

22. Benton, W.J. and Miller, C.A. (1983) *J. Phys. Chem.* **87**, 4981.

Benton, W.J. and C.A. Miller (1984) "A New Optically Isotropic Phase in the Dilute Region of the Sodium Octanoate-Decanol-Water-System" in *Surfactants in Solution*, K.L. Mittal and B. Lindman (ed). Plenum Press, New York, p. 205.

23. Israelachvili, J.N., Mitchell, D.J., and Ninham, B.W. (1976) *J. Chem. Soc. Faraday Trans. II* **72**, 1525.

Mitchell, D.J., Tiddy, G.J.T., Waring, L., Bostock, T., and Macdonald, M.P. (1983) *J. Chem. Soc. Faraday Trans. I* **79**, 975.

24. Fontell, K. (1981) *Mol. Cryst. Liq. Cryst.*, **63**, 59.

25. Tiddy, G.J.T. (1980) *Phys. Rep.*, **57**, 1.

26. Gaines, G.L. (1966) *Insoluble Monolayers at Liquid-Gas Interfaces*, Wiley, New York.

27. Ries, H.E. and Kimball, W.A. (1957) *Proc. 2nd Intl Cong. Surface Activity*, **1**, 75.

28. Harris, W.F. (1964) "A Study of the Mechanism of Collapse of Monomolecular Films at Fluid/Fluid Interfaces". M.S. Thesis, University of Minnesota.

29. Smith, R.D. and Berg, J.C. (1980) *J. Colloid Interface Sci.* **74**, 273.

30. Alexander, A.E. and Johnson, P. (1949) *Colloid Science*, **vol. 2**, Clarendon Press, Oxford, p. 686.

31. Saito, H. and Shinoda, K. (1967) *J. Colloid Interface Sci.* **24**, 10.

32. Hoar, T.P. and Schulman, J.H. (1943) *Nature* **152**, 102.

Bowcott, J.E. and Schulman, J.H. (1955) *Z. Elektrochem.* **59**(4), 283.

Schulman, J.H., Stoeckenius, W. and Prince, L.M. (1959) *J. Phys. Chem.* **63**, 1677.

33. Rehbinder, P.A. (1957) *Proc. 2nd Intl. Cong. Surface Activity* **1**, 476.

Shchukin, E.D. and Rehbinder, P.A. (1958) *Colloid J. USSR* **20**, 645.

34. Ruckenstein, E. and Chi, J.C. (1975) *J. Chem. Soc. Faraday Trans. II* **71**, 1690.

35. Ruckenstein, E. (1978) *Chem. Phys. Lett.* **57**, 517.

36. Murphy, C.L. (1966) "Thermodynamics of Low Tension and Highly Curved Surfaces", Ph.D. Thesis, University of Minnesota. See also Ref. (54) below.

37. Miller, C.A. and Neogi, P. (1980) *AIChE J.* **26**, 212.

Mukherjee, S., Miller, C.A., and Fort, T. (1983) *J. Colloid Interface Sci.* **91**, 223.

38. Hill, T.L. (1963) *Thermodynamics of Small Systems*, **Vol. 1**, W.A. Benjamin, New York. See also **Vol. 2** (1964).

39. Huh, C. (1982) *Soc. Petrol. Eng. J.* **23**, 829.

40. Mitchell, D.J. and Ninham, B.W. (1981) *J. Chem. Soc. Faraday Trans. II* **77**, 601.

41. Ruckenstein, E. (1983) *Chem. Phys. Lett.* **98**, 573.

42. Safran, S.A. and Turkevich, L.A. (1983) *Phys. Rev. Lett.* **50**, 1930.

43. Reed, R.L. and Healy, R.N. (1977) "Some Physico-chemical Aspects of Microemulsion flooding: A Review", in *Improved Oil Recovery by Surfactant and Polymer Flooding*, D.O. Shah and R.S. Schechter (eds), Academic Press, New York. p. 383.

44. Hwan, R., Miller, C.A., and Fort, T. (1979) *J. Colloid Interface Sci.* **68**, 221.

45. Miller, C.A., Hwan, R., Benton, W.J., and Fort, T. (1977) *J. Colloid Interface Sci.* **61**, 554.

46. Scriven, L.E. (1976) *Nature* **267**, 333.

47. Talmon, Y. and Prager, S. (1978) *J. Chem. Phys.* **69**, 2984.

Talmon, Y. and Prager, S. (1982) *J. Chem. Phys.* **76**, 1535.

Jouffroy, J., Levinson, P. and deGennes, P. (1982) *J. Physique* **43**, 1241.

48. Shinoda, K. and Friberg, S. (1975) *Adv. Colloid Interface Sci.* **4**, 281.

49. Huh, C. (1979) *J. Colloid Interface Sci.* **71**, 408.

50. Fleming, P.D. and Vinatieri, J.E. (1979) *AIChE J.* **25**, 493.

51. Fleming, P.D., Vinatieri, J.E., and Glinsmann, G.R. (1980) *J. Phys. Chem.* **84**, 1526.

52. Cazabat, A.M., Langevin, D., Meunier, J., and Pouchelar, A. (1982) *Adv. Colloid Interface Sci.* **16**, 175.

53. Huang, J.S. and Kim, M.W. (1982) "Critical Scaling Behavior of Microemulsions", SPE/DOE Preprint 10787 presented at Symposium on Improved Oil Recovery, Tulsa.

54. Robbins, M.L. (1976) "Theory of the Phase Behavior of Microemulsions", in *Micellization, Solubilization, and Microemulsions*, K.L. Mittal (ed), **Vol. 2**, Plenum, New York. p. 713.

55. Salter, S.J. (1977) "The Influence of Type and Amount of Alcohol on Surfactant-Oil-Brine Phase Behavior and Properties", SPE Preprint 6843, presented at Annual Fall Meeting, Denver.

56. Baviere, M., Wade, W.H., and Schechter, R.S. (1981) "The Effect of Salt, Alcohol, and Surfacant on Optimum Middle Phase Composition", in *Surface Phenomena in Enhanced Oil Recovery*, D.O. Shah (ed), Plenum, New York, p. 117.

57. Salager, J.L., Bourrel, M., Schechter, R.S., and Wade, W.H. (1979) *Soc. Petrol. Eng. J.* **19**, 271.

58. Cash, R.L., Cayias, J.L., Fournier, G., MacAllister, D.J., Scharer, T. Schechter, R.S., and Wade, W.H. (1977), *J. Colloid Interface Sci.* **59**, 39.

59. Cayias, J.L., Schechter, R.S., and Wade, W.H. (1976) *Soc. Petrol. Eng. J.* **16**, 351.

60. Puerto, M.C. and Reed, R.L. (1983), *Soc. Petrol. Eng. J.* **23**, 669.

61. Qutubuddin, S., Miller, C.A., and Fort, Jr., T. (1984) "Phase Behavior of pH-Dependent Microemulsions". *J. Colloid Interface Sci.*, in press.

62. Shinoda, K. and Kunieda, H. (1973) *J. Colloid Interface Sci.* **42**, 381.

63. Friberg, S., Mandell, L., and Larsson, M. (1969) *J. Colloid Interface Sci.* **29**, 155.

Friberg, S. (1971) *J. Colloid Interface Sci.* **37**, 291.

64. Manev, E.D., Sazdanova. S.V., Rao, A.A., and Wasan, D.T. (1982) *J. Disp. Sci. Technol.* **3**, 435.

65. Bourrel, M., Graciaa, A., Schechter, R.S., and Wade, W.H. (1979) *J. Colloid Interface Sci.* **72**, 161.

Vinatieri, J.E. (1980) *Soc. Petrol. Eng. J.* **20**, 402.

Salager, J.L., Loaiza-Maldonado, I., Minana-Perez, M., and Silva, F. (1982) *J. Disp. Sci. Tech.* **3**, 279.

66. Griffin, J. (1949) *J. Soc. Cosmet. Chem.* **1**, 311.

67. Davies, J.T. and Rideal, E.K. (1963) *Interfacial Phenomena*, 2nd ed. Academic Press, New York, Ch. 8.

68. Stevenson, D.G. (1961) "Ancillary Effects in Detergent Action". in *Surface Activity and Detergency*, K. Durham (ed), MacMillan, London. p. 146.

69. Chan, A.F., Evans, D.F., and Cussler, E.L. (1976) *A.I.Ch.E. J.* **22**, 1006.

Shaeiwitz, J.A., Chan, A.F., Evans, D.F. and Cussler, E.L. (1981) *J. Colloid Interface Sci.* **84**, 471.

70. Broome, F.K., Hoerr, C.W., and Harwood, H.J. (1951) *J. Am. Chem. Soc.* **73**, 3350.

71. Friman, R., Danielsson, I., and Stenius, P. (1982) *J. Colloid Interface Sci.* **86**, 501.

72. Adamson, A.W. (1976) *Physical Chemistry of Surfaces*, 3rd ed., Wiley, New York.

73. Hiemenz, P. (1977) *Principles of Colloid and Surface Chemistry*, Marcel Dekker, New York.

PROBLEMS

IV-1. If ΔG^0 is known as a function of temperature, show that $\Delta H^0 = -T^2\, \partial/\partial T\, (\Delta G^0/T)$ and $\Delta S^0 = (\Delta H^0 - \Delta G^0)/T$. The high temperature data of Ref. (14) are

T(°C)	10^3CMC (mole/ℓ)	$-\Delta\bar{G}^0$ (kcal/mole)	$-\Delta\bar{H}^0$ (kcal/mole)	$\Delta\bar{S}^0$ (cal/deg-mole)
25.2	3.79	5.79	4.99	2.7
25.2	3.82	5.78	4.99	2.6
40.3	4.22	5.86	6.21	- 1.1
54.2	4.94	5.78	7.50	- 5.3
76.3	6.69	5.61	10.25	-13.6
95.5	9.83	5.35	11.34	-16.3
95.5	9.91	--	--	--
114.0	13.56	4.96	12.42	-19.2
134.9	21.38	4.52	13.23	-21.4
166.0	39.35	3.81	13.47	-22.0

where the actual values of α have been used for the calculations and the overbars denote that these quantities have been divided by $\bar{n}$. Calculate

ΔG^0, ΔH^0, and ΔS^0 using $\alpha = 0$ and a modified form for Eq. [IV-6] with concentrations instead of mole fractions. Note that enthalpic effects are responsible for micelle formation in this case.

More conventional data from J.E. Adderson and H. Taylor, *J. Coll. Sci.* **19**, 495 (1964) for dodecylpyridinium bromide in water are

T(°C)	10^3 CMC (mole/ℓ)
5	11.5
10	11.2
15	11.0
20	11.2
25	11.4
30	11.8
35	12.2
40	12.8
45	13.5
50	14.0
55	14.8
60	15.4
65	16.3
70	17.2

Repeat the previous calculations. Note that CMCs pass through a minimum which is not an uncommon feature.

IV-2. Obtain an expression for $\Delta \bar{G}_N^0$ for the data in Figure IV-22 from Ref. (5). What is the significance of the fact that the lines have almost the same slope? (Obtain first $\Delta G_N^0/\bar{N}$ as a function of the carbon number n.)

IV-3. A model of micelle formation based on the mass action law is given by Eqs. [IV-1] - [IV-3]. Since micelle aggregation numbers are large and since micelle formation occurs over a very narrow concentration range, micelle formation can also be modeled with reasonable success by a phase separation model. Assume that the following expressions give the chemical potential of a surfactant species in the aqueous and micellar phases respectively

$$\mu_i = \mu_i^o + RT \ln x_i$$

$$\mu_{im} = \mu_{im}^o + RT \ln x_{im}$$

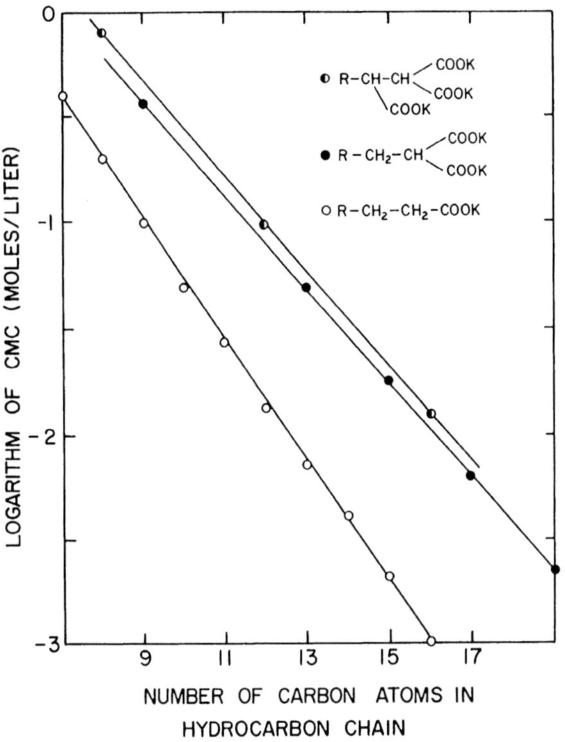

Figure IV-22. Data for Problem IV-2 from Ref. (5) with permission.

The phase separation model is convenient for studying micelle formation in surfactant mixtures.

(a) Consider a surfactant mixture with a ratio of species 2 to species 1 of n on a mole basis. Show that if this mixture is added to water, micelles first form when

$$\frac{x_1}{x_{1c}} = \frac{x_{2c}/x_{1c}}{n+(x_{2c}/x_{1c})}$$

where x_{1c} and x_{2c} are the critical micelle concentrations of surfactants 1 and 2 respectively. Calculate (x_1/x_{1c}) and x_{1m} at initial micelle formation for $(x_{2c}/x_{1c}) = 10$ and n = 0.1, 1, 10.

(b) Find (x_2/x_1) in the aqueous solution when enough of the same mixture has been added that almost all of the surfactant is present in the micelles.

IV-4. Use the thermodynamic approach of Section 8 of Chapter I for an insoluble monolayer to show that the mole fraction of surfactant at the surface is given by

$$x_s^s = 1 - \exp\left[-\frac{\pi a}{RT}\right]$$

where a is the area per molecule of both surfactant and water at the surface.

IV-5. Consider a lamellar liquid crystalline phase (Figure IV-9) whose surfactant bilayers and water layers have thicknesses h_s and h_w respectively. The repeat spacing $(h_s + h_w)$ can be obained experimentally using low angle X-ray diffraction techniques. Suppose that the spacing is measured over a range of values of surfactant-to-water ratio in the liquid crystal.

(a) If h_s is constant, show that a straight line is obtained when the spacing is plotted as a function of $(1/\phi_s)$, where ϕ_s is the volume fraction of surfactant in the liquid crystal. Also explain how h_s can be obtained from the slope of this line or by extrapolating the line to $(1/\phi_s) = 1$.

(b) If v_s is the volume of a surfactant molecule, derive an expression for area σ_s per surfactant molecule in the bilayers.

Thus, both h_s and σ_s can be obtained from suitable X-ray data.

IV-6. The differential dG_m of the Gibbs free energy of a microemulsion can be written in the following form, using the thermodynamics of small systems (35):

$$dG_m = -S_m\,dT + V_m dp + \sum_i \mu_{im} dn_{im} + EdN$$

where p is the overall pressure, N is the number of drops, and E is $(\partial G_m/\partial N)_{T,p,n_i}$. For simplicity we suppose that temperature and external pressure are maintained constant and that the microemulsion contains drops of pure oil in pure water with the surfactant entirely at the drop surfaces. Then we write dG_m as the sum of contributions dG_{mw}, dG_{mo}, and dG_{mi} from the water, oil and interface and dG_{md} from the configurational entropy of the drops and the energy of interaction among them. With this framework we have for the equilibrium condition for a single microemulsion phase

$$E = 0 = \left(\frac{\partial G_{mi}}{\partial N}\right)_{T,p,n_{im}} + \left(\frac{\partial G_{md}}{\partial N}\right)_{T,p,n_{im}}$$

Note that there is no term in G_{mo} in this equation even though the pressure p_d of oil inside the drops may vary since G_m is given by $(U_m + pV_m - TS_m)$, i.e., the external pressure p must be used in the pV term.

(a) Use Eq. [I-9] supplemented by a term (C d(2/a)) for dG_{mi} (see Problem I-4) to show that the above equilibrium condition simplifies to

$$-\frac{3}{a^2} V_d \gamma - \frac{2C}{a^2} + \left(\frac{\partial G_{md}}{\partial a}\right)_{T,p,n_{im}} = 0$$

where V_d is the (constant) total volume of the microemulsion drops. Derive an expression for the last term when G_{md} is given by the following equation used in the statistical thermodynamics of hard spheres:

$$G_{md} = NkT\left(\ell n\, \phi - 1 + \phi \frac{4-3\phi}{(1-\phi)^2} + \ell n \frac{V_w}{V_{Hs}}\right)$$

where ϕ is the volume fraction of drops, V_{Hs} the volume of a drop, and V_w the volume of a solvent molecule.

(b) Now suppose that enough oil is present that an oil-in-water microemulsion is in equilibrium with an excess oil phase. Show that minimization of total system free energy leads to a second equilibrium equation

$$8\pi a N\gamma - \frac{2C}{a^2} + \left(\frac{\partial G_{md}}{\partial a}\right)_{T,p,n_{wm},n_{sm},N} = 0$$

Again evaluate the last term using the above expression for G_{md}.

(c) To demonstrate that bending effects are important when an excess oil phase is present, evaluate γ, (C/A), $(2\gamma/a)$ and $(2(C/A)/a^2)$ for a = 3, 10, and 50 nm and ϕ = 0.01, 0.05, and 0.30. The significance of the last two terms may be seen from Eq. (i) of Problem I-4. Take V_w = 0.015 nm^3 and T = 300°K in your calculations.

V

Interfaces in Motion-Stability and Wave Motion

1. BACKGROUND

So far the discussion of interfaces has dealt mainly with their equilibrium properties. Such topics as interfacial thermodynamics, equilibrium shapes of fluid interfaces, equilibrium contact angles, and equilibrium phase behavior in systems containing surfactants are central to the understanding of interfacial phenomena and are discussed to a greater or a lesser extent in most textbooks on the subject.

Interfaces in motion or otherwise not in equilibrium occur frequently, however, in situations of practical importance. Nonequilibrium interfaces are common in mass transfer processes such as distillation, gas absorption, and liquid-liquid extraction which are ubiquitous and of great importance in the chemical processing industry. Nonequilibrium behavior in the form of interfacial motion is also decisive in both formation and destruction of disperse systems. Whether energy is supplied by mechanical stirring, ultrasonic vibration, electric fields, or other means, processes for forming emulsions, foams or aerosols starting from the two bulk phases necessarily involve interfacial motion, instability, and breakup. On the other hand, destruction of a disperse fluid system is the result of coalescence of many pairs of drops (or bubbles). Before coalescence occurs, the film between two drops must thin by drainage. Coalescence itself is initiated by an instability of the draining film.

In the remainder of this book we present information on phenomena where dynamic interfacial effects are important. We begin in this chapter with interfacial stability and the closely related subject of interfacial oscillation or wave motion. It is frequently of great interest to know the conditions for interfacial instability. We may ask, for instance, how far a fluid jet leaving a circular orifice travels before it breaks up into drops. Or when we can expect spontaneous convection to arise near an interface across which one or more species diffuse.

Suitable stability analyses can provide answers to these and similar questions. They basically establish whether perturbations from some initial state grow or diminish with time. In the former case the system is unstable and never returns to its initial state. In the latter case the system is stable and the initial state is regained although sometimes via oscillations of ever decreasing amplitude.

If only small perturbations are considered, terms involving the squares of or products of the various perturbed quantities may be neglected. Under these conditions the stability analysis is linear in nature and hence much simplified from the general case.

Nonlinear analysis is required for larger perturbations. Though more complex mathematically, it provides additional information, e.g., whether or not an unstable system eventually approaches some new steady state and, if so, the nature of that steady state. It also shows whether a system stable with respect to small perturbations can be unstable with respect to some large perturbations as well. In this book we shall consider only small perturbations and the techniques of linear stability analysis.

2. LINEAR ANALYSIS OF INTERFACIAL STABILITY

2.a Differential Equations

We consider stability of the interface between two incompressible fluids A and B, as illustrated in Figure V-1. When the upper fluid is the denser of the two, a gravitationally produced instability arises. It is the prototype of various instabilities which are produced by acceleration of fluids in contact and which are employed in forming emulsions, aerosols and foams by various shaking processes and ultrasonic vibration. While important in its own right, it also serves here as a vehicle for introducing the techniques of linear stability analysis which will be employed throughout this chapter and the next.

Initially, the interface in Figure V-1 occupies the horizontal plane $z = 0$, and there is no motion in either fluid. At time $t = 0$ the interface is deformed slightly to a new configuration $\bar{z}(x)$. Whatever the functional form of $\bar{z}(x)$, it can be expressed as a Fourier integral

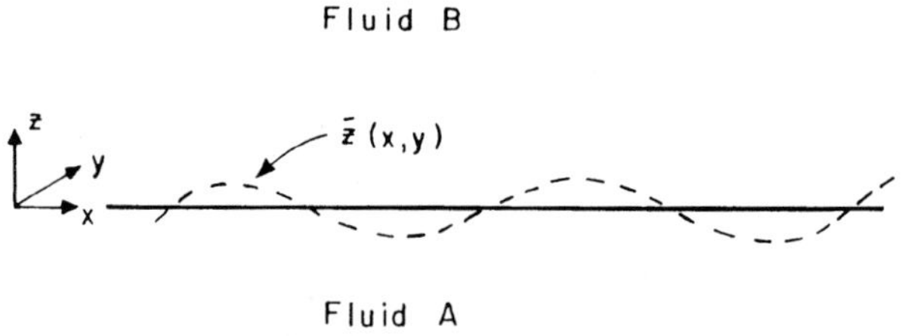

Figure V-1. Wavy perturbation of the interface between superposed fluids. Gravity acts in the negative z-direction.

$$\bar{z}(x) \quad = \quad \int_0^\infty [B_1(\alpha) \cos \alpha x + B_2(\alpha) \sin \alpha x] \, d\alpha \qquad [V-1]$$

The analysis given below focuses on determining the time-dependent interfacial position $z^*_{1\alpha}(x,t)$ (or $z^*_{2\alpha}(x,t)$) when the initial deformation is that produced by a single Fourier component $\bar{z}_{1\alpha} = \cos \alpha x$ (or $\bar{z}_{2\alpha} = \sin \alpha x$). For the linear nature of the governing equations for small perturbations implies that interfacial position $z^*(x,t)$ for a general deformation $\bar{z}(x)$ may be obtained by superposition of these results:

$$z^*(x,t) \quad = \quad \int_0^\infty [B_1(\alpha) \, z^*_{1\alpha}(x,t) + B_2(\alpha) \, z^*_{2\alpha}(x,t)] \, d\alpha \qquad [V-2]$$

The underlying idea behind the stability analysis is that all possible initial perturbations $\bar{z}(x)$ consistent with system boundary conditions may occur. If all of these perturbations diminish in amplitude with time, the system is stable. But if any perturbation grows, instability occurs and the interface never returns to its initial flat configuration. It is clear from Eq. [V-2] that if the system is stable with respect to all initial interfacial deformations having the forms $\bar{z}_{1\alpha}$ and $\bar{z}_{2\alpha}$ of individual Fourier components, it is stable with respect to a general deformation. But if it is unstable with respect to even one Fourier component, interfacial deformation can be expected to increase continuously with time, and there is no return to the initial state.

Accordingly, we concentrate on solving the equations of change for an initial deformation given by the single Fourier component B cos αx (or B sin αx) which has a wavelength λ equal to $(2\pi/\alpha)$. Actually, for reasons explained below, we choose to write the deformation in the form $e^{i\alpha x}$, it being understood that only the real (or imaginary) part of this expression is of physical significance. For simplicity two-dimensional initial deformations of the form $\bar{z}(x,y)$ are not considered here. The basic method of solution is the same, however, and is described elsewhere(1).

For small interfacial deformations the velocity components v_x and v_z are small. If we neglect terms involving the squares or products of these components, a valid procedure when disturbance amplitude is much smaller than the wavelength, the equations of motion in either fluid A or fluid B simplify to

$$\rho \frac{\partial v_x}{\partial t} = - \frac{\partial p}{\partial x} + \mu \nabla^2 v_x \qquad [V-3]$$

$$\rho \frac{\partial v_z}{\partial t} = - \frac{\partial p}{\partial z} - \rho g \mu \nabla^2 v_z \qquad [V-4]$$

Here p is the pressure, ρ and μ are fluid density and viscosity, t is time, g is the gravitational acceleration and ∇^2 is the operator $(\partial^2/\partial x^2 + \partial^2/\partial z^2)$. The pressure may be eliminated by differentiating Eq. [V-3] with respect to z and Eq. [V-4] with respect to x and subtracting:

$$\rho \frac{\partial \omega}{\partial t} = \mu \nabla^2 \omega \qquad [V-5]$$

$$\omega = \frac{\partial v_x}{\partial z} - \frac{\partial v_z}{\partial x} \qquad [V-6]$$

We note that ω in these equations is the y component of the vorticity $\nabla \times \mathbf{v}$.

If Eq. [V-5] is differentiated with respect to z, the result can be written as

$$\rho \frac{\partial}{\partial t} \left[\frac{\partial^2 v_x}{\partial z^2} - \frac{\partial}{\partial x} \left(\frac{\partial v_z}{\partial z} \right) \right] = \mu \nabla^2 \left[\frac{\partial^2 v_x}{\partial z^2} - \frac{\partial}{\partial x} \left(\frac{\partial v_z}{\partial z} \right) \right] \qquad [V-7]$$

Moreover, local continuity for an incompressible fluid requires that

$$\frac{\partial v_x}{\partial x} + \frac{\partial v_z}{\partial z} = 0 \qquad [V-8]$$

Substituting Eq. [V-8] into Eq. [V-7], we obtain

$$\left(\frac{\partial}{\partial t} - \nu \nabla^2 \right) \nabla^2 v_x = 0 \qquad [V-9]$$

where ν is the kinematic viscosity (μ/ρ). A similar manipulation leads to

$$\left(\frac{\partial}{\partial t} - \nu \nabla^2 \right) \nabla^2 v_z = 0 \qquad [V-10]$$

Of course, these equations can also be readily obtained using vector calculus (see (1)).

Let us suppose that v_z has the same periodic behavior as the interfacial deformation itself, so that

$$v_z(x,z,t) = u(z,t)\, e^{i\alpha x} \quad [V\text{-}11]$$

The validity of this procedure will be confirmed if we are successful in finding a solution which satisfies the governing equations and boundary conditions. Substitution of Eq. [V-11] into Eq. [V-10] yields

$$\left(\frac{\partial}{\partial t} - \nu \frac{\partial^2}{\partial z^2} - \nu \alpha^2 \right)\left(\frac{\partial^2}{\partial z^2} - \alpha^2 \right) u = 0 \quad [V\text{-}12]$$

Next, we try a further separation of variables and suppose that

$$u\,(z,t) = W(z)\, \delta\,(t) \quad [V\text{-}13]$$

When this expression for u is substituted into Eq. [V-12], the result may be written as

$$\frac{\delta'(t)}{\delta(t)} = \nu \frac{\left(\frac{\partial^2}{\partial z^2} - \alpha^2 \right)\left(\frac{\partial^2}{\partial z^2} - \alpha^2 \right) W(z)}{\left(\frac{\partial^2}{\partial z^2} - \alpha^2 \right) W(z)} = \beta \quad [V\text{-}14]$$

Here β is a separation constant. Since the first two terms in Eq. [V-14] are functions only of t and z respectively and since they are equal, both terms must be equal to the same constant β.

Eq. [V-14] yields two ordinary differential equations for $\delta(t)$ and $W(z)$. The solutions in fluids A and B, subject to the condition that v_z becomes vanishingly small at large distances from the interface are

$$\delta_A(t) = \delta_{oA} e^{\beta_A t} \quad [V\text{-}15]$$

$$\delta_B(t) = \delta_{oB} e^{\beta_B t} \quad [V\text{-}16]$$

$$W_A(z) = D_1 e^{\alpha z} + D_2 e^{\alpha r_A z} \quad [V\text{-}17]$$

$$W_B(z) = D_3 e^{-\alpha z} + D_4 e^{-\alpha r_B z} \quad [V\text{-}18]$$

$$r_A^2 = 1 + \frac{\beta_A}{\nu_A \alpha^2} \qquad ; \qquad r_B^2 = 1 + \frac{\beta_B}{\nu_B \alpha^2} \qquad \text{[V-19]}$$

The tangential velocity v_x is given by the real (or imaginary) part of

$$v_x = \frac{i}{\alpha} W'(z) \; e^{i\alpha x} \; \delta(t) \qquad \text{[V-20]}$$

As is easily shown, this expression satisfies the differential equation [V-9].

In employing certain boundary conditions, we also require an expression for the pressure. Differentiating Eqs. [V-3] and [V-4] with respect to x and z respectively, adding the resulting expressions, and invoking Eq. [V-8], we find

$$\frac{\partial^2 p}{\partial x^2} + \frac{\partial^2 p}{\partial z^2} = 0 \qquad \text{[V-21]}$$

Before perturbation the initial pressure distribution is hydrostatic

$$p_{iA} = p_o - \rho_A g z$$
$$p_{iB} = p_o - \rho_B g z \qquad \text{[V-22]}$$

where p_o is the initial pressure at the interface. Let us assume that the perturbation p_p from this initial pressure distribution has the following form in each fluid:

$$p_p = P(z) \; e^{i\alpha x} \; \delta(t) \qquad \text{[V-23]}$$

Combining this equation with Eq. [V-21], we find that

$$\frac{d^2 P}{dz^2} = \alpha^2 P \qquad \text{[V-24]}$$

Similarly, the derivative of [V-4] with respect to z yields

$$\frac{d^2 P}{dz^2} = -\rho \beta \frac{dW}{dz} + \mu \left(\frac{d^3 W}{dz^3} - \alpha^2 \frac{dW}{dz} \right) \qquad \text{[V-25]}$$

Finally, equating the right hand members of Eqs. [V-24] and [V-25] leads to the following expression for P:

$$P(z) = \frac{\mu}{\alpha^2}\frac{d^3W}{dz^3} - \mu r^2 \frac{dW}{dz} \qquad [V-26]$$

Hence, the pressure distributions in the two fluids are readily found using Eqs. [V-17] and [V-18].

The basic procedure employed here, i.e., writing each dependent variable as the sum of an initial contribution and a small perturbation, substituting these expressions into the governing differential equations, neglecting squares and products of perturbations, and solving the resulting partial differential equations by separation of variables, is applicable in most cases where a linear stability analysis is to be performed.

2b. Boundary Conditions

The above solution of the equations of motion for a given value of the wave number α contains several unknown constants : D_1, D_2, D_3, D_4, B, β_A, and β_B (the constants δ_{0A} and δ_{0B} can be incorporated into the constants D_i when the overall expressions for v_{zA} and v_{zB} are written in accordance with Eqs. [V-11] and [V-13]). Suitable boundary conditions along the wavy interface and the initial condition must be used to determine these constants. It is through these conditions that interfacial properties such as interfacial tension enter the analysis.

Some boundary conditions are basically requirements that quantities such as overall mass, momentum, energy, and electrical charge be conserved. We shall derive expressions for the first two of these in this chapter although in forms more general than required for solution of the simple problem of superposed fluids considered in the preceeding section. Other conservation equations including individual component material balances are treated in the following chapter which deals with transport processes.

We begin with conservation of mass for a "pillbox" control volume such as that shown in Figure I-3. We take the control volume itself as fixed in space but permit the reference surface S to move normal to itself with velocity $\dot{a}_n$. Employing a procedure similar to that used in Chapter I, we first write the mass balance for the overall control volume

$$\int_{S_A} \rho_A \mathbf{v}_A \bullet \mathbf{n}\, dS - \int_{S_B} \rho_B \mathbf{v}_B \bullet \mathbf{n}\, dS - \int_{S_0} \rho\mathbf{v} \bullet \mathbf{M}\, dS = \frac{d}{dt}\int_V \rho\, dV \qquad [V-27]$$

We then subtract the following expressions which would apply if the regions below and above S were occupied by bulk phases A and B respectively:

$$\int_{S_A} \rho_A \mathbf{v}_A \bullet \mathbf{n}\, dS - \int_S \rho_A(\mathbf{v}_A \bullet \mathbf{n} - \dot{a}_n)\, dS - \int_{S_{oA}} \rho_A \mathbf{v}_A \bullet \mathbf{M}\, dS = \frac{d}{dt}\int_{V_A} \rho_A dV \qquad [V-28]$$

$$\int_S \rho_B(\mathbf{v}_B \bullet \mathbf{n} - \dot{a}_n)\, dS - \int_{S_B} \rho_B \mathbf{v}_B \bullet \mathbf{n}\, dS - \int_{S_{oB}} \rho_B \mathbf{v}_B \bullet \mathbf{M}\, dS = \frac{d}{dt}\int_{V_B} \rho_B dV \qquad [V-29]$$

The result is

$$\int_S [\rho_A(v_{An} - \dot{a}_n) - \rho_B(v_{Bn} - \dot{a}_n)]\, dS + \int_{S_o} \Delta(\rho \mathbf{v}) \bullet \mathbf{M}\, dS = \frac{d}{dt}\int_V \Delta\rho\, dV \qquad [V-30]$$

$$\text{where} \quad \Delta(\rho\mathbf{v}) = \begin{cases} \rho\mathbf{v} - \rho_A \mathbf{v}_A & \text{in } V_A \\ \rho\mathbf{v} - \rho_B \mathbf{v}_B & \text{in } V_B \end{cases}$$

Note that use of these equations implies that even in the nonequilibrium situation, a method can be found for extrapolating the bulk phase densities and velocities into the interfacial region. In many situations of interest, this extrapolation presents no difficulties.

If we make the same small curvature approximation as in Chapter I, the second and third integrals in Eq. [V-30] become

$$\int_C \left[\int_{\lambda_A}^{\lambda_B} \Delta(\rho\mathbf{v})\, d\lambda \right] \bullet \mathbf{m}\, ds \qquad [V-30a]$$

with m the outward pointing normal to C (see Figure I-3)

$$\frac{d}{dt}\int_S \Gamma\, dS = \int_S \frac{\partial\Gamma}{\partial t} dS + \int_C \Gamma\, \mathbf{v} \bullet \mathbf{m}\, ds + \int_S (-2H)\, \dot{a}_n\, \Gamma\, dS \qquad [V-30b]$$

It is clear that the integral given by Eq. [V-30a] becomes vanishingly small as the pillbox thickness $(\lambda_B - \lambda_A)$ decreases. Hence, it can be neglected in comparison with the first integral in Eq. [V-30]. The time derivative of the surface integral can be written as the sum of three integrals as shown in Eq. [V-30b]. The second and third terms of Eq. [V-30b] represent the rate of change of the area of S due to (a) tangential motion of the curve C which forms the boundary of S and (b) translation of S perpendicular to itself (cf. Eq. [I-19]). Substituting Eq. [V-30b] into [V-29] and invoking Eq. [I-36], we obtain

$$\int_S [\rho_A(v_{An} - \dot{a}_n) - \rho_B(v_{Bn} - \dot{a}_n) - \frac{\partial\Gamma}{\partial t} - \nabla_s \bullet \Gamma\, \mathbf{v}]\, dS = 0 \qquad [V-31]$$

As the extent of S is arbitrary, the integrand itself must vanish, and we obtain the differential interfacial mass balance

$$\frac{\partial \Gamma}{\partial t} = -\nabla_s \cdot \Gamma \underline{v} + \rho_A(v_{An} - \dot{a}_n) - \rho_B(v_{Bn} - \dot{a}_n) \qquad [V\text{-}32]$$

According to this general equation, the local surface excess mass per unit area Γ can change due to inflow or outflow of mass by surface convection or by inflow from the bulk fluids.

For the case of superposed fluids considered in Section 2a we neglect the terms in Eq. [V-32] involving the surface excess mass Γ. Moreover we note that the term $(\rho_A(v_{An} - \dot{a}_n))$ represents the rate at which material from phase A crosses the interface. Since no transfer of either A or B across the interface occurs in this case, we conclude that

$$v_{An} = v_{Bn} = \dot{a}_n \qquad [V\text{-}33]$$

In terms of Eqs. [V-15] - [V-18] we have

$$W_A(o)\ \delta_A(t) = W_B(o)\ \delta_B(t) \qquad [V\text{-}34]$$

This equation implies that the two exponential time factors β_A and β_B are equal. We shall therefore drop the subscripts and simply use β. Note that W_A and W_B are evaluated at $z = 0$. Although the interface is slightly perturbed from this position, the small correction is of higher order and hence neglected in our linear analysis because the quantities W_A and W_B are themselves small perturbations.

A similar procedure can be used to derive a general interfacial momentum balance. The result, a generalization of the static momentum balance derived in Section 4 of Chapter I, is

$$\frac{\partial \mathbf{T}}{\partial t} = -\nabla_s \cdot \mathrm{T}\,\mathbf{v} + \Gamma \hat{F} + (p_A + \tau_{An} - p_B - \tau_{Bn})\,\mathbf{n} + 2H\gamma\,\mathbf{n} + \nabla_s \gamma$$

$$+ (\boldsymbol{\tau}_{At} - \boldsymbol{\tau}_{Bt}) + \rho_A \mathbf{v}_A (v_{An} - \dot{a}_n) - \rho_B\,\mathbf{v}_B\,(v_{Bn} - \dot{a}_n) \qquad [V\text{-}35]$$

Here $\mathbf{T}$ is the surface excess momentum defined by

$$\mathbf{T} = \int_{\lambda_A}^{\lambda_B} \Delta\,(\rho\mathbf{v})\,d\lambda \qquad [V\text{-}36]$$

with $\Delta(\rho\mathbf{v})$ defined as at Eq. [V-30]. Also τ_{An} and τ_{Bn} are the normal viscous stresses at the interface and $\boldsymbol{\tau}_{At}$ and $\boldsymbol{\tau}_{Bt}$ are the shear stresses at the interface in fluids A and B respectively. The sign convention used for the normal viscous stresses and shear stresses is that employed by Bird, Stewart, and Lightfoot (2). We note that no excess stresses along the interface caused by interfacial flow itself have been considered. Some discussion of such stresses produced by an effective interfacial viscosity is given in Chapter VII.

The last two terms in Eq. [V-35] represent the net momentum input to the interface by material crossing the interface, e.g., material which changes from liquid to vapor during a boiling process. Normally these terms are small and their effect on the interfacial momentum balance is negligible. We shall consider briefly, however, in Chapter VI an interesting exception where such momentum effects are the source of one type of interfacial instability.

For the superposed fluids of Section 2a, we neglect terms in Eq. [V-35] involving surface excess mass and momentum, assume that interfacial tension is uniform, and take the bulk fluids to be Newtonian with viscosities μ_A and μ_B in evaluating the normal viscous and shear stresses. The general vector equation then simplifies to the following two scalar equations reflecting the normal and tangential components:

$$(p_A - 2\mu_A \frac{\partial v_{zA}}{\partial z}) - (p_B - 2\mu_B \frac{\partial v_{zB}}{\partial z}) + 2H\gamma = 0 \qquad [V\text{-}37]$$

$$\mu_A \left(\frac{\partial v_{xA}}{\partial z} + \frac{\partial v_{zA}}{\partial x} \right) = \mu_B \left(\frac{\partial v_{xB}}{\partial z} + \frac{\partial v_{zB}}{\partial x} \right) \qquad [V\text{-}38]$$

The pressures p_A and p_B include both the initial pressures of Eq. [V-22] and the perturbations given by Eq. [V-23]. They must be evaluated at the actual position z^* of the interface, but with all terms of second and higher order in small perturbation quantities neglected. The result for the pressure in fluid A is, for instance,

$$p_A = p_o - \rho_A g z^* + P_A(o)\, e^{i\alpha x}\, \delta(t) \qquad [V\text{-}39]$$

Eq. [V-37] also contains a term involving curvature. There is only one nonzero radius of curvature in this case. It can be evaluated using Eq. [I-42], neglecting terms of second and higher order as before. The result is that the mean curvature H is given by $(\frac{1}{2})(\partial^2 z^*/\partial x^2)$.

In addition to the boundary conditions based on conservation of mass and momentum, another condition required for the case of superposed fluids is continuity of tangential velocity at the interface:

$$v_{xA} = v_{xB} \quad \text{at} \quad z = 0 \qquad \text{[V-40]}$$

Finally, the initial condition that the interface is at position B cos αx (or B sin αx) at $t = 0$ must be imposed. Since the interface moves with the fluids, the time derivative (dz*/dt) of interfacial position is given by v_{zA} at the interface

$$\frac{dz^*}{dt} = W_A(o)\, e^{i\alpha x}\delta(t) = (D_1 + D_2)\, e^{i\alpha x}\, e^{\beta t} \qquad \text{[V-41]}$$

It is clear from this expression that the initial condition is satisfied, provided that

$$B = \frac{D_1 + D_2}{\beta} \qquad \text{[V-42]}$$

2.c. Stability Condition and Wave Motion for Superposed Fluids

Let us summarize our progress to this point in analyzing stability of the interface between the superposed fluids of Figure V-1. We solved the equations of continuity and motion in Section 2a to obtain the perturbations in the velocity and pressure distributions in the two fluids (Eqs. [V-17]-[V-20], [V-26]). With Eq. [V-42] and the equality of β_A and β_B implied by Eq. [V-34], the list of unknown constants in these solutions given at the beginning of Section 2b has been reduced to D_1, D_2, D_3, D_4, and β. Moreover, we derived four applicable boundary conditions in Section 2b (Eqs. [V-34], [V-37], [V-38], and [V-40]). When the velocity and pressure distributions are substituted into these equations, we find the following relationships among the unknown constants:

$$D_1 + D_2 - D_3 - D_4 = 0 \qquad \text{[V-43]}$$

$$D_1\mu_A\left(1 + r_A^2 + \frac{(\gamma\alpha^2 + (\rho_A - \rho_B)\, g)}{\alpha\beta\mu_A}\right) + D_2\mu_A\left(2r_A + \frac{(\gamma\alpha^2 + (\rho_A - \rho_B) g)}{\alpha\beta\mu_A}\right)$$

$$+ D_3\mu_B(1 + r_B^2) + 2\, D_4\mu_B r_B = 0 \qquad \text{[V-44]}$$

$$2\mu_A D_1 + \mu_A(1 + r_A^2) D_2 - 2\mu_B D_3 - \mu_B(1 + r_B^2) D_4 = 0 \qquad [V-45]$$

$$D_1 + r_A D_2 + D_3 + r_B D_4 = 0 \qquad [V-46]$$

Note that these are four linear, homogeneous equations in the four unknowns D_1 through D_4. An obvious solution but an uninteresting one because it implies no interfacial deformation and no flow is the trivial solution $D_1 = D_2 = D_3 = D_4 = 0$. According to the theory of linear equations, the condition for a nontrivial solution to exist is that the determinant of coefficients vanish for Eqs. [V-43] - [V-46]. This relationship, which, of course, does not contain any of the D_i's, may be solved to obtain the time factor β as a function of the wave number α and the physical properties of the fluids. If the real part of β is positive, the perturbation grows and the interface is unstable. But if the real part of β is negative, the perturbation diminishes and the interface ultimately returns to the flat configuration. A nonzero imaginary part of β implies that oscillation can be expected.

It is instructive to consider some special cases. First let fluid A be a gas of negligible density and viscosity. Then Eqs. [V-43] and [V-46] which require continuity of normal and tangential velocity at the interface are not needed. If the derivation leading to Eq. [V-42] is repeated for fluid B, the quantity (D_1+D_2) in that equation is replaced by (D_3+D_4). Thus, with A a gas, Eq. [V-44] becomes

$$D_3 \left(1 + r_B^2 + \frac{\gamma\alpha^2 - \rho_B g}{\alpha\beta\mu_B}\right) + D_4 \left(2r_B + \frac{\gamma\alpha^2 - \rho_B g}{\alpha\beta\mu_B}\right) = 0 \qquad [V-47]$$

Noting that terms in D_1 and D_2 also disappear from Eq. [V-45], we find, after some manipulation, that the determinant of coefficients of Eqs. [V-45] and [V-47] simplifies to

$$r_B^4 + 2r_B^2 - 4r_B + 1 = \frac{-\beta_B^{*2}}{\nu_B^2\alpha^4} \qquad [V-48]$$

where

$$\beta_B^{*2} = \frac{1}{\rho_B}(\gamma\alpha^3 - \rho_B g\alpha) \qquad [V-49]$$

When the liquid B has a low viscosity, we anticipate that the dimensionless group $|\beta/\nu_B\alpha^2| \gg 1$. In this case if we retain only the leading term of the left side of Eq. [V-48], we obtain

$$\beta^2 = -\beta_B^{*2} = g\alpha - \frac{\gamma\alpha^3}{\rho_B} \qquad \text{[V-50]}$$

Thus for small wavenumbers α, i.e., long wavelengths $(2\pi/\alpha)$, β is positive and the perturbation grows. Such gravitationally-produced instability is often called the Rayleigh-Taylor instability and is expected in this case where the liquid is placed above the gas. However, for sufficiently large α, i.e., for short wavelengths, interfacial tension is able to stabilize the interface in spite of the unfavorable positions of the fluids relative to gravity. For then we have

$$\beta = \pm i \beta_B^* \quad , \quad \beta_B^* > 0 \qquad \text{[V-51]}$$

and the interface oscillates with frequency β_B^*. (There is no damping of the oscillations since viscous dissipation has been neglected in going from Eq. [V-48] to Eq. [V-50].)

The critical wavenumber α_c for which β vanishes is given by

$$\alpha_c = [\rho_B g/\gamma]^{1/2} \qquad \text{[V-52]}$$

For the air-water system this critical wavelength $(2\pi/\alpha_c)$ is about 1.7 cm. Only perturbations having wavelengths greater than this value will grow. Hence the instability can be suppressed by limiting the lateral extent of the interface between the fluids.

We expect the rate of perturbation growth to be zero for $\alpha = \alpha_c$ at the transition between stable and unstable regions and also to be small for long wavelengths (small α) where interfacial deformation and hence the gravitational driving force for the instability vary slowly along the interface. Between these extremes there should be a fastest growing perturbation. By differentiating Eq. [V-50] with respect to α, we can find the wavenumber α_M which has the fastest rate of growth. The result is

$$\alpha_M = [\rho_B g/3\gamma]^{1/2} \qquad \text{[V-53]}$$

$$\beta_M = [4\rho_B g^3/27\gamma]^{1/4} \qquad \text{[V-54]}$$

For the air-water system the fastest-growing disturbance has a wavelength

$(2\pi/\alpha_M)$ of about 2.9 cm and its time constant β_M^{-1} for growth is about 0.027 sec.

We expect disturbances with wavenumbers near α_M to dominate the early stages of the instability where the linear theory applies. In practice the dominant disturbance observed experimentally in many cases of instability frequently has a wavenumber near α_M even though the linear analysis is not applicable for the rather large perturbations which are usually required to detect the instability.

As the liquid viscosity μ_B increases, the condition Eq. [V-52] for the critical wavenumber remains unchanged, as it is basically a thermodynamic condition (see Section 7 below). But the rates of growth of unstable disturbances decrease. The slowing of growth is greatest for short wavelengths where velocity gradients are greatest. As a result the fastest growing disturbance shifts to longer wavelengths (smaller α).

Another limiting case is reached when μ_B becomes very large. Now we have $|\beta/\nu_B\alpha^2| \ll 1$ and expansion of Eq. [V-48] in terms of this quantity leads to

$$\beta = \frac{-\beta_B^{*2}}{2\nu_B\alpha^2} \qquad \text{[V-55]}$$

Under these conditions there is no oscillation and the disturbance either grows or decays exponentially, depending on the sign of β_B^{*2}. The critical wavenumber α_c is still given by Eq. [V-52], but there is no finite α_M. For the smaller the value of α (consistent with $|\beta/\nu_B\alpha^2| \ll 1$), the greater the rate of growth in the unstable region.

If we consider instead the situation where A is a liquid and B a gas, we obtain instead of Eq. [V-48]:

$$r_A^4 + 2r_A^2 - 4r_A + 1 = \frac{-\beta_A^{*2}}{\nu_A^2\alpha^4} \qquad \text{[V-56]}$$

$$\beta_A^{*2} = \frac{1}{\rho_A}(\gamma\alpha^3 + \rho_A g\alpha) \qquad \text{[V-57]}$$

We expand as before for the low viscosity limit but this time retain the two leading terms of the left side of Eq. [V-56]. The result is

$$\beta^2\left(1 + \frac{4\nu_A\alpha^2}{\beta}\right) = -\beta_A^{*2} \qquad \text{[V-58]}$$

In the case of an inviscid liquid ($\nu_A = 0$), we obtain $\beta = \pm i\beta_A^*$ as before and the interface oscillates without damping. There is no instability because with the liquid below the gas, both gravity and interfacial tension act to restore the deformed interface to its initial planar configuration. But with no viscous resistance to flow, the momentum developed during the return carries the interface beyond the flat configuration to a new deformation of opposite sign to the original one.

When ν_A is small, we anticipate that β has the form $(\pm i\beta_A^* + \beta_1)$, where β_1 is a small damping factor produced by viscous dissipation such that $(\beta_1/\beta^*) << 1$. Substituting this expression into Eq. [V-58] and keeping only the leading terms, we find

$$\beta = \pm i\beta_A^* - 2\nu_A\alpha^2 \qquad [V-59]$$

The second term in this expression is the rate at which the oscillations are damped. As might be expected, higher viscosities lead to more rapid damping in this case where the flow is primarily determined by the oscillatory motion.

A second limiting case of Eq. [V-56] is the high viscosity situation for which $|\beta/\nu_A\alpha^2| << 1$. If Eq. [V-56] is expanded in powers of $(\beta/\nu_A\alpha^2)$ and only the leading nonzero term is kept, we find in a manner analogous to that used in deriving Eq. [V-55] that

$$\beta = \frac{-\beta_A^{*2}}{2\nu_A\alpha^2} \qquad [V-60]$$

Here the resistance to flow is so great that the interface does not continually overshoot its equilibrium position as above, i.e., no oscillatory motion develops. Instead the deformed interface simply returns gradually to the flat configuration. Higher viscosities produce slower damping because they reduce the velocities and velocity gradients.

In the liquid-liquid case, it can be shown by similar procedures (3) that in the limit of low fluid viscosities

$$\beta^2 = -\beta^{*2} = -\left[\frac{\gamma\alpha^3 + (\rho_A - \rho_B)g\alpha}{\rho_A + \rho_B}\right] \qquad [V-61]$$

Thus β is real and positive and the perturbation grows whenever β^{*2} is negative. The expressions for the critical and fastest growing wavenumbers α_c and α_M can be obtained by replacing ρ_B by $(\rho_B - \rho_A)$ in Eqs. [V-52] and [V-53]. But the growth factor β_M for the fastest growing disturbance is given by

$$\beta_M = \left[\frac{4(\rho_B - \rho_A)^3 g^3}{27\,\gamma(\rho_A + \rho_B)^2} \right]^{1/4} \qquad [V\text{-}62]$$

When the interface is stable, the time factor β in the low viscosity limit has the form (4)

$$\beta = \pm i\,\beta^* - (1 \pm i) \left[\frac{(2\beta^* \mu_A \mu_B \rho_A \rho_B)^{1/2}}{(\rho_A + \rho_B)[(\mu_A \rho_A)^{1/2} + (\mu_B \rho_B)^{1/2}]} \right] \qquad [V\text{-}63]$$

Thus, the frequency of oscillation is slowed slightly from the inviscid value β^* by viscous effects and the damping factor has an entirely different form than in Eq. [V-59], being, for instance, dependent on the frequency β^*. The reason is that the requirement Eq. [V-40] for continuity of tangential velocity at the interface leads to formation of an oscillating boundary layer near the interface in which viscous dissipation is much greater than that for the liquid-gas case.

In the high viscosity limit, the result analogous to Eqs. [V-55] and [V-60] is

$$\beta = \frac{-\beta^{*2}(\rho_A + \rho_B)}{2\alpha^2(\mu_A + \mu_B)} \qquad [V\text{-}64]$$

As before, this equation applies to both stable and unstable interfaces.

Example V-1. Characteristics of Wave Motion for Free Interfaces

Calculate the frequency of oscillation and the rate of damping at an air-water interface for wavelengths λ ($=2\pi/\alpha$) of 100, 10, 1, and 0.1 cm. Repeat for an oil-water interface with ρ_B = 0.85 gm/cm^3, μ_B = 1 cp for the oil and with interfacial tension γ_{AB} = 50 dyne/cm.

Solution For the air-water interface, we take ρ_A = 1 gm/cm^3, μ_A = 1 cp, γ = 72 dyne/cm. The frequency β_A^* can be calculated from Eq. [V-57] and the damping factor β_1 from Eq. [V-59] in the low viscosity limit. Results are

λ (cm)	α(cm^{-1})	β_A^*(sec^{-1})	$-\beta_1$(sec^{-1})
100	0.0628	7.85	7.9×10^{-5}
10	0.628	25.2	7.9×10^{-3}
1	6.28	155	0.79
0.1	62.8	4230	79

Similarly for the oil-water interface Eqs. [V-61] and [V-63] may be used. The corrected frequency is given by β_c^*

λ (cm)	$\beta^*(\mathrm{sec}^{-1})$	$\beta_c^*(\mathrm{sec}^{-1})$	$-\beta_1(\mathrm{sec}^{-1})$
100	2.24	2.23	4.14×10^{-3}
10	7.52	7.45	7.59×10^{-2}
1	84.9	82.3	2.55
0.1	2590	2450	141

In both cases damping increases with decreasing wavelength. This behavior is due partly to the increased frequency and partly to the increased velocity gradient for short wavelengths.

3. DAMPING OF CAPILLARY WAVE MOTION BY INSOLUBLE SURFACTANTS

Mass transfer across a fluid interface is enhanced by convection in the vicinity of the interface. One source of such convection is wave motion. An increase in the rate of damping of waves can thus be expected to reduce mass transfer rates. As we shall now show, surfactants can cause a significant increase in damping of capillary waves at a liquid-gas interface.

We consider the simplest case of an insoluble surfactant. In the initial motionless state with a flat interface, the surfactant is uniformly distributed and interfacial tension is uniform. But during wave motion the local concentration of surfactant varies with position along the interface with the result that interfacial tension gradients arise. Because these gradients influence the interfacial momentum balance (through the term $\mathbf{\nabla}_s \gamma$ in Eq. [V-35]), the velocity distribution is altered and with it the damping rate.

The solution of the equations of motion and application of the boundary conditions proceeds as before except for an additional term $(d\gamma/dx)$ on the right side of Eq. [V-38]. If fluid B is a gas, we obtain instead of Eqs. [V-44] and [V-45]:

$$D_1\left(1 + r_A^2 + \frac{\beta_A^{*2}}{\beta\alpha^2 \nu_A}\right) + D_2\left(2 r_A + \frac{\beta_A^{*2}}{\beta\alpha^2 \nu_A}\right) = 0 \qquad \text{[V-65]}$$

$$2\mu_A D_1 + \mu_A(1 + r_A^2)\, D_2 + \frac{(d\gamma/d\Gamma_s)(\partial\Gamma_s/\partial x)}{(-i\alpha)\, e^{i\alpha x} e^{\beta t}} = 0 \qquad \text{[V-66]}$$

where Γ_s is the surface concentration of surfactant. Now $(d\gamma/d\Gamma_s)$ is a property of the surfactant film related to its compressibility. Here we shall assume that its value, which normally has a negative sign, is constant at least over the small range of Γ_s involved in the perturbation.

The general boundary condition for conservation of mass of some species in the interfacial region is derived in Chapter VI. For the present case of an insoluble surfactant and in the absence of surface diffusion, we anticipate that the first two terms of Eq. [V-32] should suffice with Γ replaced by Γ_s:

$$\frac{\partial \Gamma_s}{\partial t} = - \frac{\partial(\Gamma_s v_x)}{\partial x} \qquad [V\text{-}67]$$

If we assume that Γ_s is the sum of an initial concentration Γ_{os} and a perturbation term, we may write

$$\Gamma_s = \Gamma_{os} + \Gamma_{1s} e^{i\alpha x} e^{\beta t} \qquad [V\text{-}68]$$

Substituting this expression into Eq. [V-67], we find

$$\Gamma_{1s} = \frac{\Gamma_{os}}{\beta} W_A'(o) = \frac{\Gamma_{os}\alpha}{\beta} (D_1 + r_A D_2) \qquad [V\text{-}69]$$

Thus, Eq. [V-66] becomes

$$D_1 \left[2 + \frac{\Gamma_{os}\, \alpha\, (-d\gamma/d\Gamma_s)}{\beta\, \mu_A} \right] + D_2\left[1 + r_A^2 + r_A \frac{\Gamma_{os}\, \alpha\, (-d\gamma/d\Gamma_s)}{\beta\, \mu_A}\right] = 0 \qquad [V\text{-}70]$$

The determinant of coefficients of Eqs. [V-65] and [V-70] yields the following dispersion equation for periodic motion of a liquid pool covered by an insoluble monolayer:

$$-(r_A - 1)^2 + \left[r_A^2 + r_A + \frac{\beta_A^{*2}}{\beta \nu_A \alpha^2}\right]\left[1 + r_A + \left(\frac{\nu_A \alpha^2}{\beta}\right)\left(\frac{\Gamma_{os}(-d\gamma/d\Gamma_s)}{\mu_A \nu_A \alpha}\right)\right] = 0 \qquad [V\text{-}71]$$

Two limiting cases are of interest. In one the expression in the first set of brackets in this equation is very large, e.g. owing to a large value of interfacial tension and hence β_A^{*2}. As a result, the expression in the second set of brackets is nearly zero. Under these conditions "longitudinal" wave motion ensues (5) where the flow near the surface is dominated by the effects of surface tension variation rather than by surface deflection. Longitudinal waves, which are an important tool for measuring elastic properties of soluble surfactants, are considered further in Problem V-3 and in Chapter VI.

In the second limiting case the surface elasticity $\Gamma_{os}(-d\gamma/d\Gamma_s)$, or more precisely the dimensionless surface elasticity number $[\Gamma_{os}(-d\gamma/d\Gamma_s)/\mu_A\nu_A\alpha]$, becomes so large that the first expression in brackets must be nearly zero. Lateral flow along the interface is virtually precluded and effects of surface deflection are important in this "inextensible" case which might be expected for a concentrated, nearly incompressible monolayer.

Setting the first expression in brackets in Eq. [V-71] equal to zero and taking the low viscosity limit in the usual way, we find

$$\beta = \pm i\, \beta_A^* - (1 \pm i)\, \frac{\alpha(\beta_A^* \nu_A)^{1/2}}{2^{3/2}} \qquad [V\text{-}72]$$

The expression for the damping factor here is very different from that found at Eq. [V-59] for the free surface case and indeed resembles that of Eq. [V-63] for a free liquid-liquid interface. We conclude that a surfactant film causes an oscillating boundary layer to develop and thus increases the damping rate. The boundary layer arises because the film prevents all lateral motion at the surface in this inextensible limit. Indeed, Eq. [V-70] reduces to a requirement that $(D_1 + r_A D_2)$ and hence v_x must vanish at the surface. That the damping rate is higher with the surfactant film is indicated by its proportionality to $(\mu_A)^{1/2}$ in Eq. [V-72], in contrast to μ_A in Eq. [V-59], which of course, is the smaller quantity for low viscosities. For intermediate surfactant concentrations the time factor β is still calculated by requiring that the determinant of coefficients of Eqs. [V-65] and [V-70] vanish, but with all terms of the latter equation retained. It can be shown that for low fluid viscosities the damping factor passes through a maximum with increasing surfactant concentration Γ_{os} (see (6) and Problem V-1). The maximum damping factor is twice that given by Eq. [V-72]. Evidently velocity gradients in the oscillating boundary layer can achieve values greater than those of the inextensible case because tangential velocities at the surface and near the edge of the boundary layer are out of phase and thus have opposite directions during some parts of the oscillation cycle.

For an inextensible liquid-liquid interface, the time factor β is given by the following expression (7):

$$\beta = \pm i\, \beta^* - (1 \pm i)\, \frac{\alpha\, \beta^{*1/2}[(\mu_A \rho_A)^{1/2} + (\mu_B \rho_B)^{1/2}]}{2^{3/2}(\rho_A + \rho_B)} \qquad [V\text{-}73]$$

Damping is predominantly due to an oscillating boundary layer as for a surfactant-free interface (cf. Eq. [V-63]). Because surfactants do not change the basic nature of the flow, their damping effect is smaller at a liquid-liquid than at a liquid-gas interface.

Experimentally, it is not convenient to measure the damping with time of standing waves of a fixed wavelength, the situation considered so far. Instead, oscillations of a fixed frequency β^* and fixed amplitude are imposed on a surface by, for instance, vibration of a bar. For liquids of low viscosity in air, waves having a definite wavelength develop which move away from the bar, their amplitude diminishing with increasing distance of travel. Both the wavelength and the variation of amplitude with position can be measured by optical or acousto-mechanical methods (see (8)).

In our previous discussion the "dispersion equation" obtained by setting the determinant of coefficients equal to zero was used to solve for the time factor β, which could be a complex number, in terms of a known real wavenumber α. For the situation of the experiments just described, the same equation is solved for α, which may be complex, for a given imaginary value of β corresponding to the imposed frequency. Since interfacial configuration varies as $e^{i\alpha x}$, according to Eq. [V-41], we see that the real part α_R is the wavenumber, i.e., the wavelength λ is $(2\pi/\alpha_R)$. The imaginary part α_I is a damping factor characterizing the exponential decay of wave amplitude with distance from the source of vibrations. We note that it was this situation involving a complex α which was the motivation for using $e^{i\alpha x}$ instead of cos αx or sin αx in equations such as Eq. [V-11].

In the low viscosity limit the analysis shows that α_R is given by Eq. [V-57] with β_A^* the imposed frequency. The waves travel outward at a speed of (β_A^*/α_R) or $(\beta_A^* \gamma/\rho_A)^{1/3}$ for short wavelengths where the effect of gravity is unimportant. For a free surface the damping factor α_I is found under the same circumstances to be

$$\alpha_I = \frac{4\ \mu_A \beta_A^*}{3\ \gamma} \qquad \text{[V-74]}$$

For high surfactant concentrations the interface is inextensible, an oscillating boundary layer develops, and we obtain

$$\alpha_I = \frac{2^{1/2}}{6} \left(\beta_A^{*5} \rho_A \mu_A^3 / \gamma^4\right)^{1/6} \qquad \text{[V-75]}$$

At intermediate surfactant concentrations a maximum in α_I is predicted, just as was a maximum in the damping factor β_1 of the earlier discussion.

Figure V-2 shows experimental damping factors α_I obtained by the experimental technique described previously. The data are for monolayers of various fatty acids on water and the frequency of oscillations is 200 Hz (9). As the theory predicts, adding surfactant produces a significant increase in damping. Moreover, there is a maximum damping rate at an intermediate surfactant concentration. We note that only the long chain acids are described by the present theory. The short chain acids have appreciable solubility in water and adsorption-desorption and bulk diffusion effects must be considered. Because diffusion provides a means of surfactant transport from regions of high to low surfactant concentration along the interface, interfacial tension gradients are smaller and less damping occurs, as the figure shows. This effect is considered in Chapter VI.

Experiments of this type can be used to determine the surface tension γ and surface compressibility Γ_{os} $(-d\gamma/d\Gamma_s)$. The former is obtained from the measured value of wavelength using Eq. [V-57], the latter from the damping factor and the general dispersion equation found by setting equal to zero the determinant of coefficients of Eqs. [V-65] and [V-70]. A comparsion of surface properties obtained in this way with those obtained independently from film balance measurements is shown in Figure V-3 for monolayers of a particular anionic surfactant. Agreement is good, a confirmation of the validity of the capillary wave motion analysis presented above.

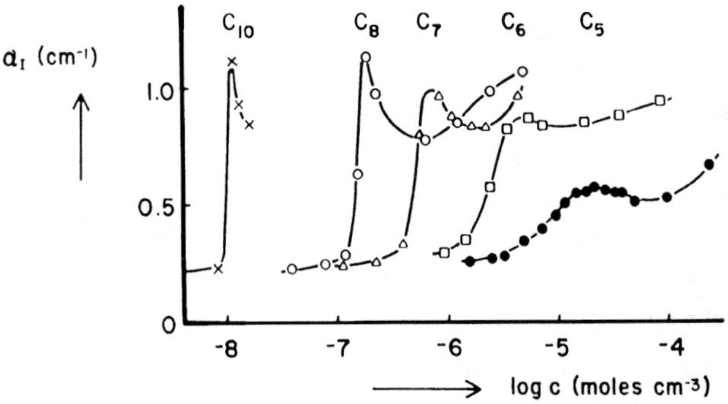

Figure V-2. Damping on mololayers of fatty acids at 200 Hz. Reprinted with permission from ref. (38).

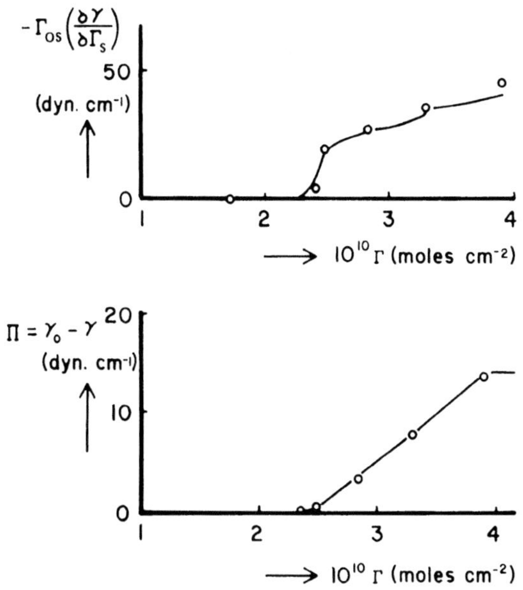

Figure V-3. Comparison of measured surface pressure and elasticity coefficient with values calculated from ripple properties. Monolayers of dodecyl-p-toluenesulphonate. (--) Measured; (◦) calculated. Reprinted with permission from ref. (8).

In a similar manner, the frequency of oscillation and the rate of damping of surface waves on an oscillating drop or bubble can be used to determine interfacial tension and compressibility. A derivation of pertinent equations and tabulation of key results may be found in the paper by Miller and Scriven (7). Some useful formulas are given in Problem V-4.

The above discussion has centered on wave motion imposed on a surface by, for instance, an oscillating bar. But thermal fluctuations cause wave motion of small amplitude even on interfaces that are not disturbed by external means. With developments in laser light scattering techniques in recent years it is possible to measure interfacial tension from analysis of surface fluctuations. This method has been applied to measurement of ultralow interfacial tensions between liquid phases (10). Presumably, it could also be used to determine surface compressibility or other rheological properties.

Example V-2. Characteristics of Wave Motion for Inextensible Interfaces.

Repeat the calculations of Example V-1 for an inextensible interface.

Solution Eqs. [V-72] and [V-73] apply to the liquid-gas and liquid-liquid cases respectively.

	Air-Water		Oil-Water	
λ (cm)	β_c^* (sec^{-1})	$-\beta_1$ (sec^{-1})	β_c^*(sec^{-1})	$-\beta_1$(sec^{-1})
100	7.84	6.22×10^{-3}	2.23	4.24×10^{-3}
10	25.1	0.112	7.45	7.77×10^{-2}
1	152	2.77	82.3	2.61
0.1	4080	144	2450	144

As expected the high surface elasticity greatly increases damping for the liquid-gas case over the free surface values of Example V-1. But it has only a small effect for the liquid-liquid case where boundary layer flow exists even with a free surface.

Surface elasticity is very effective in damping ripples of short wavelengths on a liquid surface. Although the damping rate is increased for long wavelengths, too, as the above results show, these waves are damped very slowly with or without a surfactant film. The increased damping of the ripples causes them to disappear rapidly and leave the surface with a smooth appearance. The same is true when a second immiscible liquid is added. A combination of these two effects is responsible for the ability of added oil to produce an apparent calming of the sea.

4. INSTABILITY OF FLUID CYLINDERS OR JETS

A slow-moving, thin cylindrical stream of water issuing from a faucet can often be seen to become nonuniform in diameter at some distance below the faucet and eventually to break up into drops. In a similar manner the beads on a spider's web are formed by instability of a cylinder of sticky liquid which surrounds the thread leaving the spider's body. This instability is sometimes used in industrial processes to disperse one fluid in another by injecting it in the form of a jet.

We consider a cylinder of an incompressible, inviscid liquid A in gas B (see Figure V-4). Initially there is no flow and the cross section is everywhere circular with radius R. At time t = 0 the fluid interface is deformed to the position given by the real part of the following expression:

$$\bar{r} = R + B \; e^{i(m\theta + \alpha z)} \qquad \text{[V-76]}$$

$$m = 0,1,2, \ldots .$$

If effects of gravity may be neglected, the equation of motion simplifies to

$$\rho \frac{\partial \mathbf{v}}{\partial t} = -\nabla p \qquad \text{[V-77]}$$

Taking the divergence of this equation and invoking the continuity equation $\nabla \cdot \mathbf{v} = 0$, we find

$$-\rho \frac{\partial}{\partial t} (\nabla \cdot \mathbf{v}) = 0 = \nabla^2 p \qquad \text{[V-78]}$$

We look for a solution of this equation which has the form

$$p(r,\theta,z,t) = P(r) \; e^{i(m\theta + \alpha z)} \; \delta(t) \qquad \text{[V-79]}$$

Substitution of Eq. [V-79] into Eq. [V-78] yields

$$r^2 \frac{d^2P}{dr^2} + r \frac{dP}{dr} - (m^2 + \alpha^2 r^2) P = 0 \qquad \text{[V-80]}$$

The solution of this equation which remains finite at r = 0 is given by

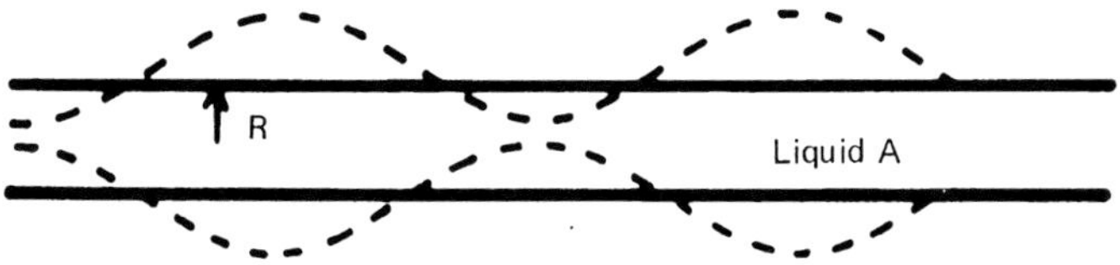

Figure V-4. Wavy deformation of liquid jet.

$$P(r) = D\, I_m(\alpha r) = D\, i^{-m} J_m(i\alpha r) \qquad [V\text{-}81]$$

where I_m is a modified Bessel function of the first kind.

In a similar manner we take the radial velocity v_r to have the form

$$v_r = W(r)\, e^{i\,(m\theta + \alpha z)}\, \delta(t) \qquad [V\text{-}82]$$

Substituting this expression into the radial component of Eq. [V-77], we obtain

$$W[r]\, \delta'(t) = -\frac{1}{\rho} \frac{dP(r)}{dr}\, \delta(t) \qquad [V\text{-}83]$$

Separation of variables yields

$$\frac{\delta'(t)}{\delta(t)} = -\frac{dP(r)/dr}{\rho W(r)} = \beta \qquad [V\text{-}84]$$

where β is a separation constant. The solutions are

$$\delta(t) = \delta_o\, e^{\beta t} \qquad [V\text{-}85]$$

$$W(r) = -\frac{1}{\rho\beta} \frac{dP(r)}{dr} = -\frac{D\alpha}{\rho\beta} \left(\frac{m}{\alpha r} I_m(\alpha r) + I_{m+1}(\alpha r) \right) \qquad [V\text{-}86]$$

We next invoke the requirement that the time derivative of the interfacial perturbation r^* must equal the radial velocity v_r at the interface. Clearly this condition and the initial condition [V-76] can be satisfied if the following relationships hold

$$r^* - R = B\, e^{i\,(m\theta + \alpha z)}\, e^{\beta t} \qquad [V\text{-}87]$$

$$B = W(R)/\beta \qquad [V\text{-}88]$$

The normal component of the interfacial momentum balance Eq. [V-35] must be satisfied at the fluid interface. The tangential component cannot be imposed because neglect of viscous effects has reduced the order of the equation of motion and thus diminished the number of boundary conditions which can be satisfied. Since there are no viscous stresses here, Eq.[V-35] simplifies to the Young-Laplace equation

$$p_A - p_B = -2H\gamma \qquad [V-89]$$

Before the interface is perturbed this equation simply states that p_A exceeds p_B by an amount (γ/R). For the case of an axisymmetric perturbation (m = 0) the curvature of the interface in a plane passing through the axis can be calculated from Eq. [I-42] by a procedure similar to that outlined following Eq. [V-39] for a nearly plane interface. The second radius of curvature is the local radius of the cylinder. Hence,

$$-2H = \frac{\partial^2(r^* - R)}{\partial z^2} + \frac{1}{r^*} \qquad [V-90]$$

$$= \frac{1}{R} + (\alpha^2 - \frac{1}{R^2})(r^* - R) \quad ; \quad m = 0 \qquad [V-91]$$

For $m \neq 0$ it can be shown that this expression must be augmented as follows:

$$-2H = \frac{1}{R} + (\alpha^2 + \frac{m^2 - 1}{R^2})(r^* - R) \qquad [V-92]$$

Substituting Eqs. [V-81], [V-87], [V-88], and [V-92] into Eq. [V-89], we obtain the following equation for the time factor β :

$$\beta^2 = -\frac{\gamma}{\rho}(\alpha^3 + \frac{\alpha(m^2 - 1)}{R^2})(\frac{m}{\alpha R} + \frac{I_{m+1}(\alpha R)}{I_m(\alpha R)}) \qquad [V-93]$$

It is clear that for nonaxisymmetric disturbances ($m \neq 0$), the right side of Eq. [V-93] is always negative. That is, β is an imaginary number, and oscillations exist without growth or decay, the expected behavior for a stable, inviscid system. But for the axisymmetric case (m = 0), instability does occur whenever the wave number is less than a critical value α_c given by

$$\alpha_c = \frac{1}{R} \qquad [V-94]$$

That is, a fluid cylinder is unstable with respect to disturbances having wavelengths greater than $(2\pi/\alpha_c)$ or the circumference $2\pi R$. The reason is that interfacial area and hence interfacial free energy actually decrease for such disturbances.

As before, the wavenumber α_M of the fastest growing disturbance can be obtained by differentiating Eq. [V-93] with respect to α (with m = 0) and requiring the resulting expression to vanish. This calculation, which was first carried out by Rayleigh, yields the following numerical results:

$$\alpha_M = 0.697/R \qquad \text{[V-95]}$$

$$\beta_M = 0.3433\ (\gamma/\rho R^3)^{1/2} \qquad \text{[V-96]}$$

Thus, the fastest growing disturbance has a wavelength λ_M approximately equal to 9.0 R. For a jet of water in air with a radius R of 0.2 cm, the corresponding time factor β_M is about 32.6 sec^{-1}.

Comparing Eq. [V-96] with Eq. [V-54], we see that β_M increases with increasing surface tension for instability of a cylinder but that the opposite is true for superposed fluids. This result is not surprising since the change in interfacial free energy is the source of the instability here, whereas it is a stabilizing effect for superposed fluids.

While the above analysis applies to breakup of a stationary cylinder, the results can be used to obtain a rough estimate of the length L_B required for breakup of a jet injected with uniform velocity V. In particular $L_B = Vt_B$, where t_B is the time required for the amplitude of a perturbation with wavelength λ_M to increase from its initial value B to a value comparable to the radius R. Unfortunately, the initial amplitude is unknown. However, the exponential nature of the growth causes t_B to be somewhat insensitive to the value of B. For instance, t_B for the water jet considered above is 0.14 sec if the amplitude must increase by a factor of 100 before breakage and 0.21 sec if it must increase by a factor of 1000. Using these values, we would estimate a jet length L_B of several centimeters.

Another procedure for determining L_B is to recast the stability analysis to recognize explicitly that the instability of a jet is a spatial rather than a temporal one, i.e., that perturbation amplitude increases with distance from the point of injection rather than with time at a fixed location as in the stability analysis developed so far. For the spatial analysis one proceeds much as in Section 3 and assumes that a perturbation with frequency β^* is imposed on the moving jet as it leaves the orifice. The time factor β is taken as $i\beta^*$ and α is obtained from the stability analysis. Its real part is related to perturbation wavelength and its imaginary part to the rate of growth (or decay) of perturbation amplitude with distance from the orifice. Perturbations of all frequencies would be deemed possible, and L_B would be the breakup length for the frequency whose amplitude increases most rapidly with distance from the orifice. Such an analysis has been given by Keller *et al* (11).

5. OSCILLATING JET

As indicated above all perturbations of a fluid cylinder for which $m \neq 0$ are stable. For $m = 1$ consideration of Eq. [V-87] shows that the cross section remains circular with radius R_0 at all points but that the axis becomes sinuous. For $m = 2$ the axis remains straight but the cross section is perturbed to a somewhat elliptical shape. It is this latter case which is of interest in this section.

If a jet issues from an orifice of circular cross section, we suppose that it can experience perturbations of the type specified by Eq. [V-76] with all possible values of m and α. As shown in the previous section an instability develops characterized by wavenumber α_M and growth factor β_M. But if the orifice has a rather elliptical cross section, a perturbation characterized by $m = 2$ and $\alpha = 0$ is imposed. Since Eq. [V-93] is not convenient to use for $\alpha = 0$, we return to Eq. [V-80] and solve directly for P(r) in this limit. The solution which remains finite for $r = 0$ is given by

$$P(r) = E\, r^m \qquad [V\text{-}97]$$

The remainder of the analysis proceeds in the same manner as before, the final result being

$$\beta^2 = -\frac{\gamma m(m+1)(m-1)}{\rho R^3} \qquad [V\text{-}98]$$

For $m = 2$, the behavior is a sustained oscillation with frequency β^*

$$\beta = \pm i\, \beta^* = \pm i\, (6\gamma/\rho R^3)^{1/2} \qquad [V\text{-}99]$$

If the liquid moves along the jet with velocity V, the distance Λ traveled during one cycle of oscillation is given by

$$\Lambda = \frac{2\pi V}{\beta^*} = \pi V\, (2R^3\rho/3\gamma)^{1/2} \qquad [V\text{-}100]$$

If the wavelength Λ of jet oscillation is measured, for instance from a photograph of the jet, and if ρ, R, and V are known, the surface tension γ can be calculated from Eq. [V-100]. In practice, the oscillating jet is used to measure variation of surface tension with time due to diffusion of surfactant from the bulk liquid to the surface and adsorption there, a subject considered further in Chapter VI. The procedure is simply to measure Λ at various distances from the orifice corresponding to different times available for surfactant diffusion and adsorption.

The above equation is based on a linear analysis which applies only for infinitesimal perturbations. Perturbations of finite amplitude are required, however, for a practical experiment. An early paper of Niels Bohr (12) dealt with the necessary extension. He also included the effect of liquid viscosity. A corrected equation suitable for perturbations whose amplitude b is finite but still smaller than the mean radius R is

$$\gamma = \frac{2\rho V^2 \pi^2 R^3}{3\ \Lambda^2} \left[\frac{1 + \dfrac{37\ b^2}{24\ R^2}}{1 + \dfrac{5}{3} \dfrac{\pi R^2}{\Lambda^2}} \right] \left[1 + 2 \left(\frac{\mu\Lambda}{\rho V \pi R^2} \right)^{3/2} + 3 \left(\frac{\mu\Lambda}{\rho V \pi R^2} \right)^2 \right] \qquad \text{[V-101]}$$

Other corrections which are sometimes required are discussed by Defay and Petre (13).

Example V-3. Surface Tension of Oscillating Jet

A jet of an aqueous solution has a mean radius R of 0.2 cm and travels with a velocity V of 60 cm/sec. Density and viscosity may be taken as 1 gm/cm^3 and 1 cp respectively. Calculate the surface tension using the simple and corrected equations [V-100] and [V-101] if the wavelength Λ for an oscillation is 2.40 cm and if (b/R) is 0.25.

Solution: Solving Eq. [V-100] for γ and substituting the above values, we find γ = 32.9 dynes/cm. With Eq. [V-101] the corresponding result is γ = 34.8 dynes/cm. In this particular case the viscosity correction is negligible, so that the chief correction is for the finite amplitude.

6. STABILITY AND WAVE MOTION OF THIN LIQUID FILMS; FOAMS

Coalescence of drops or bubbles is a phenomenon of fundamental importance to the understanding of disperse fluid systems such as foams and emulsions. As two drops or bubbles approach, a film forms between them. It drains and, if coalescence occurs, ultimately breaks. We are concerned here with this last step which involves film instability.

Careful experiments by Sheludko on thin liquid films have shown that instability often occurs when film thickness is a few tens of nanometers, i.e., a few hundred Angstrom units (14). For these very thin films the effects of London - van der Waals forces and of electrical double layer interaction discussed in Chapter III must be included in the analysis of film stability. Each of them contributes a local "body force" which acts on each volume element of liquid within the film and influences flow much as does the familiar gravitational body force per unit volume ρg. As indicated in Chapter III (Eq. [III-18]), the electrical body force is $(-\rho_e \nabla\psi)$, where ρ_e is the local free charge density and ψ the electrical potential. Now it is well known that the gravitational body force can be written as $(-\nabla\phi_g)$, where ϕ_g is the usual gravitational potential energy. Similarly, the electrical body force has the form $(-\nabla\phi_e)$ provided that flow is sufficiently slow that diffusional equilibrium of the ions is maintained at all times (15). The potential is given by

$$\phi_e = \frac{-\varepsilon\kappa^2}{4\pi} \left(\frac{kT}{\nu e_0}\right)^2 (\cosh u - 1) \qquad \text{[V-102]}$$

Finally, the van der Waals body force can be written as $(-\nabla\phi_{vw})$, where the van der Waals energy ϕ_{vw} can be calculated, for instance, by integration as in Section 1 of Chapter III.

If we neglect gravity in a thin film of liquid A and assume that it is of uniform thickness h_0 with no flow (see Figure V-5), the equation of motion simplifies to

$$\frac{d\Phi}{dz} = 0 \qquad \text{[V-103]}$$

with

$$\Phi = p_A + \phi_e + \phi_{vw} \qquad \text{[V-104]}$$

Thus, the total potential Φ is uniform in such a film, but the initial pressure p_{Ai} varies.

If the film surfaces are given a small wavy perturbation, the equations of motion now become Eqs. [V-3] and [V-4] with p replaced by Φ. Clearly Φ can be eliminated from these equations in the same manner as was p at Eq. [V-5]. Hence, the basic differential equations [V-9] and [V-10]

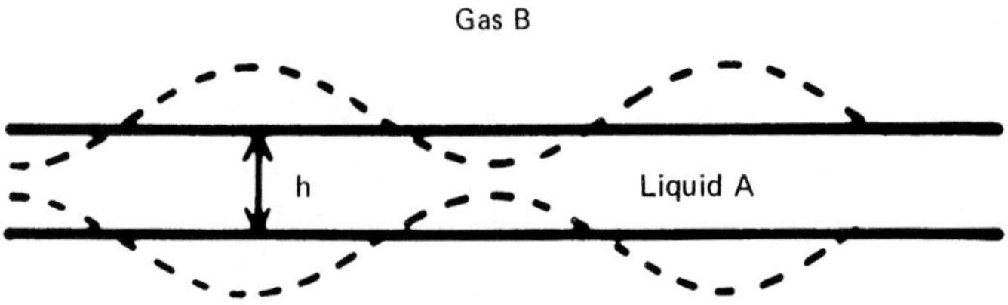

Figure V-5. Wavy deformation of thin liquid film.

for the velocity components, and naturally also the solutions to these equations such as Eq. [V-17], show no explicit dependence on van der Waals and electrical double layer forces. As we shall see shortly, these forces do make their appearance in the boundary conditions. Note that the use of Eqs. [V-3] and [V-4] involves an assumption that the film is at rest before perturbation, i.e., flow due to film drainage (see below) is neglected. This assumption amounts to requiring that the instability develop rapidly in comparison with changes in film thickness due to drainage.

One feature of the thin film analysis is that two interfaces exist, each of which can receive a general deformation such as that given by Eq. [V-1]. For each wavenumber α let us consider a general perturbation in which the two interfaces are displaced to $\bar{z}_1 = B_{11} e^{i\alpha x}$ and $\bar{z}_2 = B_{12} e^{i\alpha x}$. Such a perturbation can be expressed as a linear combination of two special perturbations: a symmetric one about the film centerline for which $\bar{z}_{1s} = -\bar{z}_{2s} = B_s e^{i\alpha x}$ and an antisymmetric one for which $\bar{z}_{1a} = \bar{z}_{2a} = B_a e^{i\alpha x}$. In particular, B_{11} can be written as the sum of B_a and B_s and B_{12} as their difference. Thus, it suffices to analyze symmetric and antisymmetric perturbations for all values of α.

Of primary interest here are symmetric perturbations because these lead to variations in film thickness and eventual film breakup if they are unstable. Since the resulting velocity distribution must be antisymmetric, we obtain instead of Eq. [V-17]

$$W_A(z) = D_1 \sinh \alpha z + D_2 \sinh \alpha r_A z \qquad \text{[V-105]}$$

The tangential velocity and the perturbation Φ_p in the total potential Φ defined at Eq. [V-103] can be found from Eqs. [V-20] and [V-26] respectively using this expression for W_A. If we assume that the film surfaces contain sufficient surfactant to prevent lateral motion, a common situation in practice, we have the inextensible case considered previously. Hence, the tangential component of the momentum balance becomes a requirement that v_x vanish. A derivation similar to that of Section 2b leads to the following counterpart of Eq. [V-42]:

$$\beta B_s = D_1 \sinh \frac{\alpha h_0}{2} + D_2 \sinh \frac{\alpha r_A h_0}{2} \qquad \text{[V-106]}$$

The key to the analysis is the normal component of the momentum balance. Because the liquid film is surrounded by gas, the terms for fluid B in Eq. [V-37] may be neglected. But another term must be added as at Eq. [III-29] to account for electrical stresses acting on the interface (15,16). We thus obtain the following form for the boundary condition

$$p_A - \frac{\varepsilon}{8\pi} \left(\frac{\partial \psi}{\partial z} \right)^2 - 2\mu_A \frac{\partial v_{Az}}{\partial z} + 2H\gamma = 0 \qquad \text{[V-107]}$$

It remains to evaluate the first two terms of this equation or, more precisely, the perturbation

$$\Delta \left(p_A - \frac{\varepsilon}{8\pi} \left(\frac{\partial \psi}{\partial z} \right)^2\right)$$

which occurs at the film surface when it is deformed. Expressing this quantity in terms of the total potential Φ, we find that

$$\Delta \left(p_A - \frac{\varepsilon}{8\pi} \left(\frac{\partial \psi}{\partial z} \right)^2\right) = \Delta\Phi - \Delta\phi_{vw} - \Delta \left(\phi_e + \frac{\varepsilon}{8\pi} \left(\frac{\partial \psi}{\partial z} \right)^2\right) \qquad \text{[V-108]}$$

The quantity $\Delta\, \Phi$ in Eq. [V-108] can be evaluated by making use of Eq. [V-26] for the initial and perturbation contributions to Φ

$$\Delta\Phi = \left. \frac{d\Phi_i}{dz} \right|_{\frac{h_0}{2}} + \Phi_p \left(\frac{h_0}{2} \right) \quad e^{i\alpha x}\, e^{\beta t}$$

$$= \left(D_1\, \mu\alpha \cosh \frac{\alpha h_0}{2} + D_2 \mu \alpha r^3 \cosh \frac{\alpha r h_0}{2} \right) e^{i\alpha x}\, e^{\beta t} \qquad \text{[V-109]}$$

Also $\Delta\phi_{vw}$ can be readily evaluated if we limit consideration to perturbations for which $(\alpha\, h_0) \ll 1$, i.e., those having wavelengths much greater than initial film thickness h_0. For then, the potential energy at the surface of a film with local thickness h relative to that at the interface of a bulk fluid can be found by integration

$$\phi_{vwo} = \int_0^{\infty}\int_h^{\infty} \frac{2\pi n^2 \beta dr dz}{(r^2 + z^2)^3} \qquad \text{[V-110]}$$

The integration can be carried out using the procedure described following Eq. [III-2]. In this manner we obtain

$$-\Delta\phi_{vw} = \frac{d}{dh}\left(\frac{-A_H}{6\pi h^3}\right) 2\, B_s\, e^{i\alpha x}\, e^{\beta t} \qquad \text{[V-111]}$$

The quantity in parentheses in this equation has the dimensions of pressure and has been termed the van der Waals contribution Π_{vw} to the film "disjoining pressure" by Deryagin. We note that $(-\Pi_{vw})$ is the derivative with respect to h of the film potential energy given by Eq. [III-14].

Again restricting consideration to the case $(\alpha h_o) \ll 1$, we can use Eq. [V-102] to write the last term of Eq. [V-108] in terms of an electrical contribution Π_{el} to the disjoining pressure:

$$-\Delta\left[\phi_e + \frac{\varepsilon}{8\pi}\left(\frac{\partial\psi}{\partial z}\right)^2\right] = \frac{d\Pi_{el}}{dh}\, 2B_s\, e^{i\alpha x}\, e^{\beta t} \qquad \text{[V-112]}$$

$$\Pi_{el} = \frac{\varepsilon\kappa^2}{4\pi}\left[\left(\frac{kT}{\nu e_o}\right)^2(\cosh u_o - 1) - \frac{\varepsilon}{8\pi}\left(\frac{\partial\psi^2}{\partial z}\right)\right]_{\text{surface}} \qquad \text{[V-113]}$$

Comparing with Eq. [III-28] and [III-29] we see that Π_{el} is equal to the electrical force per unit area F_e which acts on the surface of a film of uniform thickness. Substituting Eqs. [V-108], [V-109], [V-111], and [V-112] into Eq. [V-107] and evaluating the mean curvature H as in Section 2b, we obtain

$$D_1\left[\mu_A\alpha \cosh\frac{\alpha h_o}{2} + \frac{1}{\beta}\sinh\frac{\alpha h_o}{2}\left(2\frac{d}{dh}(\Pi_{vw} + \Pi_{el}) - \gamma\alpha^2\right)\right]$$
$$+ D_2\left[\mu_A\alpha\, r_A^3 \cosh\frac{\alpha r_A h_o}{2} + \frac{1}{\beta}\sinh\frac{\alpha r_A h_o}{2}\left(2\frac{d}{dh}(\Pi_{vw} + \Pi_{el}) - \gamma\alpha^2\right)\right] = 0 \qquad \text{[V-114]}$$

The condition for no tangential flow discussed above implies that

$$D_1 \cosh\frac{\alpha h_o}{2} + r_A D_2 \cosh\frac{\alpha r_A h_o}{2} = 0 \qquad \text{[V-115]}$$

The determinant of coefficients of these two equations yields, as before, a relation which can be used to calculate the time factor β. We have already assumed $(\alpha h_o) \ll 1$ in writing Eq. [V-110]. If we further assume $|\alpha r_A h_0| << 1$, we can expand the hyperbolic functions and obtain a result first given by Vrij *et al* (17):

$$\beta = \frac{-\left[\frac{\gamma\alpha^3}{\rho_A} - \frac{2\alpha}{\rho_A}\frac{d(\Pi_{vw} + \Pi_{el})}{dh}\right](\alpha h_0)^3}{24\,\nu_A\alpha^2} \qquad [V\text{-}116]$$

That is, the perturbation either grows or decays without oscillation. The form of Eq. [V-116] is similar to that found above for a single interface in the high viscosity limit (e.g., at Eq. [V-60]), yet no assumption of high viscosity has been made here. Basically, the inextensible interfaces so inhibit flow that a high viscosity is not required to prevent oscillatory motion in a thin film.

For the hypothetical case of an inviscid liquid film with free surfaces, it is readily shown that for small values of (αh_0), oscillatory motion is predicted by the analysis with an angular frequency β_F^* given by

$$\beta_F^{*2} = (\alpha h_0)\left[\frac{\gamma\alpha^3}{\rho_A} - \frac{2\alpha}{\rho_A}\frac{d(\Pi_{vw} + \Pi_{el})}{dh}\right] \qquad [V\text{-}117]$$

Hence, we can write Eq. [V-116] in a form even more similar to Eq. [V-60]

$$\beta = -\frac{\beta_F^{*2}(\alpha h_0)^2}{24\,\nu_A\alpha^2} \qquad [V\text{-}118]$$

Since $(d\Pi_{vw}/dh) > 0$, according to Eq. [V-111], for a liquid film in air where $A_H > 0$, we see from Eq. [V-116] that van der Waals forces favor instability. With film thickness less than the effective range of intermolecular forces, the potential energy of a molecule in a thin region of the film is greater than that of a molecule in a thick region which is surrounded by more molecules with which it can interact. Hence, the body force due to van der Waals interactions promotes flow from thin to thick regions of the film. This flow increases perturbation amplitude and leads eventually to film breakage. We note that the destabilizing effect of van der Waals forces is entirely consistent with Eq. [III-14], which states that total film energy per unit area (as opposed to the energy per molecule of the above argument) decreases with decreasing film thickness.

As indicated, above, the electrical disjoining pressure Π_{el} is equal to the electrical force F_e discussed in Chapter III. Examining the various expressions for F_e given there, we conclude that (dF_e/dh) and hence $(d\Pi_{el}/dh)$ are negative. With this information and Eq. [V-116] we see that repulsion between the double layers of the two film surfaces has a stabilizing influence on the film, the expected result. Also stabilizing is the tension γ of the film surfaces, especially for short wavelengths. Thus, perturbations having short wavelengths comparable to film thickness, which are not adequately described by the present analysis, are of little interest for film stability.

For situations where the film is unstable, Eq. [V-116] can be solved for the critical wavenumber α_c for which β vanishes

$$\alpha_c^2 = \frac{2}{\gamma} \frac{d}{dh} (\Pi_{vw} + \Pi_{el}) \qquad \text{[V-119]}$$

Only wavelengths exceeding $(2\pi/\alpha_c)$ are unstable. Similarly, the fastest growing disturbance can be found. It is characterized by

$$\alpha_M^2 = \frac{1}{\gamma} \frac{d}{dh} (\Pi_{vw} + \Pi_{el}) \qquad \text{[V-120]}$$

$$\beta_M = \frac{(\alpha h_0)^3}{24\mu_A \alpha} \frac{d}{dh} (\Pi_{vw} + \Pi_{el}) \qquad \text{[V-121]}$$

Of course, not all thin films are unstable. If the stabilizing effect of electrical forces is sufficiently large, it can outweigh the destabilizing influence of van der Waals forces. In this case β_F^*, as calculated from Eq. [V-117] is positive, and the film is stable with respect to disturbances of all wavelengths.

Also of significance is that initial instability of a thin flim in accordance with the above mechanism does not inevitably lead to film rupture. The analysis, like all others in this chapter, is based on linear stability theory and hence is valid only for small amplitude perturbations. It has been observed experimentally that at low surfactant concentrations instability of a film some tens of nanometers in thickness does produce rupture. But for some surfactants it is found that above a critical concentration the instability leads to formation of "black" films which are only slightly thicker than the total length of two surfactant molecules (14). These black films can be very stable and are a major factor in foam stability.

Indeed, an important application of knowledge of thin film behavior is to foams. Many foams of practical interest contain large proportions of the gas phase (over 90% by volume). The gas cells are polyhedral since such high volume fractions are impossible in dispersions of spheres having moderate drop size distributions. The cells are separated by thin liquid films of uniform thickness. The junctions of the films are called Plateau borders. A balance of forces requires that three films meet at a Plateau border and that the angle between adjacent films be 120°, as illustrated in Figure V-6a.

In view of the interfacial curvature in the Plateau border region (see Figure V-6a), it is clear from the Young-Laplace equation [I-22] that the pressure there is below that in the film, i.e., a driving force exists for film drainage. Thus, when a foam forms above a bulk liquid phase, liquid drains from the individual films into the Plateau borders and from the latter into the bulk liquid. Note that since hydrostatics requires liquid pressure to decrease with increasing height, the radius of curvature of the Plateau borders must decrease as well. As a result, the liquid holdup in the Plateau borders is less at greater elevations above the liquid phase.

An interesting feature of Plateau borders is their nonzero contact angles with the individual films. Owing to the effects of disjoining pressure, the effective tension γ_F of a thin film differs from the sum 2γ of the tensions of its two interfaces with the surrounding gas. A balance of forces at the junction between Plateau border and thin film thus requires existence of a finite contact angle λ (see Figure V-6b). Indeed, γ_F exceeds by 2γ the excess energy E_T of the film per unit area which is

(a)

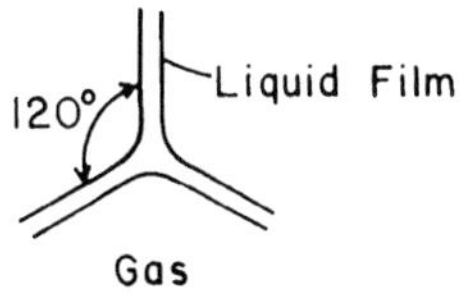

(b)

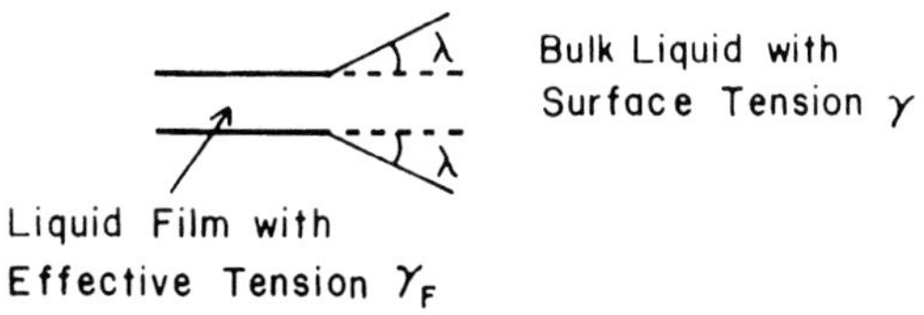

Figure V-6. (a) Plateau border in a foam. (b) Detailed view showing finite contact angle between thin film and bulk liquid in Plateau border region.

given by Eq. [III-50]. Hence, in view of the geometry of Figure V-6b, we have

$$\gamma_F = 2\gamma + E_T = 2\gamma \cos \lambda \qquad [V-122]$$

Values of such contact angles are typically small, ranging from somewhat less than one degree to a few degrees (18,19). Simultaneous measurement of contact angle and film thickness provides a method of determining disjoining pressure as a function of thickness and hence of obtaining information on the magnitude of intermolecular forces, e.g., values of the Hamaker constant A_H.

Variation of film thickness with time as a result of drainage is of great interest for foams and indeed for the coalescence process in general. As indicated previously, the driving force for flow in a thin film is the gradient of the total potential energy per unit volume Φ given by Eq. [V-104]. For a small circular film of radius R and uniform thickness h(t) and for situations with sufficient surfactant present to eliminate lateral flow at the film surfaces, the equation of motion can be solved to obtain (14):

$$\frac{d}{dt}(1/h^2) = \frac{4}{3\mu R^2}(\Delta p_c - \Pi_{vw} - \Pi_{el}) \qquad [V-123]$$

where Δp_c is the pressure difference mentioned above between film and Plateau border. Since Π_{vw} is negative, van der Waals forces promote thinning, the expected result since thinning transfers molecules to the bulk liquid where their potential energy due to interaction with surrounding molecules is lower than in the film. On the other hand, Π_{el} is positive and electrical repulsion opposes drainage. Indeed, if Π_{el} becomes equal to $(\Delta p_c - \Pi_{vw})$, an equilibrium film thickness is reached and drainage stops. In situations where the surface potential ψ_0 is known values of equilibrium film thickness can be used to estimate the Hamaker constant A_H.

Experiments on drainage of individual films can also be used to estimate A_H (14). In aqueous films with high electrolyte contents and in films of organic liquids, Π_{el} is negligible and $\Pi_{vw}(h)$ can be otained from Eq. [V-123] and data on the time dependence of h. The results confirm that Π_{vw} varies inversely with h^3 (cf. Eq. [V-111]) and provide values of A_H. Thin liquid films thus provide three possibilities for experimentally obtaining information on intermolecular forces (A_H): measurement of contact angles with adjacent bulk liquids, measurement of equilibrium film thicknesses, and measurement of film drainage rates.

Drainage and instability of individual films are very important in foam behavior, but other factors must be considered as well. For example, the lateral extent of individual films may increase during deformation of a foam, the additional liquid being drawn form the surrounding Plateau borders (20,21). The new portions of the film are normally thicker than the old. By such behavior a foam may remain stable in spite of mechanical perturbations.

The polyhedral cells in a foam usually have a range of sizes. As the pressure is largest for gas in the smallest cells, a driving force exists for transfer of gas from small to large cells by diffusion across the thin liquid films. By this mechanism the number of cells diminishes with time even in the absence of rupture of individual films. Lemlich (22,23) has presented a model of gas diffusion in foams.

To close this section, we briefly mention some applications of foams. Consumers prefer detergents and shampoos which produce some foam during use (but not too much). Formulators must design their products accordingly. Foams are also used in fire fighting, mineral flotation, food products, and "foam fractionation" processes which can separate materials having different degrees of surface activity. And, of course, as with emulsions, foams often develop where they are unwanted and hinder process performance. In this case they must be broken or steps taken to prevent their initial formation.

Example V-4. Stability of a Liquid Film

Consider a thin liquid film in air for which the Hamaker constant A_H is 5×10^{-13} erg and electrical double layer effects are negligible. If viscosity is 1 cp, density is 1 gm/cm^3, film thickness is 50 nm and surface tension is 40 mN/m, find the wavenumbers α_C and α_M of the critical and fastest growing disturbances and the time factor β_M for growth of the latter using the inextensible interface results given above. Compare with the corresponding results for a free interface in the inviscid approximation.

Solution The disjoining pressure Π_{VW} is $(-A_H/6\pi h^3)$, according to Eq. [V-111]. Hence, from Eqs. [V-119] - [V-121], we have for inextensible film surfaces

$$\alpha_c = \left(\frac{A_H}{\pi\gamma h_o^4}\right)^{1/2} = 2.52 \times 10^3 cm^{-1} \quad ; \quad \lambda_c = (2\pi/\alpha_c) = 2.49 \times 10^{-3} cm$$

$$\alpha_M = (\alpha_c / 2^{1/2}) = 1.78 \times 10^3 cm^{-1} \quad , \quad \lambda_M = 3.53 \times 10^{-3} cm$$

$$\beta_M = \frac{\alpha_M^2 A_H}{48\pi\mu_A h_o} = 0.422 \text{ sec}^{-1} \quad ; \quad \text{time constant } \beta_M^{-1} = 2.37 \text{ sec}$$

The inviscid result is simply

$$\beta^2 = - \beta_F^{*2}$$

where β_F^{*2} is given by Eq. [V-117]. It is clear from inspection of this equation and Eq. [V-117] that α_c is the same as before, the expected result since α_c is determined by thermodynamics alone. Upon differentiating Eq. [V-117], we find that α_M is again given by Eq. [V-120], so that it also has the same value as above. But β_M now has a value of $4.50 \times 10^4 sec^{-1}$, so that the time constant for growth is 2.22×10^{-5}sec. Hence, a surfactant-free film breaks very rapidly because of the rapid development of the instability. As a result, stable foams cannot be produced in the absence of surfactants.

7. ENERGY AND FORCE METHODS FOR THERMODYNAMIC STABILITY OF INTERFACES

In the examples of interfacial stability considered thus far the systems have been at rest in their initial states. Hence, the predictions of when instability can be expected are, in fact, conditions for thermodynamic stability. We have chosen not to emphasize this point and to carry out the analyses in terms of perturbations of the general equations of change because we obtain in this way not only the stability condition but also the rates of growth of unstable perturbations and the appropriate frequencies of oscillation and/or damping factors for stable perturbations. Also the basic method of analysis used above is applicable to systems not initially in equilibrium, as we shall see later in this chapter and in Chapter VI.

Nevertheless, there may be situations where information on growth rates and the characteristics of the fastest-growing disturbance is not of prime importance. In this case the energy and force methods, when they are applicable, can be used to determine the condition for thermodynamic stability with less effort than required for the more general analysis presented above (15).

The energy method may be used whenever an energy function such as the free energy exists which is known to have a local minimum at any state of stable equilibrium. Basically, the change in this function is calculated for all possible perturbations from the equilibrium state. If all such changes are positive, the initial state is stable. But if any possible peturbation produces a decrease in the energy function, instability can be expected.

The force method is often the easiest method of stability analysis to use because it requires only that the local normal force acting on the interface after deformation be calculated. If for all possible deformations this force acts to return the interface to its initial configuration, the system is stable. But if the force for any possible deformation is in a direction to increase deformation amplitude, the interface is unstable. The force method applies to fewer situations than the other methods, however. It requires not only that an energy function exist but also that body forces be irrotational, that fluids be incompressible, and that normal displacement at the system boundaries vanish except for the interface being analyzed (15).

Tyuptsov (24) and Huh (25) independently derived a general thermodynamic stability condition for an interface between fluids A and B using the energy method. If the fluids are of infinite extent or are bounded by rigid solid surfaces except at their interface, if gravity is the only external force acting on the system, and if any contact lines at the periphery of the interface are either fixed or move with no change in contact angle from the initial equilibrium values, the stability condition simplifies to

$$\int_S \{\gamma[-\nabla_s^2 \eta + (2K - 4H^2)\,\eta] - (\rho_A - \rho_B)g_n\eta\}\,\frac{\eta}{2}\,dS > 0 \qquad \text{[V-124]}$$

Here η is the local displacement of the interface in the normal direction (toward fluid B), K is the local Gaussian curvature (product of the principal curvatures $(1/r_1)$ and $(1/r_2)$), g_n is the local normal component of the gravitational acceleration, and the integration is over the entire fluid interface.

Eq. [V-124] simply states that stability requires the system's free energy to increase for all possible perturbations. The last term of the integrand clearly represents the local change in gravitational potential energy due to interfacial deformation. The first term, however, is only part of the local change in interfacial free energy, that is, the part due to *changes* in curvature during the deformation. For were the interface deformed at constant curvature, the change in interfacial free energy would, in view of the Young-Laplace equation [I-22], exactly balance the work done at the interface by the pressures in the two fluids.

According to Eq. [V-124], interfacial curvature changes for two reasons. The first, represented by the term $(-\nabla_s^2 \eta)$ is the nonuniformity, e.g., waviness, of the perturbation itself. The second, represented by the term $(2K - 4H^2)\,\eta$, is displacement of an initially curved surface normal to itself. For instance, uniform outward displacement of a spherical surface produces an increase in the radii of curvature and hence a decrease in magnitude in mean curvature H.

Eq. [V-124] can be applied to some of the situations we have already analyzed such as gravitationally-produced instability of superposed fluids and capillary instability of a fluid cylinder. It has also been applied to more complex situations such as stability of a pendant drop (25,26). We remark that the expression in braces in Eq. [V-124] is the local force acting in the normal direction on the deformed interface to restore it to its initial configuration. Hence application of the force method amounts to a requirement that this expression be positive for stability.

We remark that energy methods can also be applied to systems not initially at equilibrium. A *Liapunov function* must be found which (a) vanishes for the initial state, (b) is positive for all other states, and (c) decreases in value as perturbation amplitude decreases. The initial state is stable if the function's value decreases continuously during system response to all possible perturbations. As with the thermodynamic energy method, the Liapunov method is generally easier to use than perturbation of the governing equations, especially for perturbations of large amplitude. However, a Liapunov function must first be found. Further information on this approach may be found in references (27)-(29).

Example V-5. Energy Method for Stability of Superposed Fluids

Use Eq. [V-124] to determine the stability condition for superposed fluids.

Solution: Suppose that the interface in Figure V-1 is displaced from the plane $z = 0$ to the position $\bar{z} = B \cos \alpha x$. In this case $\eta = \bar{z}$, $H = K = 0$ and Eq. [V-124] becomes

$$\int_S \left[\gamma\alpha^2 \frac{\bar{z}^2}{2} - (\rho_A - \rho_B)\, g\, \frac{\bar{z}^2}{2}\right] dS > 0$$

Hence, at marginal stability

$$\alpha_c^2 = \frac{(\rho_B - \rho_A)g}{\gamma}$$

This result agrees with that obtained by the general analysis above (see remark following Eq [V-61]).

8. INTERFACIAL STABILITY FOR FLUIDS IN MOTION - KELVIN-HELMHOLTZ INSTABILITY

We now turn to interfaces in systems not initially at rest. From the manifold possible situations of this type we choose two for detailed study. One is the so-called Kelvin-Helmholtz instability at the interface between two fluids initially moving in a direction parallel to the interface but at different velocities. The other is wave motion on a falling liquid film, a situation of great practical interest.

We begin with analysis of stability of the interface between two inviscid, incompressible fluids which initially have uniform velocities V_A and V_B respectively in the x-direction (see Figure V-7). Kelvin's interest in this problem was as a possible source of wind-generated wave motion on the sea. But the instability also occurs in various cases of layered or annular flow of fluids, for instance with high-speed liquid jets. In the atmosphere it causes cloud surfaces to be irregular. Large vertical gradients in wind velocity may even be the source of much "clear air turbulence" (30).

After initial deformation of the interface from its initial planar state to the position $\bar{z} = Be^{i\alpha x}$, the equation of motion, to first order in perturbation terms, takes the form

$$\rho\left(\frac{\partial \mathbf{v}}{\partial t} + V \frac{\partial \mathbf{v}}{\partial x}\right) = -\nabla p + \rho \mathbf{g} \qquad \text{[V-125]}$$

Taking the divergence of this equation, we obtain

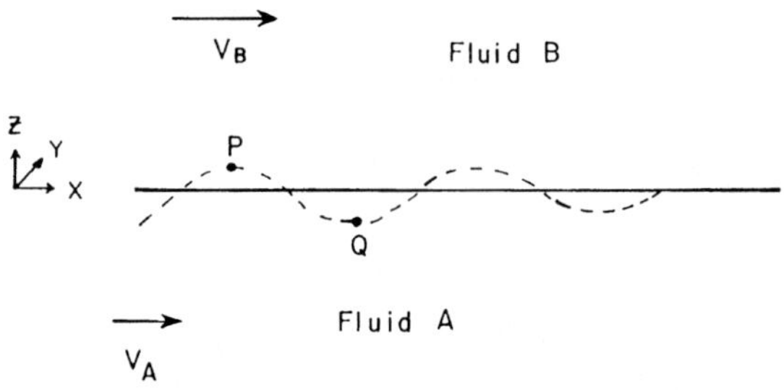

Figure V-7. Wavy perturbation of the interface between fluids moving with different velocities in the direction parallel to the interface.

$$\nabla^2 p = 0 \qquad [V-126]$$

If we take the perturbation in pressure to have the form $P(z)\, e^{(i\alpha x + \beta t)}$, we can easily solve Eq. [V-126] using procedures similar to those of Section 2a. The solutions which vanish far from the interface are

$$P_A(z) = A_1 e^{\alpha z} \qquad [V-127]$$

$$P_B(z) = A_2 e^{-\alpha z} \qquad [V-128]$$

Now with the z-component v_z of velocity given by $W(z)\, e^{(i\alpha x + \beta t)}$, we find from the z-component of Eq. [V-125] that

$$W(z) = -\frac{1}{\rho(\beta + i\alpha V)} \frac{dP}{dz} \qquad [V-129]$$

As before the perturbation in the tangential velocity v_x can be found from Eq. [V-20].

Now the rate of interfacial motion is equal to the normal component of velocity at the interface. The latter includes not only v_z at the interface as before, but also the component normal to the deformed interface of the initial velocity V. This condition requires that

$$\beta B = W_A(o) - i\alpha V_A B \qquad [V-130]$$

$$\beta B = W_B(o) - i\alpha V_B B \qquad [V-131]$$

These equations can be solved for A_1 and A_2 in terms of B.

$$A_1 = -\frac{\rho_A}{\alpha}(\beta + i\alpha V_A)^2 B \qquad \text{[V-132]}$$

$$A_2 = \frac{\rho_B}{\alpha}(\beta + i\alpha V_B)^2 B \qquad \text{[V-133]}$$

As before the tangential component of the interfacial momentum balance cannot be imposed for inviscid fluids. Nor can continuity of tangential velocity at the interface. But the normal component of the momentum balance is, as usual, a key equation. For the present situation Eq. [V-37] simplifies to

$$A_1 - A_2 = (\gamma\alpha^2 + (\rho_A - \rho_B)g)B \qquad \text{[V-134]}$$

Substituting Eqs. [V-132] and [V-133] into this expression amd solving for β, we obtain

$$\beta = \frac{-i\alpha(\rho_A V_A + \rho_B V_B)}{\rho_A + \rho_B} \pm i\left[\frac{\gamma\alpha^3 + (\rho_A - \rho_B)g\alpha}{\rho_A + \rho_B} - \frac{\alpha^2\rho_A\rho_B(V_A - V_B)^2}{(\rho_A + \rho_B)^2}\right]^{1/2} \qquad \text{[V-135]}$$

According to Eq. [V-135], instability occurs whenever the expression inside the brackets is negative. That β always has an imaginary part implies that wave motion exists, whether perturbation amplitude grows or not. The wave travels in the x-direction, its speed being $(\rho_A V_A + \rho_B V_B)/(\rho_A + \rho_B)$ and thus intermediate between V_A and V_B for unstable disturbances. The terms involving gravity and interfacial tension indicate that these effects act to stabilize the interface just as they do in the absence of tangential flow. As the destabilizing term inside the brackets is proportional to $(V_A - V_B)^2$, occurrence of the instability does not depend on which fluid has the greater velocity. When the two tangential velocities are equal, Eq. [V-135] reduces to the result Eq. [V-61] found previously for superposed inviscid fluids.

One way of viewing the basic instability mechanism is to recognize that for the situation illustrated in Figure V-7 where $V_B > V_A$, tangential velocity increases for the portion of fluid A displaced into the region near point P which was initially occupied by B. Similarly, tangential velocity decreases for the portion of fluid B displaced into the region near point Q. Because kinetic energy is proportional to the square of the velocity, more kinetic energy is extracted from the flow during the latter process than is added during the former. Thus, energy is made available to overcome gravity and interfacial tension and deform the interface.

The Kelvin-Helmholtz instability has been confirmed experimentally by Francis (31) for a situation where air was blown over a viscous oil. At a critical air speed small ripples appeared and grew rapidly. The critical speed was in good agreement with that predicted by Eq. [V-135].

Example V-6. Kelvin-Helmholtz Instability for Air-Water System

Calculate the critical velocity and wavenumber for air at 1 atm and 25°C blowing over water.

Solution: We have $\gamma = 72$ mN/m, $\rho_A = 1$ gm/cm^3, $\rho_B = (pM/RT)$ by the ideal gas law = 0.00119 gm/cm^3. At marginal stability the expression in brackets in Eq. [V-135] vanishes. Inspection of Eq. [V-135] shows that for a given value of $|V_A - V_B|$, this condition occurs whenever the stabilizing quantity $(\gamma\alpha + (\rho_A - \rho_B)g/\alpha)$ is minimized. Thus, at marginal stability we have

$$\alpha^2 = \alpha_c^2 = \frac{(\rho_A - \rho_B)g}{\gamma}$$

Hence, $\alpha_c = 3.69$ cm^{-1} ($\lambda_c = 1.70$ cm) when the instability first develops. The critical value of the velocity difference is given by

$$|V_A - V_B| = \left[\frac{(\gamma\alpha_c + (\rho_B - \rho_A)g/\alpha_c)(\rho_A + \rho_B)}{\rho_A\rho_B}\right]^{1/2} = 669\ \frac{\text{cm}}{\text{sec}}$$ or about 15 mph.

The speed c of the traveling wave, can be found from the first term of Eq. [V-135]

$$c = -\frac{\beta}{i\alpha} = \frac{\rho_A V_A + \rho_B V_B}{\rho_A + \rho_B} = 0.794\ \text{cm/sec}$$

Thus, air blowing over a lake or the sea should be able to generate waves when wind velocity exceeds about 15 miles per hour (about 13 nautical miles per hour).

9. WAVES ON A FALLING LIQUID FILM

It is found that when a film of liquid flows down an inclined surface as in Figure V-8, waves often develop along the fluid interface. The enhanced convection near the interface can lead to appreciable increases in heat and mass transfer rates between the liquid and gas phases. We note

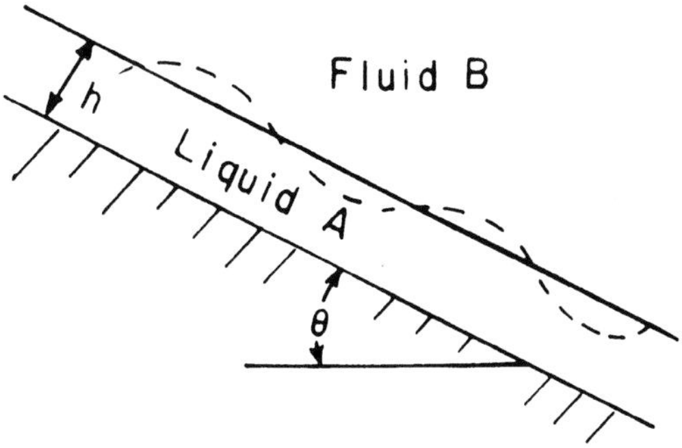

Figure V-8. Flow of a liquid film down an inclined plane.

that the films of interest here are much thicker than those considered in Section 6, so that effects of disjoining pressure are negligible.

The velocity distribution in a film of uniform thickness h where the flow is laminar is well known (2)

$$v_x = \frac{\rho\, g \sin\theta}{2\mu}\,(2z\, h - z^2) \qquad [V\text{-}136]$$

It can be written in terms of the average velocity $\bar{v}_x$ in the film as

$$v_x = \frac{3\bar{v}_x}{2}\left(2\,\frac{z}{h} - \frac{z^2}{h^2}\right) \qquad [V\text{-}137]$$

We seek information about characteristics of the traveling waves which develop when the interface is deformed. Because of the initial velocity, a direct stability analysis based on the differential equations of change is more complex mathematically than the analyses presented above although it has been carried out (32 - 34). We present here instead a simpler, but slightly less accurate, analysis based on the integral momentum equation. It follows for the most part the procedure of Kapitza (35) which has been described by Levich (36).

The x-component of the equation of motion for the falling film of Figure V-8 has the general form

$$\frac{\partial v_x}{\partial t} + v_x\,\frac{\partial v_x}{\partial x} + v_z\,\frac{\partial v_x}{\partial z} = -\frac{1}{\rho}\,\frac{\partial p}{\partial x} + g\sin\theta + \nu\left(\frac{\partial^2 v_x}{\partial x^2} + \frac{\partial^2 v_x}{\partial z^2}\right) \qquad [V\text{-}138]$$

We can simplify this equation for wavelengths long in comparison with film thickness by using the usual boundary layer approximations:

(a) $$\frac{\partial^2 v_x}{\partial x^2} \ll \frac{\partial^2 v_x}{\partial z^2}$$

(b) pressure variation in the z direction is hydrostatic,

(c) velocity distribution at each position x is given by Eq. [V-137] using local values of $\bar{v}_x$ and h. From condition (b) and the Young-Laplace equation at the fluid interface we obtain

$$p = \rho g(h-z)\cos\theta - \gamma\frac{\partial^2 h}{\partial x^2} \qquad \text{[V-139]}$$

With this equation and condition (a), Eq. [V-138] becomes

$$\frac{\partial v_x}{\partial t} + v_x\frac{\partial v_x}{\partial x} + v_z\frac{\partial v_x}{\partial z} = \frac{\gamma}{\rho}\frac{\partial^3 h}{\partial x^3} - g\cos\theta\frac{\partial h}{\partial x} + g\sin\theta + \nu\frac{\partial^2 v_x}{\partial z^2} \qquad \text{[V-140]}$$

We now integrate Eq. [V-140] over the entire film thickness for a given value of x using condition (c). Integrating by parts and invoking continuity, we write for the integral of the third term

$$\int_0^h v_z\frac{\partial v_x}{\partial z}\,dz = v_x v_z\Big|_0^h - \int_0^h v_x\frac{\partial v_z}{\partial z}\,dz$$

$$= -v_x(h)\int_0^h\frac{\partial v_x}{\partial x}\,dz + \int_0^h v_x\frac{\partial v_x}{\partial x}\,dz \qquad \text{[V-141]}$$

With this expression and with Eq. [V-136] for v_x and its derivatives, the integral momentum equation becomes, to first order in perturbation quantities,

$$\frac{3}{2}\frac{v_o}{h_0}\frac{\partial h}{\partial t} + \frac{3}{2}\frac{v_o^2}{h_0}\frac{\partial h}{\partial x} = \frac{\gamma}{\rho}\frac{\partial^3 h}{\partial x^3} - g\cos\theta\frac{\partial h}{\partial x} \qquad \text{[V-142]}$$

where v_o and h_0 are the average velocity and thickness of the initial unperturbed film of uniform thickness. An overall mass balance at each position x is also required. Noting from Eqs. [V-136] and [V-137] that $\bar{v}_x$ is proportional to h^2 and again keeping only first order terms, we find

$$\frac{\partial h}{\partial t} = -\frac{\partial}{\partial x}(\bar{v}_x h) = -3v_o\frac{\partial h}{\partial x} \qquad \text{[V-143]}$$

Let us now suppose that the waves travel down the film with a velocity c and with no increase or decrease in amplitude. That is, we focus on the condition of marginal stability at the boundary between stable and unstable regions. Then the thickness h should have the form

$$h = h_0 (1 + h_1 e^{i\alpha(x-ct)}) \qquad [V-144]$$

where α and c are both real. Substitution of this expression into Eq. [V-143] yields

$$c = 3 v_o \qquad [V-145]$$

The speed of the traveling waves is thus three times the average velocity or twice the surface velocity for a uniform film.

Similarily, substitution of Eqs. [V-144] and [V-145] into Eq. [V-142] provides an equation for the wavenumber α

$$\alpha^2 = \frac{\rho}{\gamma} (3 \frac{v_o^2}{h_0} - g \cos \theta) \qquad [V-146]$$

$$= \frac{\rho}{\gamma} (\frac{\rho^2 g^2 h_0^3 \sin^2 \theta}{3\mu^2} - g \cos \theta)$$

For a film with a given inclination θ and with a fluid of given properties, a real solution for α exists only when h_0 exceeds some critical value. We interpret this result to mean that the film is stable for sufficiently small values of h_0, i.e., for sufficiently small velocities v_o.

If we define a film Reynolds number N_{Re} by $(v_o h_0 \rho/\mu)$, we find that the condition for $\alpha = 0$ in Eq. [V-146] becomes

$$N_{Re} = \cot \theta \qquad [V-147]$$

That is, disturbances of all wavelengths are stable for $N_{Re} < \cot \theta$. The more rigorous analysis (22) predicts stability for $N_{Re} < (5 \cot \theta)/6$. In any case a film flowing vertically ($\theta = \pi/2$) is always unstable with respect to perturbations with sufficiently long wavelengths, no matter how small the Reynolds number. In practical situations, however, the rate of growth for small Reynolds numbers is so slow that the instability is not usually detected.

For fixed values of film inclination and Reynolds number, Eq. [V-146] gives the value of α for which wave amplitude remains constant. We might expect that disturbances of shorter wavelength (smaller α) would be stabilized by interfacial tension while disturbances of longer wavelength (larger α) would be unstable. The more detailed analysis confirms this expectation although the value of α at marginal stability differs slightly from that given by Eq. [V-146].

Example V-7. Wave Motion on Falling Water Film

A film of water flows down a surface with an inclination θ from the horizontal of 60°.

(a) Find the Reynolds number, average velocity and film thickness when the film first becomes unstable with respect to perturbations having very long wavelengths.

(b) For a Reynolds number of 5.0, find the speed of propagation of the waves and their wavelength for the situation where no growth or damping occurs.

Solution

(a) Eq. [V-147] gives a critical Reynolds number of 0.577. Using Eqs. [V-136] and [V-137] to obtain the average velocity $\bar{v}_x$, we find

$$N_{Re} = 0.577 = \frac{\rho \bar{v}_x h}{\mu} = \frac{\rho^2 g \sin\theta \, h^3}{3\mu^2}$$

Solving for h (with μ = 1 cp and ρ = 1 gm/cm^3), we find h = 5.89×10^{-3} cm. The average velocity in the film is $(\rho g h^2 \sin\theta / 3\mu)$ or 0.981 cm/sec. Note that the more rigorous analysis gives a critical Reynolds number of 0.480 and even smaller values of film thickness and velocity.

(b) For N_{Re} = 5 we find by a similar calculation that h = 1.21×10^{-2} cm and $\bar{v}_x$ = 4.14 cm/sec. From Eqs. [V-145] and [V-146] we obtain a speed of wave propagation of 3x4.14 or 12.4 cm/sec and a wavenumber α of 7.22 cm^{-1}.

The corresponding wavelength is $(2\pi/\alpha)$ or 0.869 cm.

REFERENCES

General References on Interfacial Stability and Wave Motion

Chandrasekhar, S. (1961) *Hydrodynamic and Hydromagnetic Stability.* Clarendon Press, Oxford.

Levich, V.G. (1962) *Physicochemical Hydrodynamics*. Prentice-Hall, Englewood Cliffs, N.J.

Lucassen-Reynders, E.H. and Lucassen, J. (1970) *Adv. Colloid Interface Sci.*, **2**, 347.

Miller, C.A. (1978) "Stability of Interfaces," in *Surface and Colloid Science*, E. Matijevic (ed), **vol. 10**. Plenum, New York, p. 227.

Wehausen, J.V. and Laitone, E.V. (1960) "Surface Waves," in *Handbuch der Physik*, **vol. 9**, Springer-Verlag, Berlin, p. 446.

Textual References

1. Miller, C.A. (1978) "Stability of Interfaces," in *Surface and Colloid Science*, E. Matijevic (ed), **vol. 10**. Plenum, New York, p. 227.

2. Bird, R.B., Stewart, W.E., and Lightfoot, E.N. (1960) *Transport Phenomena*, Wiley, New York.

3. Chandrasekhar, S. (1961) *Hydrodynamic and Hydromagnetic Stability*. Clarendon Press, Oxford.

4. Wehausen, J.V. and Laitone, E.V. (1960) "Surface Waves," in *Handbuch der Physik*, **vol. 9**, Springer-Verlag, Berlin, p. 446.

5. Lucassen, J. (1968) *Trans. Faraday Soc.* **64**, 2221, 2230.

6. Dorrestein, R. (1954) *Proc. Konikl. Ned. Akad. Wetenschap.* **B54**, 260, 350.

7. Miller, C.A. and Scriven, L.E. (1968) *J. Fluid Mech.* **32**, 417.

8. Lucassen-Reynders, E.H. and Lucassen, J. (1970) *Adv. Colloid Interface Sci.* **2**, 347.

9. Lucassen, J. and Hansen, R.S. (1967) *J. Colloid Interface Sci.* **23**, 319.

10. Bouchiat, M.A. and Meunier, J. (1972) *J. de Physique*, Colloque C1, Supplement au no. 2-3, **33**, C1-141.
Zollweg, J., Hawkins, G., Smith, I.W., Giglio, M., and Benedek, G.B., ibid, p. C1-135.
Cazabat, A.M., Langevin, D., Meunier, J., and Pouchelar, A. (1983) *Adv. Colloid Interface Sci.* **16**, 175.

11. Keller, J.B. Rubinow, S.I., and Tu, Y.O. (1973) *Phys. Fluids* **16** 2052.

12. Bohr, N. (1909) *Phil. Trans. Roy. Soc. (London)* **A209**, 281

13. Defay, R. and Petre, G. (1971) "Dynamic Surface Tension" in *Surface and Colloid Science*, E. Matijevic (ed), **vol. 3**, Wiley, New York, p.27.

14. Sheludko, A. (1967) *Adv. Colloid Interface Sci.* **1**, 391.

15. Miller, C.A. and Scriven, L.E. (1970) *J. Colloid Interface Sci.* **33**, 360.
16. Sanfeld, A. (1968) *Introduction to the Thermodynamics of Charged and Polarized Layers*, Wiley, New York.

17. Vrij, A., Hesselink, F. Th., Lucassen, J., and van den Tempel, M. (1970) *Proc. Konikl. Ned. Akad. Wetenschap.* **B73**, 124.

18. Huisman, H.F. and Mysels, K.J. (1969) *J. Phys. Chem.* **73**, 489.

19. Princen, H.M. and Frankel, S. (1971) *J. Colloid Interface Sci.* **35**, 386.

20. Mysels, K.J., Shinoda, K., and Frankel, S. (1959) *Soap Films; Studies of Their Thinning and Bibliography*. Pergamon, New York.

21. Lucassen, J. (1981) "Free Liquid Films and Foams," in *Anionic Surfactants*, E.H. Lucassen-Reynders (ed), Marcel Dekker, New York, p.217.

22. Lemlich, R. (1978) *Ind. Eng. Chem. Fundamen.* **17**, 89.

23. Ranadive, A.Y. and Lemlich, R. (1979) *J. Colloid Interface Sci.* **70**, 392.

24. Tyuptsov, A.D. (1966) *Fluid Dynam.* **1** (2), 51.

25. Huh, C. (1969) "Capillary Hydrodynamics: Interfacial Instability and the Solid/Fluid/Fluid Contact Line," Ph.D. Thesis, University of Minnesota.

26. Pitts, E. (1974) *J. Fluid Mech.* **63**, 487.

27. Denn, M.M. (1975) *Stability of Reaction and Transport Processes*. Prentice-Hall, Englewood Cliffs, N.J.

28. Dussan V., E.B. (1975) *Arch. Rational Mech. Anal.* **57**, 363.

29. Joseph, D.D. (1976) *Stability of Fluid Motions* **vol. 2** Springer-Verlag, Berlin, Chapter XIV.

30. Dutton, J.A. and Panofsky, H.A. (1970) *Science* **167**, 937.

31. Francis, J.R.D. (1954) *Philos. Mag. Ser 7* **45**, 695.

32. Benjamin, T.B. (1957) *J. Fluid Mech.* **2**, 554.

33. Yih, C.S. (1963) *Phys. Fluids* **6**, 321.

34. Krantz, W.B. and Goren, S.L. (1970) *Ind. Eng. Chem. Fundam.* **9**, 107.

35. Kapitsa, P.L. (1948) *Zh. Ekper. Teoret. Fiz.* **18**, 3.

36. Levich, V.G. (1962) *Physicochemical Hydrodynamics*. Prentice-Hall, Englewood Cliffs, N.J.

37. Miller, C.A. (1977) *AIChE J.* **23**, 959.

38. Lucassen, J. and Hansen, R.S. (1967) *J. Colloid Interface Sci.* 23. 310.

PROBLEMS

V-1. Show that for a liquid of low viscosity in contact with a gas the damping factor, i.e., the real part of the time factor β, passes through a maximum with increasing surfactant concentration and that it approaches the value given by Eq. [V-72] for large surfactant concentrations.

(a) Form the determinant of coefficients of Eqs. [V-65] and [V-70] and show that the dispersion equation has the form

$$r_A^4 + 2r_A^2 - 4r_A + 1 + r_A(r_A^2-1)\, N = -\frac{\beta_A^{*2}}{\nu_A^2\alpha^4}\left[1 + \frac{\nu_A\alpha^2}{\beta}(r_A - 1)\, N\right]$$

$$\text{where} \quad N = \frac{\Gamma_{os}\,\alpha\left(-\frac{d\gamma}{d\Gamma_s}\right)}{\beta\,\mu_A}$$

(b) In the low viscosity limit, i.e., for $|\beta/\nu_A\alpha^2| >> 1$, show that the above equation may be approximated by its leading terms to obtain

$$\beta^2\left[1 + N\left(\frac{\nu_A\alpha^2}{\beta}\right)^{1/2} + 4\,\frac{\nu_A\alpha^2}{\beta}\right] = -\beta^{*2}\left[1 + N\left(\frac{\nu_A\alpha^2}{\beta}\right)^{1/2} - N\left(\frac{\nu_A\alpha^2}{\beta}\right)\right]$$

(c) Assume that the order of magnitude of N is at least as large as $(\beta/\nu_A\alpha^2)^{1/2}$. Then set β equal to $(i\beta^* + \beta_1)$ with $|\beta_1/\beta^*| << 1$. Neglecting terms in β_1^2, show that the result obtained in (b) becomes

$$\beta_1 \simeq \frac{\alpha(\beta^*\nu_A)^{1/2}}{2i}\left(\frac{\beta^*}{\beta}\right)^{1/2}\frac{1}{1 + \frac{1}{N}\left(\frac{\beta}{\nu_A\alpha^2}\right)^{1/2}}$$

(d) Show that $\beta^{1/2} \simeq \beta^{*\,1/2}(1+i)/(2)^{1/2}$. Substitute this expression into the result of (c) and show that

$$\beta_1 \simeq -\frac{\alpha(\beta^*\nu_A)^{1/2}}{2^{3/2}} \times \frac{1+i(1-2E)}{1-2E+2E^2}$$

$$\text{where} \quad E = \frac{\beta^{*\,3/2}(\mu_A\rho_A)^{1/2}}{2^{1/2}\,\Gamma_{os}\,\alpha^2\left(\frac{-d\gamma}{d\Gamma_s}\right)}$$

(e) Show that in the limit $E \to 0$, the inextensible result Eq. [V-72] is obtained. In addition, show that the real part of β_1 has its greatest magnitude for $E = 1/2$ and that the damping factor in this case is twice that for an inextensible surface. Finally, show that if the wavelength is short enough for gravitational effects to be negligible, the surface compressibility Γ_{os} $(-d\gamma/d\Gamma_s)$ at maximum damping is given by

$$\Gamma_{os}\left(\frac{-d\gamma}{d\Gamma_s}\right) = 2^{1/2}\gamma^{3/4}\alpha^{1/4}\nu_A^{1/2}\rho_A^{1/4}$$

V-2. Use Eq. [V-71] to derive the expressions for damping of capillary waves given by Eqs. [V-74] and [V-75]. Also find the speed at which waves travel away from the source of vibrations.

(a) Show that in the free surface case the dispersion equation is given by

$$r^4 + 2r^2 - 4r + 1 + \frac{\beta^{*2}}{\nu^2\alpha^4} = 0$$

(b) Show that if $\beta = -i\bar{\beta}$, where $\bar{\beta}$ is the imposed frequency, the speed i of the traveling waves is $(\bar{\beta}/\alpha_R)$, where α_R is the real part of α, i.e., $\alpha = \alpha_R + i\,\alpha_I$. If $|\alpha_I/\alpha_R| \ll 1$, substitute these expressions for β and α into the dispersion equation and keep only the lead terms of the real and imaginary parts of this equation in the low viscosity approximation $|\beta/\nu\alpha^2| \gg 1$. Then solve for α_R and α_I. Show that for short wavelengths where gravity is negligible, α_I is given by [V-74] and

$$\alpha_R = \left(\frac{\rho\bar{\beta}^2}{\gamma}\right)^{1/3}$$

$$c = \left(\frac{\bar{\beta}\gamma}{\rho}\right)^{1/3}$$

(c) Repeat for the inextensible surface case. At one point in this derivation you will need to prove that the lead term of r in the low viscosity approximation is

$$(\bar{\beta}^{1/2}/\alpha_R\nu^{1/2})(1 - i)/2^{1/2}$$

V-3. In the experiments described in Section 3 "surface wave" motion was generated by small oscillations of a bar in a direction perpendicular to the surface. In another type of experiment for surfactant - containing interfaces the bar receives small oscillations in a direction parallel to the surface and "longitudinal wave" motion ensues (5). Develop an analysis of this situation for the case of an insoluble surfactant film at a liquid-gas interface which remains flat during longitudinal wave motion.

(a) Show that the dispersion equation in this case is given by

$$r^2 - 1 + N\,(r - 1) = 0$$

where N is as defined in Problem V-1

(b) With $\beta = -i\,\bar{\beta}$ and $\bar{\beta}$ the imposed frequency, simplify this equation to show that in the low viscosity limit

$$i\,\bar{\beta}^3\,\mu\rho = \alpha^4\,\Gamma_{os}^{\;2}\left(\frac{-d\gamma}{d\Gamma_s}\right)^2$$

(c) Let $(\alpha/\alpha_o) = \alpha_R + i\alpha_I$, where $\alpha_o = [\bar{\beta}^3\,\mu\rho/\Gamma_{os}^{\;2}\,(-d\gamma/d\Gamma_s)^2]^{1/4}$ Show that

$$(\alpha/\alpha_o) = e^{i\,\pi/8}$$

and hence that $\alpha_R = \cos\pi/8$ and $\alpha_I = \sin\pi/8$

(d) Calculate the real and imaginary parts of α for a surfactant film on an aqueous surface with $\bar{\beta} = 300\ \text{sec}^{-1}$, $\mu = 1$ cp, $\rho = 1\ \text{gm/cm}^3$, and $[\Gamma_{os}\,(-d\gamma/d\Gamma_s)] = 5$ dyne/cm. What are the wavelength and the speed of travel of the longitudinal waves generated in this case?

V-4. For a drop of a viscous liquid having a radius R oscillating in a gas, the time factor β for the case of a free surface is given by

$$\beta = \pm\, i\,\beta_D^{*} - \frac{\nu(\ell-1)(2\ell+1)}{R^2}$$

where $\beta_D^{*2} = \dfrac{\gamma\ell(\ell-1)(\ell+2)}{\rho R^3}$

and ℓ is a positive integer characterizing the mode of oscillation. For example, the drop is deformed into approximately an ellipsoidal shape for $\ell = 2$. Similarly, for an inextensible interface, we have

$$\beta = \pm\, i\,\beta_D^{*}\,(1 \mp i)\,\frac{(\beta_D^{*}\nu)^{1/2}(\ell-1)^2}{2^{3/2}\,R\,(\ell+1)}$$

If $\ell = 2$, $R = 1.5$ cm, $\mu = 1$ cp, $\gamma = 72$ dyne/cm, and $\rho = 1$ gm/cm^3, calculate the oscillation frequency and the damping rates in the two cases. Repeat for a bubble. The formulas are

$$\beta = \pm i\ \beta_B^* - \frac{\nu(2\ell+1)(\ell+2)}{R^2}$$

$$\beta = \pm i\ \beta_B^* - (1\mp i)\ \frac{(\beta_B^*\nu)^{1/2}(\ell+2)^2}{2^{3/2}\ \ell R}$$

where $\beta_B^{*2} = \dfrac{\gamma(\ell+1)(\ell-1)(\ell+2)}{\rho\ R^3}$

Why does the surfactant have a much larger effect on damping for a bubble? Hint: Consider the tangential velocities in the bulk fluids and at the interface as the drop or bubble becomes more ellipsoidal. See Problem V-5.

V-5. As shown in the text, there is no lateral motion along an inextensible plane interface subjected to small wavy deformations, i.e. for a surface where $\Gamma_{os}\ (-\ d\gamma/d\Gamma_s\ )$ is very large. The same is *not* true for a spherical interface because, in view of Eq. [I-19], local expansion occurs in regions where the interface moves outward and local compression in regions where it moves inward. If the surfactant film is nearly incompressible, tangential flow develops from the latter to the former regions. This situation is the one described by the formulas given in Problem V-4.

It is also conceivable that a film might be rather compressible but exhibit a very large resistance to shear. In this case where the so-called shear elasticity is large, there is little lateral motion along the interface and, the following expressions can be derived (37)

<u>Drop</u>

$$\beta = \pm i\ \beta_D^* - (1\mp i)\ \frac{(\beta_D^*\nu)^{1/2}(\ell+1)}{2^{3/2} R}$$

<u>Bubble</u>

$$\beta = \pm i\ \beta_B^* - (1\mp i)\ \frac{(\beta_B^*\nu)^{1/2}\ell}{2^{3/2}\ R}$$

Using the same values of physical properties as in Problem V-4, calculate damping factors using these formulas. Comment on the results in comparison with those of Problem V-4.

V-6. Develop an analysis for surface wave motion of a liquid having a finite depth H. Show that for wavelengths comparable in magnitude to H or larger an oscillating boundary layer develops at the solid surface in the low viscosity limit and that the time factor β is given for a free surface by

$$\beta = \pm\, i\, \beta^* - (1 \mp i)\, \frac{\alpha(\beta^* \nu)^{1/2}}{2^{3/2}} \ \text{sech}\ \alpha H\ \text{csch}\ \alpha H$$

with $\beta^{*2} = \dfrac{\gamma\, \alpha^3 + \rho_A g \alpha}{\rho_A \coth\, \alpha H}$

V-7. Find α_c, α_M, and β_M for the instability with respect to axisymmetric perturbations of a hollow fluid cylinder, i.e., a cylinder of gas in an infinite expanse of liquid.

V-8. Use the force method to investigate the stability of a cylinder of A in B when both fluids rotate at an angular velocity ω around the axis of the cylinder. Assume that interfacial tension exerts a local restoring force equal to $[-\gamma d(2H)]$, where H is the local mean curvature. If $\rho_B > \rho_A$, how large must ω be to overcome the basic capillary instability described in Section 4.

V-9. For air blowing over water under the conditions given in Example V-6, find the range of unstable wave numbers α, the wavenumber of the fastest growing disturbance α_M, its rate of growth, and its speed of wave propagation along the interface when wind velocity is 20 mph (892 cm/sec).

V-10. Repeat the analysis of wave motion on a falling film for the case where the fluid surface is immobile, i.e., it can deflect in the z-direction but tangential velocity v_x is zero. Such a flow might exist in the event of surfactant buildup along the film because some barrier far downstream prevents surfactant from leaving the film surface. Find the critical Reynolds number for instability as a function of θ and find c and α at marginal stability as a function of film thickness h.

VI

Transport Effects on Interfacial Phenomena

1. INTERFACIAL TENSION VARIATION

Experiments demonstrate that interfacial tension varies with both interfacial composition and temperature. At equilibrium both these quantities are uniform along an interface and hence so is interfacial tension. If transport processes, e.g., diffusion of surfactant to a newly created interface, cause interfacial tension to be time dependent while remaining uniform at each time, changes in interfacial shape may occur. Such processes may be followed by, for instance, monitoring the dimensions of a sessile or pendant drop as a function of time.

Flow or transport may also produce spatial variations in interfacial temperature or concentration and hence in interfacial tension. Indeed, we already saw in Chapter V that flow associated with surface wave motion causes surfactant concentration gradients to develop at an interface with an insoluble monolayer. The resulting interfacial tension gradients were found to significantly enhance damping of wave motion at liquid-gas interfaces with resulting adverse effects on transfer of heat or mass across the interface.

In this chapter we deal with flow generated by interfacial tension gradients, the so-called Marangoni effect. The development of "tears" on the inner surface of a glass of strong wine is a common example of such flow. We shall be particularly interested in cases of interfacial stability where flow develops by this mechanism for converting thermal or chemical energy to mechanical energy. Such instabilities can bring about severalfold increases in heat or mass transfer rates across interfaces.

2. INTERFACIAL SPECIES MASS BALANCE AND ENERGY BALANCE

General mass and momentum balances for an interfacial region were derived in Chapter V and used in the analyses of interfacial stability presented there. In dealing with heat and mass transport near interfaces we require additional balances for energy and individual species.

We consider the latter first because it can be derived in a manner similar to that employed previously for the overall mass balance but with additional terms to account for diffusion and chemical reaction. The starting point is the mass balance for species i for the pillbox control volume of Figure I-3

$$\int_{S_A} (\rho_{iA}\,\mathbf{v}_A + \mathbf{j}_{iA})\cdot\mathbf{n}\,dS - \int_{S_B} (\rho_{iB}\,\mathbf{v}_B + \mathbf{j}_{iB})\cdot\mathbf{n}\,dS - \int_{S_0} (\rho_i\,\mathbf{v} + \mathbf{j}_i)\cdot\mathbf{M}\,dS$$

$$+ \int_V r_i\,dV = \frac{d}{dt}\int_V \rho_i dV \qquad \text{[VI-1]}$$

Here ρ_i is the local mass density of species i, $\mathbf{j}_i$ the local diffusional flux of i relative to the mass average velocity $\mathbf{v}$, and r_i the local rate of production of i per unit volume as the result of one or more chemical reactions. As before the corresponding equations for the extrapolated bulk phases must be subtracted from this equation, various manipulations performed to convert the resulting equation to a statement that an intergral over the reference surface S must vanish, and the arbitrary extent of S invoked to argue that the integrand itself must be zero (compare Eqs. [V-28] - [V-32]).

The resulting differential interfacial mass balance is given by

$$\frac{\partial\Gamma_i}{\partial t} + \nabla_s\cdot\Gamma_i\,\mathbf{v} - \rho_{iA}(v_{An} - \dot{a}_n) + \rho_{iB}(v_{Bn} - \dot{a}_n) - j_{iAn} + j_{iBn}$$

$$- R_i + \nabla_s\cdot\mathbf{j}_{is} = 0 \qquad \text{[VI-2]}$$

where $\dot{a}_n$ is the normal velocity of S as before and

$$\Gamma_i = \int_{\lambda_A}^{\lambda_B} \Delta\rho_i d\lambda \qquad \text{[VI-3]}$$

$$R_i = \int_{\lambda_A}^{\lambda_B} \Delta r_i d\lambda \qquad \text{[VI-4]}$$

$$\mathbf{j}_{is} = \int_{\lambda_A}^{\lambda_B} \Delta \, \mathbf{j}_i d\lambda \cdot (\mathbf{I} - \mathbf{nn}) \qquad [VI-5]$$

The quantities $\Delta\rho_i$, Δr_i, and $\Delta \mathbf{j}_i$ are defined in a manner analogous to $\Delta\rho$ and Δp_T at Eq. [I-33]. The term involving the unit dyadic (or tensor) $\mathbf{I}$ employed in Eq. [VI-5] assures that $\mathbf{j}_{is}$ includes only the tangential portion of the surface excess flux. Also the convective flux in the integral over S_0 has been neglected as in the previous derivation of the overall mass balance because S_0 can be made small in comparison to S by reducing pillbox thickness.

Eq. [VI-2] states that the local interfacial concentration Γ_i changes as a result of interfacial convection, interchange with the bulk phases due to convection and diffusion, interfacial chemical reaction, and diffusion within the interface. When the last effect is important some relation is required between $\mathbf{j}_{is}$ and Γ_i. Usually, $\mathbf{j}_{is}$ is taken as $(-D_{is}\nabla_s\Gamma_i)$, where D_{is} is an interfacial diffusion coefficient.

The interfacial energy balance can be derived in a similar manner, again beginning with a balance over the pillbox control volume of Figure I-3:

$$\frac{d}{dt}\int_V \rho \hat{E} \, dV = \int_{S_A} [\mathbf{q}_A + \rho_A \hat{E}_A \mathbf{v}_A + p_A \mathbf{v}_A + \tau_A \quad \mathbf{v}_A] \cdot \mathbf{n} \, dS$$
$$- \int_{S_B} [\mathbf{q}_B + \rho_B \hat{E}_B \mathbf{v}_B + p_B \mathbf{v}_B + \tau_B \quad \mathbf{v}_B) \cdot \mathbf{n} \, dS$$
$$- \int_{S_0} [\mathbf{q} + \rho \hat{E} \mathbf{v} + p_T \mathbf{v} + \tau \quad \mathbf{v}] \cdot \mathbf{M} \, dS \qquad [VI-6]$$

In this equation $\mathbf{q}$ represents a conductive heat flux, $\hat{E}$ the total energy per unit mass including internal, kinetic, and gravitational potential energy contributions, and $\boldsymbol{\tau}$ is the viscous stress dyadic. The reference states for the various internal energies are chosen in such a way that energy effects of phase changes and chemical rections occurring in the interfacial region are accounted for in this formulation.

By the same basic procedure as used before we arrive at the following differential interfacial energy balance:

$$\frac{\partial E^s}{\partial t} + \nabla_s \cdot E^s \mathbf{v} - (\rho_A \hat{E}_A + p_A + \tau_{An})(v_{An} - \dot{a}_n) - \tau_{At} v_{At}$$
$$+ (\rho_B \hat{E} + p_B + \tau_{Bn})(v_{Bn} - \dot{a}_n) + \tau_{Bt} v_{Bt} - q_{An} + q_{Bn}$$

$$+ \nabla_s \cdot \mathbf{q}_s - \nabla_s \cdot \gamma \, \mathbf{v}_s = 0 \qquad [VI-7]$$

Here the surface excess energy E^s per unit area and the surface excess heat flux $\mathbf{q}_s$ are given by

$$E^s = \int_{\lambda_A}^{\lambda_B} \Delta(\rho \, \hat{E}) \, d\lambda \qquad [VI-8]$$

$$\mathbf{q}_s = \int_{\lambda_A}^{\lambda_B} \Delta\mathbf{q} \, d\lambda \cdot (\mathbf{I} - \mathbf{nn}) \qquad [VI-9]$$

Convective transport and work done by viscous stresses have been neglected in the integral over S_0 because of the small pillbox thickness. Finally, tangential velocity $\mathbf{v}_s$ has been assumed uniform in the interfacial region so that

$$\int_{\lambda_A}^{\lambda_B} \Delta(p_T \, \mathbf{v}_s) \, d\lambda \approx \mathbf{v}_s \int_{\lambda_A}^{\lambda_B} \Delta \, p_T d\lambda = -\gamma \, \mathbf{v}_s \qquad [VI-10]$$

This assumption should be acceptable for thin pillbox control volumes.

The general equation [VI-7] is rather lengthy, but frequently many of the terms are unimportant. Usually the terms involving E^s and $\mathbf{q}_s$ are small, for instance. The last term in Eq. [VI-7] is of interest because it indicates that interconversion between thermal and mechanical energy is possible during interfacial flow. As indicated above, such interconversion is precisely what happens during Marangoni flow. Mechanical energy release owing to contractive interfacial flow in regions of high interfacial tension exceeds mechanical energy removal by expansive flow in regions of low tension. The excess is available to replace mechanical energy lost by viscous dissipation in the bulk fluids and thus maintain the flow. In most realistic situations both these effects are small in comparison to the normal heat fluxes q_{An} and q_{Bn}.

3. INTERFACIAL INSTABILITY FOR A LIQUID HEATED FROM BELOW OR COOLED FROM ABOVE

Around the turn of the century Benard (1) observed development of a regular hexagonal pattern of convection cells in an initially stagnant, thin layer of liquid heated from below. Figure VI-1 is an example of this striking phenomenon. For many years it was assumed that the flow was

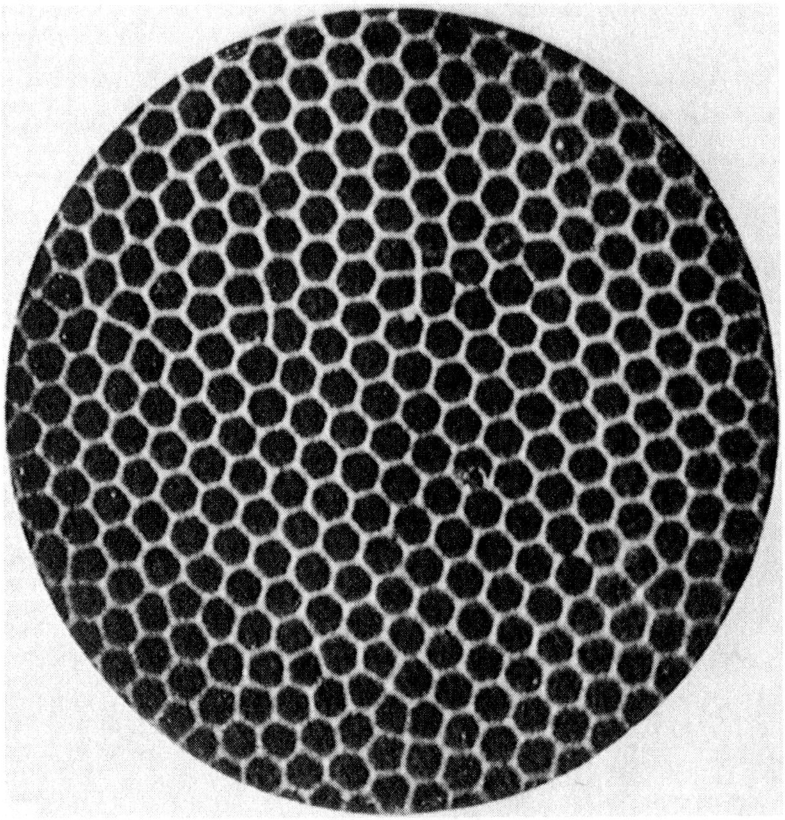

Figure VI-1. Cellular convection produced by surface tension gradients in a thin liquid layer heated from below. Reproduced from Ref. 78 with permission of the French Ministry of Defense.

generated by natural convection effects since low density fluid is situated beneath high density fluid in the stagnant film and hence has a tendency to rise. Pearson's analysis (2) demonstrated that interfacial tension gradients could produce cellular convection independent of any natural convection. His analysis explained why cellular convection was sometimes observed in thin paint films of various orientations including those on the underside of horizontal surfaces where the effect of natural convection should be stabilizing. We note that such instabilities are undesirable in paint films because they impart a cellular appearance, popularly know as "orange peel", to the finished surface. Convective instabilities in paint films have been reviewed by Hansen and Pierce (3).

We shall apply linear stability analysis to the thin layer illustrated in Figure VI-2. In order to appreciate the basic mechanism, let us suppose that, due to a small disturbance, some warm liquid from the interior of the layer reaches the interface at point P. Since interfacial tension decreases with increasing temperature, this disturbance produces a local lowering of interfacial tension at P and hence an outward directed lateral force along the interface. As a result, an outward flow arises which, in turn, draws more warm liquid to the surface, reinforcing the initial flow (see Figure VI-2). Thus, the initial stagnant layer is unstable, interfacial convection develops, and heat transfer rates increase substantially over those in the absence of flow. Note that if the layer were cooled from below the interfacial tension gradients would oppose the initial flow and the system would be stable.

The stability analysis is similar in principle to those of Chapter V but must be extended to account for heat transport effects. The differential equation describing heat transport in the liquid layer is

$$\frac{\partial T}{\partial t} + \mathbf{v} \cdot \nabla T = D_T \nabla^2 T \qquad [VI-11]$$

where T is the temperature and D_T the thermal diffusivity given by $(k/\rho c_p)$ with k the thermal conductivity, ρ the density, and c_p the specific heat of the liquid. In the initial stagnant film, Eq. [VI-11] implies that the temperature profile is linear. We take its slope to be ς. If we assume that the perturbation in temperature has the form $\Theta(z)\, e^{i\alpha x}\, e^{\beta t}$, we find upon substituting this expression into Eq. [VI-11] and retaining only terms of first order in perturbation quantities:

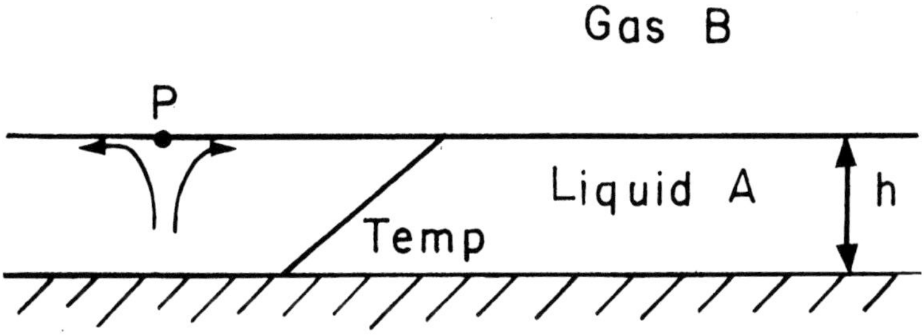

Figure VI-2. Origin of cellular convection driven by interfacial tension gradients in a thin layer of liquid heated from below or cooled from above. Warm fluid reaching the interface at P reduces interfacial tension locally.

$$\frac{d^2\Theta}{dz^2} - \alpha^2 q^2 \Theta = \frac{\zeta}{D_T} W \qquad [VI-12]$$

where W(z) is a function describing the z-dependence of the normal component of velocity as before and

$$q^2 = 1 + \frac{\beta}{D_T\alpha^2} \qquad [VI-13]$$

Now W(z) must satisfy Eq. [V-12] if the variation of density and viscosity with temperature is neglected. Thus, if we apply the operator of Eq. [V-12] to both sides of Eq. [VI-12], we obtain

$$\left(\frac{d^2}{dz^2} - \alpha^2 r^2 \right)\left(\frac{d^2}{dz^2} - \alpha^2 \right) \left(\frac{d^2}{dz^2} - \alpha^2 q^2 \right) \Theta (z) = 0 \qquad [VI-14]$$

This equation has the general solution

$$\Theta (z) = A_1' \sinh \alpha rz + A_2' \cosh \alpha rz + A_3' \sinh \alpha z + A_4' \cosh \alpha z$$

$$+ A_5' \sinh \alpha qz + A_6' \cosh \alpha qz \qquad [VI-15]$$

The normal velocity W(z) may be obtained from this equation using Eq. [VI-12] and the tangential velocity and pressure distribution can be found using Eqs. [V-20] and [V-26].

To simplify the algebra in the following discussion, we shall restrict our attention to states of marginal stability where the time factor β vanishes. As a result, we shall be able to identify conditions where the system is stable or unstable, the information of prime interest, but we shall not be able to compute growth rates. With this scheme Eqs. [VI-12], [VI-14], and [VI-15] are replaced by

$$\frac{d^2\Theta}{dz^2} - \alpha^2 \theta = \frac{\zeta}{D_T} W \qquad [VI-16]$$

$$\left(\frac{d^2}{dz^2} - \alpha^2 \right)^3 \Theta = 0 \qquad [VI-17]$$

$$\Theta (z) = A_1 \sinh \alpha z + A_2 \cosh \alpha z + A_3 z \sinh \alpha z + A_4 z \cosh \alpha z + A_5 z^2 \sinh \alpha z + A_6 z^2 \cosh \alpha z \qquad [VI-18]$$

It is easily verified by direct substitution that Eq. [VI-18] satisfies Eq. [VI-17]. The normal velocity distribution may be obtained from Eqs. [VI-16] and [VI-18]:

$$W(z) = \frac{D_T}{\zeta}[2\alpha A_3 \cosh \alpha z + 2\alpha A_4 \sinh \alpha z + A_5(2 \sinh \alpha z + 4\alpha z \cosh \alpha z) + A_6(2 \cosh \alpha z + 4\alpha z \sinh \alpha z)] \quad \text{[VI-19]}$$

The constants A_i can be found by imposition of suitable boundary conditions. At the solid surface $z = 0$ the velocity components v_z and v_x vanish and, if the solid is an excellent conductor such as a metal, the temperature remains uniform at its initial value. Using Eqs. [VI-19] and [V-20] to evaluate v_z and, v_x, we find that these three conditions lead to the following equations:

$$A_2 = 0 \quad \text{[VI-20]}$$

$$A_5 = - \alpha A_4/3 \quad \text{[VI-21]}$$

$$A_6 = - \alpha A_3 \quad \text{[VI-22]}$$

At marginal stability interfacial deflection remains constant so that the normal velocity at $v_z = H$ is zero. Making use of Eqs. [VI-21] and [VI-22], we find from this condition that

$$A_3 = \frac{A_4}{3\alpha^*}(1 - \alpha^* \coth \alpha^*) \quad \text{[VI-23]}$$

where α^* is a dimensionless wavenumber (αH).

The normal component of the interfacial momentum balance takes the form

$$\left(p - 2\mu \frac{\partial v_z}{\partial z}\right)\Big|_{H+z^*} = - 2H\gamma \quad \text{[VI-24]}$$

The interfacial tension γ in this equation is that of the initial stagnant layer with interfacial tension variations providing a second order correction which is neglected in our linear analysis. With the pressure given by Eqs. [V-39] and [V-26], z^* given by $B e^{i\alpha x}$, and A_3, A_5, and A_6 eliminated by Eqs. [VI-21]-[VI-23], Eq. [VI-24] becomes

$$0 = B + \frac{4\nu D_T \alpha^3}{3\zeta\beta^{*2}} A_4\left[\frac{(\alpha^* - 1)}{\alpha^*} \cosh \alpha^* - 2\alpha^* \sinh \alpha^* + 2\alpha^* \cosh \alpha^* \coth \alpha^*\right] \quad \text{[VI-25]}$$

where β_A^* is as defined at Eq. [V-57].

The key boundary condition of the analysis is the tangential component of the interfacial momentum balance, which becomes

$$\tau_{zx} \Big|_H + \frac{d\gamma}{dT}\frac{\partial T}{\partial x}\Big|_H = 0 \qquad \text{[VI-26]}$$

The coefficient $(d\gamma/dT)$ of interfacial tension variation with temperature is assumed to have a constant value. It is virtually always negative, a typical value being about (- 0.1) mN/m·K. According to Eq. [VI-26], development of a temperature gradient along an interface causes flow to arise having shear stresses which balance the lateral force produced by the interfacial tension gradient. In terms of the coefficients A_i, Eq. [VI-26] becomes

$$0 = -A_1 N_{Ma} \sinh \alpha^* + \frac{A_4 H}{3} 8(-\cosh \alpha^* + \alpha^* \cosh \alpha^* \coth \alpha^*$$
$$- \alpha^* \sinh \alpha^*) - N_{Ma}(\cosh \alpha^* - \alpha^* \sinh \alpha^* - \frac{\sinh \alpha^*}{\alpha^*}$$
$$+ \alpha^* \cosh \alpha^* \coth \alpha^* \qquad \text{[VI-27]}$$

where the dimensionless " Marangoni number" N_{Ma} is defined by

$$N_{Ma} = \frac{\zeta(\frac{d\gamma}{dT})H^2}{D_T \mu}$$

The final boundary condition is the energy balance at the interface. We use a very simple form of this balance which employs a heat transfer coefficient h to describe transfer into the gas phase. The latter is presumed to have some bulk temperature T_b:

$$-k \frac{\partial T}{\partial z}\Big|_{H+z^*} = h(T - T_b)\Big|_{H+z^*} \qquad \text{[VI-28]}$$

Substitution of the temperature profile into this equation yields after some manipulation

$$0 = B\, N_{Nu}\, \zeta + A_1(\alpha^* \cosh \alpha^* + N_{Nu} \sinh \alpha^*) + \frac{A_4 H}{3} \cosh \alpha^* + \frac{\sinh \alpha^*}{\alpha^*}$$
$$+ \alpha^* \cosh \alpha^* \coth \alpha^* + N_{Nu}(\cosh \alpha^* + \frac{\sinh \alpha^*}{\alpha^*} - \alpha^* \sinh \alpha^* + \qquad \text{[VI-29]}$$
$$+ \alpha^* \cosh \alpha^* \coth \alpha^*)$$

with the Nusselt number N_{Nu} given by

$$N_{Nu} = \frac{hH}{k} \qquad [VI-30]$$

Eqs. [VI-25], [VI-27], and [VI-29] are three linear, homogeneous equations in A_1, A_4, and B. For a nontrivial solution to exist the determinant of coefficients for these equations must vanish, a condition which leads to

$$N_{Ma} = \frac{8\alpha^*(\alpha^* \cosh \alpha^* + N_{Nu}\sinh \alpha^*)(\alpha^* - \sinh \alpha^* \cosh \alpha^*)}{\alpha^{*3} \cosh \alpha^* - \sinh^3 \alpha^* - \dfrac{8\, N_{Cr}\alpha^{*5} \cosh \alpha^*}{N_B + \alpha^{*2}}} \qquad [VI-31]$$

Here the "crispation number" N_{Cr} and the Bond number N_B are given by

$$N_{Cr} = \frac{\mu D_T}{\gamma H} \qquad [VI-32]$$

$$N_B = \frac{\rho g H^2}{\gamma} \qquad [VI-33]$$

Figure VI-3 shows plots of this marginal stability relationship for selected values of the dimensionless parameters. For Marangoni numbers N_{Ma} below those on a curve corresponding to a particular set of parameters, the system is stable. For Marangoni numbers above the curve instability occurs. Thus, in view of the definition of N_{Ma} following Eq. [VI-27], instability is favored by larger initial temperature gradients, larger variations of interfacial tension with temperature, and larger film thicknesses (because of the lower resistance to flow). It is opposed by higher fluid viscosities and higher fluid thermal conductivities, the latter because temperature gradients along the interface are diminished by conduction through the liquid.

Let us first examine the limit $N_{Cr} \to 0$ corresponding to high interfacial tensions or thick layers. In this case, which was the one considered in Pearson's original analysis (2), interfacial deflection is small. We see from Figure VI-3 that for low values of N_{Ma}, the system is stable, i.e., the small interfacial temperature gradient is insufficient to overcome the viscous resistance to flow in the thin layer. As N_{Ma} increases owing to, for instance, application of higher temperature gradients ζ, instability first sets in at critical values for N_{Ma_c} and α_c^* of about 80 and 2 respectively. Higher values of N_{Ma} are required

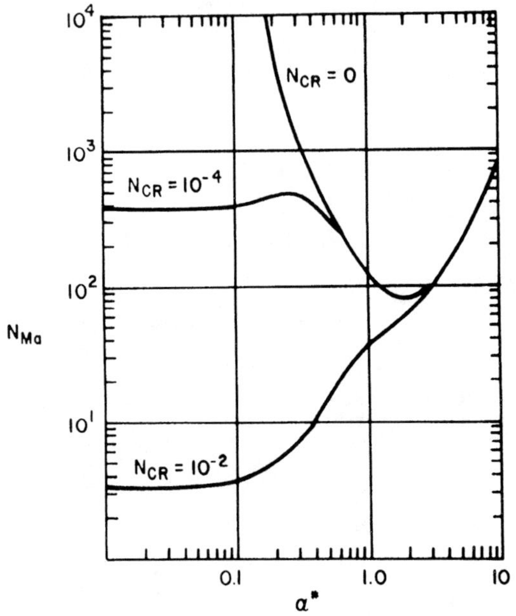

Figure VI-3. Marginal stability condition for thin layer heated from below. From (5) with permission.

to produce instability for wavelengths (cell sizes) both shorter and longer than $(2\pi/\alpha_c^*)$, the former owing to the higher velocity gradients and viscous dissipation rates in small convection cells, the latter because the interfacial temperature gradient which promotes the flow decreases faster with increasing cell size than does the rate of viscous dissipation which is determined largely by layer thickness H for long wavelengths.

Calculations show that with increasing values of the Nusselt number N_{Nu} the system becomes more stable, i.e., larger values of N_{Ma} are needed to produce instability. For instance, N_{Ma_c} in creases from about 80 to 150 N_{Nu} increases from near zero to 2. This result can be understood by recognizing that the enhanced heat loss to the gas at hot regions along the interface acts to diminish the interfacial temperature gradient. Larger values of the heat transfer coefficient h and hence of N_{Nu} increase the magnitude of this effect.

When N_{Cr} has a nonzero value, i.e., when surface deflection is permitted, the system becomes, according to Figure VI-3, much less stable to perturbations having long wavelengths (small values of α^*). The reason is that deflection promotes instability by depressing high temperature regions such as that near point P of Figure VI-4, moving them closer to the hot solid surface at the bottom of the layer (4). Similarly low temperature regions such as those near the points Q are elevated, increasing their distance from the hot solid surface. This effect is especially important when wavelengths are long, as first recognized by Smith (5), because then both flow and lateral conduction are slow and deflection is the dominant factor determining the interfacial temperature gradient. For sufficiently long wavelengths the stabilizing effect of gravity becomes important and the Marangoni number required to produce instability becomes nearly constant, as shown in Figure VI-3.

As fluid depth H increases, α^* becomes large and Eq. [VI-31] simplifies to

$$N_{Ma} \simeq 8\,\alpha^{*2}\left(1 + \frac{N_{Nu}}{\alpha^*}\right) \qquad [VI-34]$$

Rearrangement of this equation yields

$$N_{Ma}' = \frac{\zeta \frac{d\gamma}{dT}}{\mu D_T \alpha^2} = 8\left(1 + \frac{h}{\alpha k}\right) \qquad [VI-35]$$

That is, stability depends on modified Marangoni and Nusselt numbers in which α^{-1} instead of H is the characteristic length. The small magnitude of N_{Cr} for large H suggests that very small interfacial deflections can be expected. If the layer is of infinite depth, it is unstable for any nonzero value of the initial temperature gradient ζ at sufficiently long wavelengths. The reason is that viscous resistance to flow is much reduced by removal of the solid surface. Of course, wavelengths exceeding the lateral dimensions of the experimental apparatus are impossible, so that in practical cases the temperature gradient must reach some critical finite value for instability to be observed.

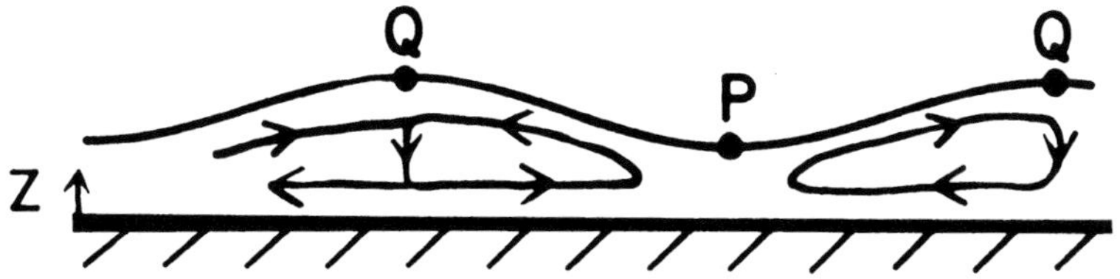

Figure VI-4. Marangoni flow with surface deflection.

Several extensions of the above analysis are possible. Frequently instability is not seen until Marangoni numbers significantly higher than the critical value predicted by the above analysis are reached (6). The enchanced stability can be attributed to small amounts of surface-active contaminant. Flow produced by interfacial temperature gradients sweeps surfactant from warm to cool regions, causing an increase in interfacial tension at the former and a decrease at the latter. This effect opposes the original interfacial tension gradient produced by temperature variations and hence, acts to diminish the flow. Relatively small amounts of an insoluble surfactant can prevent the instability from occurring altogether (7). Palmer and Berg (8) analyzed the situation for a soluble surfactant where diffusion reduces somewhat this strong stabilizing effect. They also confirmed some of their predictions experimentally.

In Benard's original experiments and in the diagram of Figure VI-2 the liquid layer is heated from below and the temperature profile is linear. If convection is generated by the cooling effect of evaporation at the fluid interface the temperature falls rapidly near the interface but more slowly in the remainder of the layer. Flow from within the layer does not, therefore, raise the surface temperature at a point such as P of Figure VI-2 as much as would be the case if the temperature gradient near the interface were present throughout the entire layer. Accordingly, the interface is much more stable than would be expected from the above analysis, as Vidal and Acrivos (9) showed.

While interfacial tension gradients are responsible for instability in thin liquid layers, natural convection due to adverse density gradients becomes an important destabilizing influence with increasing layer thickness. Nield (10) analyzed this situation, finding, as expected, that in layers of intermediate thickness, instability occurs more readily with both effects present than with either alone. His predictions regarding dependence of the stability condition on layer thickness were confirmed experimentally by Palmer and Berg (11). Generally speaking, Marangoni flow is the major destabilizing factor in layers of organic liquid less than about 1 mm thick, while natural convection is more important in thicker layers.

The linear stability analysis we have considered predicts when the instability will occur and frequently also the wavelength to be expected. However, it is unable to predict whether the instability will lead to a steady state with cellular convection as Benard observed or to ultimate break-up of the interface which is seen in many hydrodynamic instabilities such as the Rayleigh-Taylor instability for superposed fluids discussed in Chapter V. Moreover, even assuming that steady convection does develop, linear analysis yields no information on preferred cell shape. While, for simplicity, we have considered here only two-dimensional disturbances ("roll cells"), the corresponding three-dimensional linear analysis yields only a critical wave number which is consistent with various cell shapes, such as roll cells, hexagons, and rectangles. Nonlinear analysis has led to some progress in understanding the development of cellular convection and the conditions when certain cell shapes are preferred (12,13).

Example VI-1. Conditions for Development of Marangoni Instability

Determine the conditions for development of instability when N_{Nu} is very small, the initial temperature gradient $(-\zeta)$ is 1 °K/cm, and the liquid has the following properties:

ρ = 0.8 gm/cm^3

μ = 10 cp

γ = 30 mN/m (at initial surface temperature)

D_T = 1.3 x 10^{-3} cm^2/sec

$\frac{d\gamma}{dT}$ = - 0.1 mN/m K

Solution. Since the thickness H of the liquid layer is unknown, we first estimate H assuming that $N_{Cr} = 0$. According to Figure VI-2, the critical value of the Marangoni number N_{Ma} is about 80. From Eq. [VI-27] we find H = 0.32 cm. From Eq. [VI-32], $N_{cr} = 1.3 \times 10^{-5}$ which is indeed negligible. The critical wavenumber satisfies $\alpha_c^* = \alpha_c H \simeq 2$, which corresponds to a critical wavelength $\lambda_c (= 2\pi/\alpha_c)$ of 1.0 cm.

4. INTERFACIAL INSTABILITY DURING MASS TRANSFER

Spontaneous flow often known as interfacial turbulence can develop near an interface across which one or more species is being transferred. Frequently the flow patterns are less regular than the Benard cells of Figure VI-1 though a basic periodicity is usually discernable. Its source is interfacial tension gradients produced by interfacial concentration gradients. Variation of interfacial tension with concentration is typically rather strong. Hence, when both temperature and concentration vary along an interface as in evaporating multicomponent liquids, the concentraion effect usually dominates in determining whether and under what conditions interfacial turbulence develops. When it does, mass transfer rates normally increase severalfold.

Absorption of a solute from a gas phase into a thin liquid layer and the reverse process of desorption provide the closest mass transfer analogies to the analysis of heat transfer given in Section 3. Indeed, with one important exception the analysis can be used directly. When the interfacial energy balance Eq. [VI-28] is converted to a solute mass balance, an additional term must be added in the frequently encountered case of a solute having some surface activity. This term corresponds to the second term of Eq. [VI-2] and represents solute transport along the interface by convection. As might be expected from the discussion of surfactant effects in Section 3, it provides a significant stabilizing effect in the desorption case which is the mass transfer analog of a liquid layer heated from below. That is, values of N_{Ma} required for instability are much higher than predicted by the analysis of Section 3 (14), a conclusion that is in agreement with experimental findings. Absorption of a solute with some surface activity is the analog of a liquid cooled from below, and hence no instability arises.

Instability accompanying solute transfer across a liquid-liquid interface was first analyzed by Sternling and Scriven (15). Let us consider the situation shown in Figure VI-5 where solute diffuses from phase A to phase B and further suppose that interfacial tension decreases with increasing solute concentration. If a small disturbance at point P of Figures VI-5 brings some solute-rich liquid from the interior of phase A to the interface, the resulting interfacial tension gradient generates a flow which draws more liquid in A to the interface as shown, thus promoting

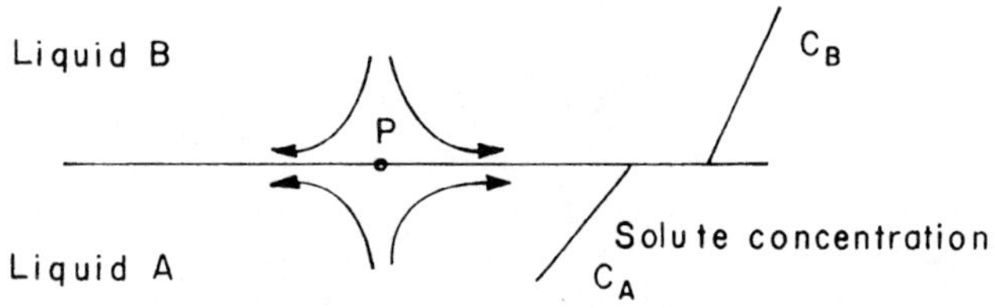

Figure VI-5. Convection due to interfacial tension gradients during solute transfer between immiscible liquid phases.

instability. But the flow also brings solute-lean fluid from B to the interface, which acts to increase interfacial tension at P and oppose the instability. Thus, whether instability develops depends on the relative magnitudes of these opposing effects.

The analysis is very similar to that of Section 3 except, of course, flow and transport in both phases must be considered. When $\alpha^* >> 1$, i.e., when both fluids are of great depth, and when solute convection along the interface can be ignored, the marginal stability condition is that obtained by Sternling and Scriven (15):

$$(N_{Ma})_A = \frac{8(m_{AB}\frac{D_B}{D_A} + 1)(\frac{\mu_B}{\mu_A} + 1)}{(1 - \frac{D_A}{D_B})} \qquad [VI-36]$$

where the Marangoni number $(N_{Ma})_A$ based on properties of phase A is given by

$$(N_{Ma})_A = \frac{(\frac{d\gamma}{dC_A})(\frac{dC_{Ai}}{dz})}{\mu_A D_A \alpha^2} \qquad [VI-37]$$

Also m_{AB} is the distribution coefficient (C_B/C_A) for solute at the interface, D_A and D_B are the solute diffusivities in A and B, $(d\gamma/dC_A)$ is the rate of change of interfacial tension with respect to interfacial concentration in phase A, and (dC_{Ai}/dz) is the initial (uniform) concentration gradient in A.

Eq. [VI-36] is closely related to the heat transfer result Eq. [VI-35] in the limit of large α^*. Indeed, setting $\mu_B << \mu_A$ and $D_A << D_B$, as might be expected in a liquid-gas system, Eq. [VI-36] becomes

$$(N_{Ma})_A = 8(1 + m_{AB}\frac{D_B}{D_A}) \qquad [VI-38]$$

This equation still differs from Eq. [VI-35], but if one recognizes that (a) the analog of the "distribution coefficient" in heat transfer is unity, (b) (D_B/D_A) is equal to the ratio of the concentration gradients $(dC_{Ai}/dz)/(dC_{Bi}/dz)$, and (c) $(h/k\alpha)$ in Eq. [VI-35] is the ratio of the temperature gradient in the liquid to the quantity $(\alpha \Delta T_{gas})$ which has the dimensions of a temperature gradient, the relationship between the two results becomes clear.

Since $(N_{Ma})_A > 0$ for the case of a surface-active solute diffusing from A to B, we see from Eq. [VI-36] that instability can occur only when $(D_A < D_B)$ and that, moreover, the value of $(N_{Ma})_A$ required for instability becomes larger as D_A becomes more nearly equal to D_B. These results can be explained in terms of the steeper concentration gradient in A than B for $D_A < D_B$. Since solute concentration increases rapidly in A with distance from the interface, the flow transports liquid quite rich in solute to point P of Figure VI-5. Thus the destabilizing effect due to the flow in A outweighs the stabilizing effect of the flow in B. We note that no matter how small the mass transfer rate, Eq. [VI-36] predicts that instability can occur for sufficiently long wavelengths (small α).

Eq. [VI-36] also indicates that the critical value of $(N_{Ma})_A$ decreases when the viscosity ratio (μ_B/μ_A) decreases. We might expect high values of μ_A to be associated with low velocity gradients in A, i.e., a slow falloff in tangential velocity with distance from the interface. Under these conditions the flow extends deeply into the bulk of liquid A and can bring to the interface material that is quite rich in solute, a destabilizing effect.

Finally, Eq. [VI-36] shows that low values of the distribution coefficient m_{AB} promote instability. The reason is that solute concentration in B is low under these conditions and the stabilizing influence of the flow in B is reduced.

For a solute which increases interfacial tension, $(N_{Ma})_A$ is negative for solute diffusion from A to B, and instability can occur only if $D_B < D_A$, according to Eq. [VI-36]. This result is entirely reasonable since in this case it is the flow in B which promotes instability and the flow in A which opposes it, just the opposite of that found for surface-active solutes.

The stability condition given by Eq. [VI-36] is, unfortunately, not the whole story. As we saw in Chapter V, β is, in general, a complex number with its real part β_R describing disturbance growth or decay and its imaginary part β_I an oscillatory motion. It is therefore possible to pass from a stable to an unstable region via a state where oscillatory motion of constant amplitude occurs with $\beta_R = 0$ and $\beta_I \neq 0$. Under these conditions the system is said to be in a state of "marginal oscillatory stability", whereas the condition Eq. [VI-36] with $\beta_R = \beta_I = 0$ describes states of "marginal stationary stability", i.e., without oscillation.

Both types of marginal stability were encountered in Chapter V. For gravitationally - produced instability of superposed fluids (Section 2.c) and the other cases where the interface was unstable from a thermodynamic point of view, no oscillatory motion was found at marginal stability. But the state of marginal stability was an oscillatory one for the Kelvin-Helmholtz instability (Section 8) with traveling waves of constant amplitude moving along the interface.

In the present situation both types of marginal stability are possible, as Sterling and Scriven demonstrated in their original paper (15). To find the states of marginal oscillatory stability, one must use the governing equations in their general forms such as Eqs. [VI-12] (adapted to the case of mass transport) and [VI-14]. The basic analysis is as before but the equation for β obtained by setting the determinant of coefficients of the boundary conditions equal to zero is rather complicated and must be solved numerically. We shall omit the details of this procedure and simply quote some key results (15).

For the case of a solute which lowers interfacial tension we have seen already that instability occurs when transport is from the phase of lower diffusivity to that higher diffusivity. With the general analysis we can draw the following conclusions:

(a) If $(D_A/D_B) < 1$ but $(\nu_A/\nu_B) > 1$, instability occurs only when solute transfer is from A to B. Both these conditions favor instability for transfer from A to B and stability for transfer from B to A by the physical arguments given above.

(b) If $(D_A/D_B) < 1$ and $(\nu_A/\nu_B) < 1$, instability occurs for transport in both directions. The diffusivity ratio is favorable to instability when transfer is from A to B, while the kinematic viscosity ratio is favorable to instability for transfer in the opposite direction. We remark that systems of type (a) are more common than those of type (b).

(c) If $D_A = D_B$, instability occurs only when transfer is from the phase of higher kinematic viscosity to that of lower kinematic viscosity, as might be expected from the above results.

Generally speaking, experimental results are in accord with these predictions although exceptions are known. As Berg (16) has pointed out, some of the exceptions may be due to factors not considered in the analysis such as natural convection effects, nonuniform transfer rates due to system geometry, and solute transport along the interface by convection. The latter effect, which can greatly enhance interfacial stability, has been incorporated in an analysis for liquid-liquid systems by Gouda and Joos (17).

The discussion thus far has for simplicity been limited to plane interfaces. Burkholder and Berg (18) studied instability of fluid cylinders and low-velocity jets when there is mass transfer between phases. They found that tangential flow produced by variations in interfacial tension modifies the stability condition from that given in Section 4 of Chapter V for a fluid cylinder in the absence of mass transfer. In particular, transfer of solute from a liquid cylinder into a surrounding gas phase has a stabilizing effect when the solute is surface-active. The reason can be seen from Figure VI-6. When the cylinder begins to deform, the usual capillary instability causes liquid rich in solute to flow from thin to thick portions of the cylinder. But the resulting increase in solute concentration at P reduces interfacial tension there and so causes a reverse flow which opposes the instability. A similar argument shows that transfer of a surface-active solute from the surrounding gas into the jet has a destabilizing effect, i.e., the distance from where the jet is formed to where it breaks up is diminished. These effects were confirmed in a qualitative manner by simple experiments.

These effects of Marangoni flow on stability apply, of course, not only to jets deliberately injected into a fluid but also to fluid cylinders formed during mixing or agitation. At high vapor rates during distillation with sieve plate columns, for instance, agitation of the liquid causes more or less cylindrical protrusions to develop as illustrated in Figure VI-7. If the more volatile component has the lower surface tension, the argument given above indicates that Marangoni flow has a stabilizing effect when a neck starts to form, thus hindering drop breakoff (Figure VI-7a). But if the less volatile component has the lower surface tension, its transfer from vapor to liquid generates a Marangoni flow which promotes neck thinning and drop breakoff (Figure VI-7b). Since the latter behavior increases interfacial area, plate efficiencies are, other factors being equal, higher in the latter case (19).

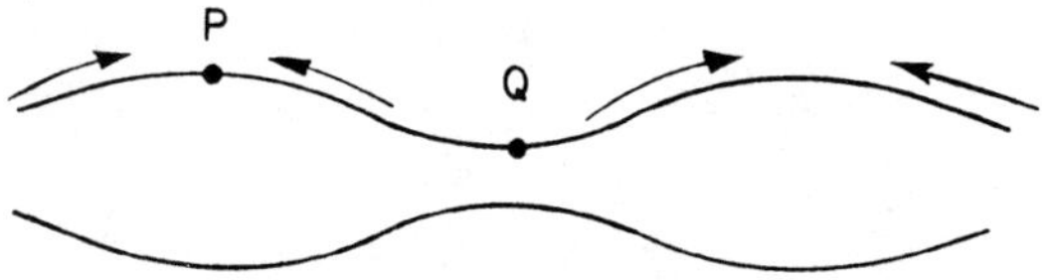

Figure VI-6. Tangential flow during axisymmetric deformation of a fluid cylinder. In the absence of interfacial tension gradients flow is from Q to P as shown.

(a) (b)

Figure VI-7. Schematic illustration of surface tension gradients on the stability of the neck in drop formation. Shaded areas denote liquid of higher surface tension, and arrows indicate the direction of interface movement. In (a) Marangoni flow hinders and in (b) it promotes drop breakoff. From ref. (19) with permission.

For a liquid jet in an immiscible liquid the situation is more complicated because the interfacial concentration is influenced by transport in both phases and, as discussed above for plane interfaces, the effects are in opposite directions. The original paper (18) and an extension incorporating relative motion between the phases due to overall jet motion (20) should be consulted for details of the theoretical analysis. Figures VI-8 and VI-9 show theoretical predictions and experimental measurements of jet length for benzene jets in water with acetone transferring into the aqueous phase (21). The major prediction of the analysis, viz, a strong destabilizing effect of the Marangoni flow, is confirmed by the experimental results.

Various other extensions of the analysis of Marangoni instability during mass transfer have been made. They incorporate such phenomena as chemical reactions in the bulk fluids (22,23) and at the interface (24,25), nonlinear initial concentration gradients (26), heat of solution effects in binary systems where interfacial concentration gradients are not possible at constant temperature (27), and spherical geometry (28). Also careful experimental observations have been made of the convection patterns generated in various systems (29,30) although many features of the observations as yet lack explanation. Further information on instability produced by Marangoni flow may be found in review articles on the subject (6,16,21,29,31,32).

5. OTHER PHENOMENA INFLUENCED BY MARANGONI FLOW

Interfacial tension gradients are responsible for other types of flow besides cellular convection. For instance a gas bubble in a liquid sub-

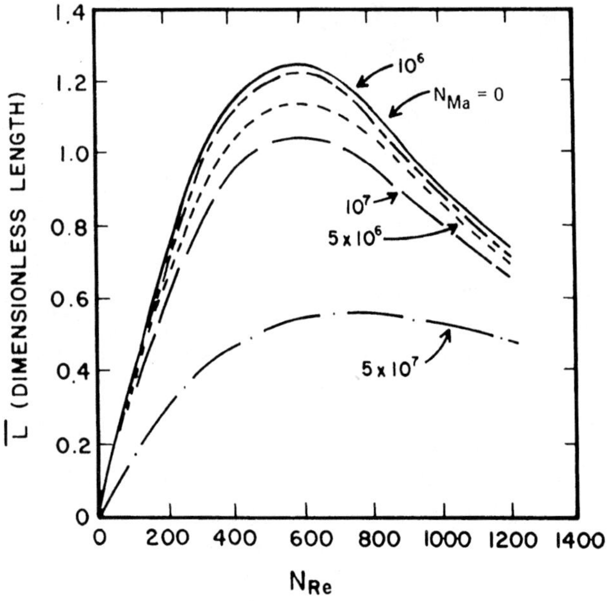

Figure VI-8. Computed linear stability results for normalized length of a benzene jet in water (undergoing outward transfer of acetone) as a function of the jet Reynolds number. Mass transfer, as represented by the Marangoni number is predicted to decrease the jet length considerably, particularly in the region of the jet length maximum. Reprinted from ref. (20) with permission.

jected to a temperature gradient moves toward the high temperature region, because by so doing it reduces its interfacial free energy. The flow which produces this motion is generated by interfacial tension gradients and, for the case of low Reynolds numbers, can be analyzed using the basic equations for flow near a drop or bubble given in Chapter VII. In a similar manner Marangoni flow can cause motion of drops or bubbles in a fluid having a concentration gradient.

The time required for coalescence to occur when two drops or bubbles approach one another is usually determined mainly by the time required for drainage of the fluid film between them. Marangoni flow can, under suitable conditions, either enhance or retard film drainage. Suppose, for example, that a solute which lowers interfacial tension transfers from the disperse to the continuous phase. We expect solute concentration to build up faster in the film than in the bulk of the continuous phase with a resulting decrease in interfacial tension of the film surfaces. The

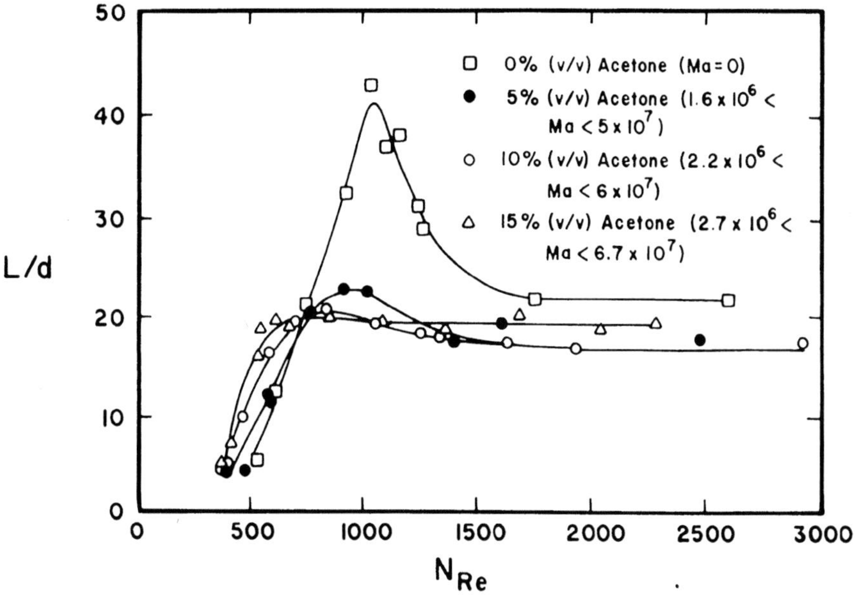

Figure VI-9. Jet breakup length data for the system of Fig. VI-8. L/d is the ratio of jet length to mean diameter. Reprinted from ref. (20) with permission.

resulting Marangoni flow speeds drainage and promotes coalescence. Such behavior can have a significant effect in mass transfer processes, being unfavorable when it is desirable to maximize the area for mass transfer and hence to minimize coalescence. Note that for transfer of the same solute from the continuous to the disperse phase, the interfacial tension gradient is reversed and its effect on flow is to retard drainage and coalescence. For a solute that increases interfacial tension a similar argument shows that transfer into the continuous phase slows and transfer into the disperse phase speeds up film drainage and coalescence.

When distillation or gas absorption is carried out in a packed column, it is important for maintaining a large area for transfer that the liquid flow as a film covering the entire solid surface and not break up into rivulets. Let us consider binary distillation for a system where the more volatile component has the higher surface tension. If a transverse wavy perturbation develops along a falling liquid film as illustrated in Figure VI-10, we would expect the concentration of the more volatile species to be greater in the thick than in the thin portions of the film at a given

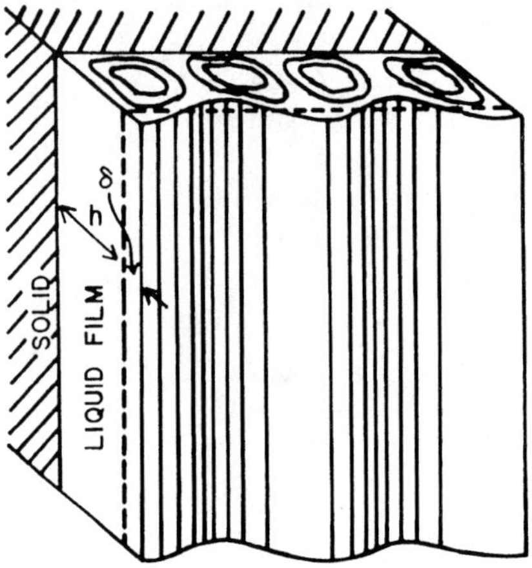

Figure VI-10. Falling film subject to disturbances leading to formation of rivulets. From ref. (35).

height. The resulting interfacial tension gradient draws liquid from thin to thick regions and promotes film breakup into rivulets. In contrast, Marangoni flow opposes ripple formation when the more volatile component has the lower surface tension. Zuiderweg and Harmens (33) found that mass transfer rates for similar conditions were about half as large in the benzene-n-heptane system where rivulet flow is favored than in the n-heptane-toluene system where it is not. Wang *et al* (34) have presented an analysis of this situation. Johnson and Berg (35) extended the analysis to include the presence of a third component which is highly surface active since such materials have been found in some cases to reduce rivulet formation.

As a final example we mention the effect of interfacial tension gradients on the rate of spreading of a liquid drop over a solid surface. We consider a drop of two components of which the more volatile has the lower surface tension. Near the drop periphery where a thin advancing layer of liquid forms (see Chapter VII for an extensive discussion of the spreading process) evaporation causes liquid composition to be leaner in the more volatile component than in the bulk drop and surface tension to be higher. An interfacial tension gradient arises which enhances flow from the bulk drop to the advancing layer and hence increases the rate of spreading, an effect which has been observed by Bascom *et al* (36). This mechanism is responsible for the formation of wine tears. To a first approximation, wine may be considered a mixture of ethanol and water with the former being more volatile and having a lower surface tension. Marangoni flow enhances the rate of spreading up the surface of the glass until enough liquid collects to form tears.

6. NONEQUILIBRIUM INTERFACIAL TENSIONS

When a system contains a soluble surfactant, diffusion, adsorption, and desorption of surfactant may cause interfacial tension to vary with both position and time. If, for instance, a fresh interface is formed on a stagnant pool of a surfactant solution, interfacial tension is found to decrease with time as surfactant diffuses to the interface and adsorbs.

Let us consider such a situation with a plane interface and with local equilibrium maintained at all times between the solution just below the interface and the adsorbed layer. If there is no convection, the diffusion equation in the bulk liquid ($z > 0$) is

$$\frac{\partial c}{\partial t} = D \frac{\partial^2 c}{\partial z^2} \qquad [VI-39]$$

Initially c has its bulk value c_b for all $z > 0$. Also at all times $c \to c_b$ far from the interface, i.e., as $z \to \infty$. Were solution concentration $c(0,t)$ at the interface known, as a function of time, Eq. [VI-39] could be solved. Some information about $c(0,t)$ can be gained by writing the interfacial mass balance and recognizing that an adsorption isotherm exists relating surface concentration $\Gamma(t)$ to $c(0,t)$:

$$\frac{\partial \Gamma}{\partial t} = \frac{d\Gamma}{dc}\frac{\partial c(0,t)}{\partial t} = D \left.\frac{\partial c(z,t)}{\partial z}\right|_{z=0} \qquad [VI-40]$$

With the additional condition $c(0,0) = 0$, Eqs. [VI-39] and [VI-40] can be solved by, for instance, the method of Laplace transforms (37). The result is

$$\Gamma(t) = 2\left(\frac{D}{\pi}\right)^{1/2}\left[c_b\, t^{1/2} - \int_0^{t^{1/2}} c(0,t-\tau)\, d(\tau^{1/2})\right] \qquad [VI-41]$$

For given values of c_b and D, $\Gamma(t)$ and hence interfacial tension $\gamma(t)$ can be calculated using numerical techniques and the known adsorption isotherm. Alternately, the measured behavior of interfacial tension $\gamma(t)$ can be used with a known surface equation of state to obtain $\Gamma(t)$ and then Eq. [VI-41] applied to calculate the diffusion coefficient D. Calculations of this type were first carried out by Ward and Tordai (38).

While the above equations were derived for a plane interface, they may also be used for curved interfaces provided that the radii of curvature are much greater than the thickness of the layer in the bulk solution where the concentration deviates significantly from c_b. This thickness is, to an order of magnitude, $(Dt)^{1/2}$ and hence only about 0.1 mm for $D = 10^{-5}$ cm^2/sec and $t = 10$ sec. Thus, interfacial tension data $\gamma(t)$ obtained with the oscillating jet technique described in Chapter V may be used with Eq. [VI-41] in many cases. Another method of measuring dynamic interfacial tension is the falling meniscus method discussed in Problem VI-4.

Diffusion also influences the damping of capillary wave motion by surfactants. It can be shown (see Problem VI-6) that when surfactant diffusion is included in the analysis, the surface elasticity term $\Gamma_{os}(-d\gamma/d\Gamma_s)$ in Eq. [V-71] must be divided by the quantity $(1 + (\alpha q D/\beta(d\Gamma_s/dc))]$, where D is the diffusion coefficient, $(d\Gamma_s/dc)$ the slope of the adsorption isotherm for the initial surface concentration, and $q^2 = [1 + (\beta/D\alpha^2)]$. If capillary waves with an angular frequency $\overline{\beta}$ are imposed by vertical oscillation of a bar at the surface as discussed in Chapter V and if the usual low viscosity assumption is made along with its diffusional counterpart $|\beta/D\alpha^2| \gg 1$, manipulation of the above expression leads to the following generalized elasticity coefficient E which governs surfactant effect on wave damping (see Problem VI-6):

$$E = \Gamma_{os}\left(-\frac{d\gamma}{d\Gamma_s}\right)\left(1 + \frac{\alpha q D}{\beta(d\Gamma_s/dc)}\right)^{-1}$$

$$\simeq \Gamma_{os}\left(-\frac{d\gamma}{d\Gamma_s}\right)\left[\frac{1 + B + iB}{1 + 2B + 2B^2}\right] \qquad [VI-42]$$

$$B = \left(\frac{d\Gamma_s}{dc}\right)^{-1}\left(\frac{D}{2\overline{\beta}}\right)^{1/2} \qquad [VI-43]$$

We see from these equations that for high frequencies $\overline{\beta}$, B is small, the elasticity is $\Gamma_{os}(-d\gamma/d\Gamma_s)$, and the behavior is that found in Chapter V for an isoluble monolayer. In this case oscillation is so fast that bulk diffusion is not able to appreciably diminish the interfacial concentration gradients produced by the oscillation. On the other hand, at low frequencies $B \gg 1$ and we find

$$E \simeq \Gamma_{os}\left(-\frac{d\gamma}{d\Gamma_s}\right)\frac{1 + i}{2B} \qquad [VI-44]$$

Clearly the real part of the elasticity is substantially reduced by diffusion effects from its value for an insoluble monolayer, an effect which was already noted for the lower molecular weight compounds of Figure V-2.

The ratio of the imaginary part of E to $\overline{\beta}$ is often called a surface viscosity although in this case the complex nature of E arises naturally from the diffusion problem and is unrelated to any relationship between interfacial stress and rate of strain. The reason for this terminology is that if the analysis of wave motion in Chapter V is carried out without any explicit consideration of surfactants but including appropriate terms in a surface elasticity Λ and a surface viscosity $(\eta + \varepsilon)$,* Eq. [V-71] is obtained with Γ_{os} $(- d\gamma/d\Gamma_s)$ replaced by a complex elasticity E given by

$$E = \Lambda + \beta\,(\eta + \varepsilon) \qquad \text{[VI-45]}$$

$$= \Lambda + i\,\overline{\beta}\,(\eta + \varepsilon)$$

with $\overline{\beta}$ again the imposed frequency. Thus, the imaginary part of E in Eq. [VI-44] does indeed play the same role in the analysis as would the product of frequency and an overall surface viscosity $(\eta + \varepsilon)$ defined in terms of interfacial stresses as in Chapter VII. Of course, it is conceivable that in some cases there would exist both an actual surface viscosity and the apparent surface viscosity involving diffusion effects which is of concern here.

In any case we see that measurement of the damping rate for capillary waves at high frequencies yields information on Γ_{os} $(- d\gamma/d\Gamma_s)$. This value can be used with similar damping measurements at low frequencies to calculate the diffusivity D if the adsorption isotherm is known. Or if D is known from separate experiments the damping rate can be used to obtain information about $(d\Gamma_s/dc)$ as a function of surfacant concentration and hence to determine the adsorption isotherm.

Another method for measuring surfactant properties by their influence on fluid motion is the use of longitudinal waves. In basic concept a bar lying along the interface is subjected to lateral oscillations instead of vertical ones as for the capillary waves considered so far. Variations in surface concentration cause variations in interfacial tension which

---

* The surface viscosity coefficients η and ε are defined and discussed in Section 6 of Chapter VII.

strongly influence the flow. Longitudinal waves are thus closely related to the Marangoni flows discussed above. As indicated in Chapter V, the dispersion relation for longitudinal waves is simply a requirement that the second term in brackets in Eq. [V-71] vanish. Diffusion is incorporated by employing the complex elasticity E given by Eq. [VI-42]. If oscillations are imposed with an angular frequency $\bar{\beta}$, the wave number α is a complex number as for capillary waves with its real part related to disturbance wave length and its imaginary part to the rate of damping of wave amplitude with distance from the oscillating bar. Longitudinal waves have some advantages over capillary waves for measuring surfactant properties since both wavelength and damping factor are influenced by elastic and diffusional phenomena, in contrast to the capillary wave case where only the damping factor varies significantly. Moreover, damping is greater than for capillary waves.

Two experimental techniques have been employed in studies of longitudinal waves. The most recently developed of these follows closely the basic concept described above with the motion of small test particles placed at various positions along the interface being used to determine the wavelength and damping rate (39). Lucassen, van der Tempel, and co-workers who originated the study of longitudinal waves (40) chose instead the arrangement illustrated in Figure VI-11. Here two parallel barriers at the surface oscillate out of phase so that the area enclosed between them undergoes a sinusoidal oscillation. Thus, at each instant of time the surface experiences a uniform expansion or compression. A Wilhelmy plate measures the interfacial tension as a function of interfacial area during the oscillation.

It can be shown that when frequency is high enough that the usual low viscosity approximation applies but low enough that disturbance wave length is much greater than the lateral dimensions of interest, the flow described by the longitudinal wave analysis does indeed correspond to a uniform expansion or contraction of the interface at each time. Moreover, the change $\Delta\gamma$ in interfacial tension from its initial value is simply the product of the fractional area change $(\Delta A/A_o)$ with the complex elasticity E. If the former is given by $A^* \sin \bar{\beta}t$ and the latter by Eq. [VI-42], we find

$$\Delta\gamma = \Gamma_{os}\left(-\frac{d\gamma}{d\Gamma_s}\right) A^*\left[\frac{1 + B}{1 + 2B + B^2}\sin \bar{\beta}t + \frac{B}{1 + 2B + B^2}\cos \bar{\beta}t\right] \qquad [VI-45]$$

$$\text{with} \quad \frac{\Delta A}{A_o} = A^* \sin \beta t \qquad [VI-46]$$

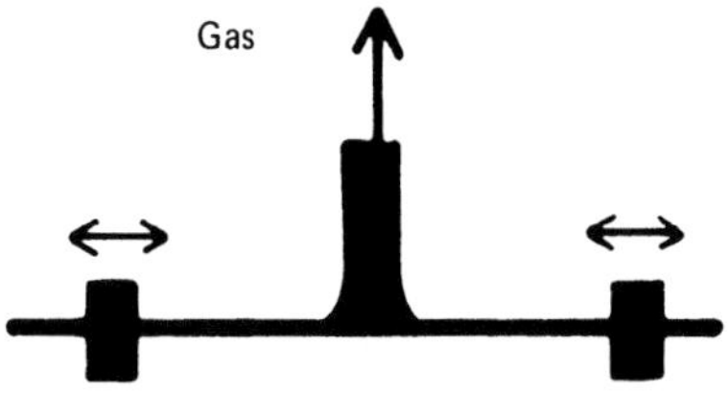

Figure VI-11. Schematic of apparatus for studying longitudinal waves (ref. 40).

In general a plot of $\Delta\gamma$ as a function of $\Delta A/A_0$ is an ellipse as illustrated in Figure VI-12 although it degenerates to a straight line in the high frequency limit where $B \to 0$ and effects of diffusion are negligible.

Figure VI-13 shows data obtained with this latter technique for two concentrations of decanoic acid in water (41). The magnitude $|E|$ of the elasticity is found to increase with increasing frequency as expected from [VI-42]. In the high frequency limit $|E|$ is Γ_0 $(- d\gamma/d\Gamma_s)$ and, as would be expected, is larger for the more concentrated solution. As indicated previously, the slope of the linear portion of the curve at low frequencies can be used to calculate $(d\Gamma_s/dc)$ at each concentration.

Yet another factor involving nonequilibrium interfaces is the possibility that adsorption equilibrium may not exist between the interface and the adjacent bulk phases, i.e., the kinetics of adsorption and desorption may be important. The terms "adsorption barrier" and "desorption barrier" are sometimes used when kinetic limitations exist for the respective processes. If a surface-active solute diffuses between phases under conditions where there is an appreciable desorption barrier, for example, interfacial concentration Γ will attain higher values than in the absence of the barrier and interfacial tension will be lower. England and Berg (42) and Rubin and Radke (43) have studied such situations. Figure VI-14 shows an example of predicted interfacial tension as a function of time for various values of a dimensionless rate constant. The low transient interfacial tension is evident. Another situation involving both flow and an interface not at adsorption equilibrium is considered in Section 7 of Chapter VII.

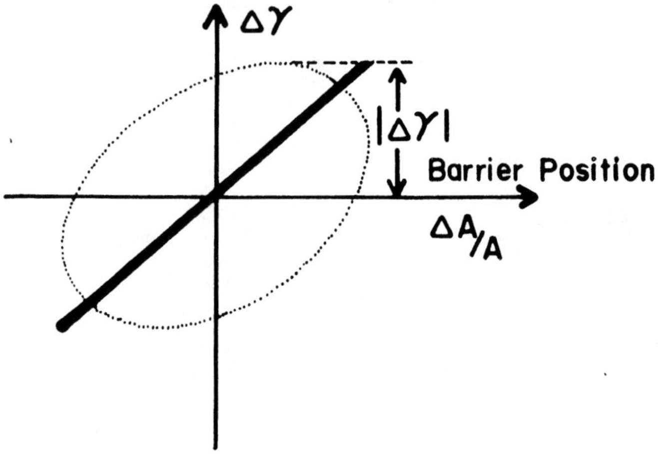

Figure VI-12. Change of surface tension with movement of oscillating barrier at given distance from starting position of barrier. Solid line-purely elastic behavior; dotted line - viscoelastic behavior. Eccentricity and tilt of the ellipse vary with distance from the barrier unless multiple reflections render the surface deformation uniform. From ref. (79) with permission.

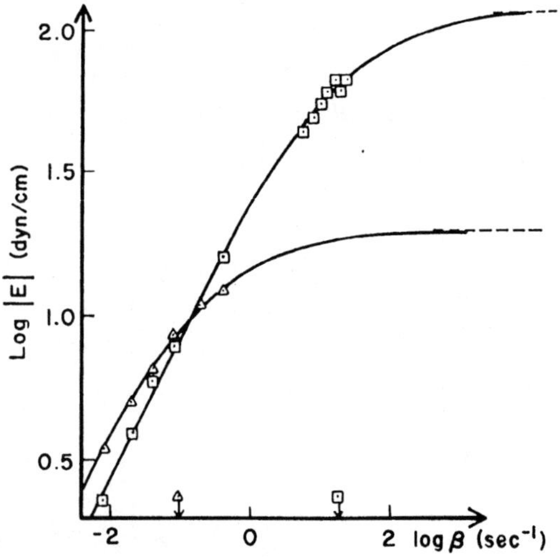

Figure VI-13. Experimental and theoretical values of |E| as a function of wave frequency $\bar{\beta}$ for decanoic acid solutions. $\Delta c = 2 \times 10^{-5}$M; $c = 10^{-4}$M. From ref. (79) with permission.

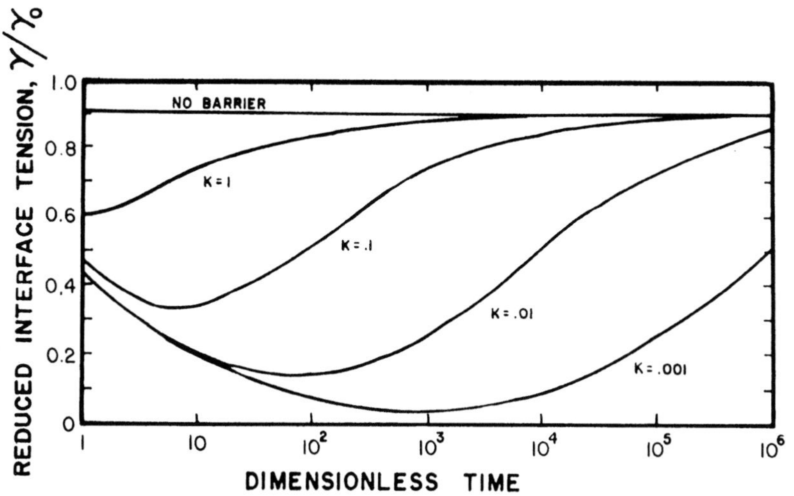

Figure VI-14. Time dependence of the interfacial tension in the presence of a desorption barrier. Smaller values of K correspond to larger desorption barriers. Reprinted from ref. (42) with permission.

7. STABILITY OF MOVING INTERFACES WITH PHASE TRANSFORMATION

We turn now to an entirely different mechanism of transport-related instability. It is important at solid-fluid interfaces where phase transformation or chemical reaction takes place and is, for instance, the basic instability giving rise to dendritic growth during solidification. We illustrate it using the simple case of pure component solidification. Figure VI-15a shows the initial moving planar interface with A and B the solid and liquid phases, respectively. Solidification is presumed to be caused by subcooling the bulk liquid below its equilibrium melting temperature T_0. The initial temperature profile is then as shown. The rate of solidification is limited by the rate at which the heat of fusion released at the interface can be transported into the bulk liquid. If the interface is perturbed as in Figure VI-15b, points such as P are exposed to more of the cooler liquid than before, the local rate of heat loss increases there, and hence so does the local solidification rate. A similar argument indicates a decrease in local solidification rate at points such as Q. Both these changes cause perturbation amplitude to grow, i.e., they are destabilizing.

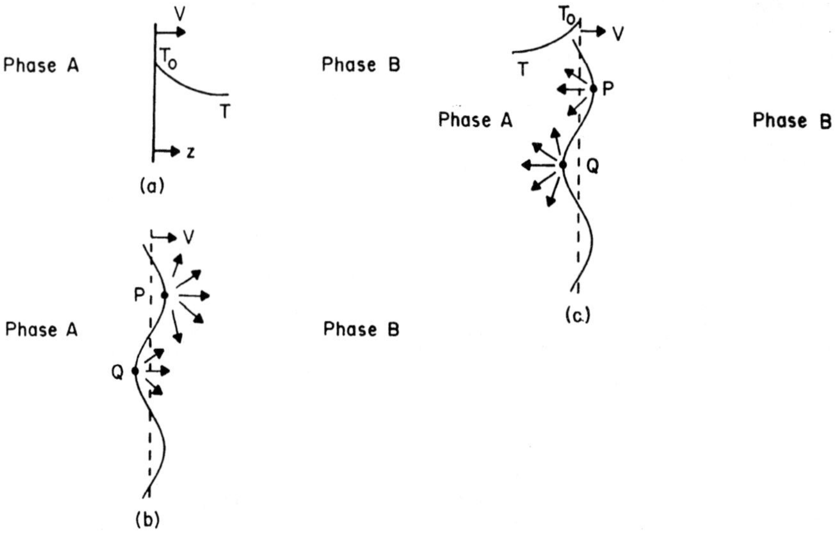

Figure VI-15. Transport effect on interfacial stability during phase transformation: (a) Phase transformation with transport into shrinking phase B - initial situation. (b) Effect of a wavy perturbation of the interface. Transport into phase B is enhanced at point P and diminished at Q. (c) Interfacial perturbation with transport into growing phase. Transport into phase A is enhanced at point Q and diminished at point P.

Suppose instead that the temperature is uniform and equal to T_0 in the liquid B but that the solid A is cooled (Figure VI-15c). Then when the wavy perturbation arises, P is exposed to less of the cooler solid than before and the local solidification rate decreases. Similarly, Q is exposed to more of the cooler solid and the local solidification rate increases. Both these effects are stabilizing since they cause perturbation amplitude to decrease.

We see from this example that heat transport is a key factor influencing interfacial stability. In solidification of binary mixtures or alloys both heat and mass transport must be considered. If the solute is preferentially soluble in the liquid phase, for example, solute must diffuse from the interface to the bulk liquid as solidification proceeds. Since diffusion coefficients are often much smaller than thermal diffusivities in liquids, the solidification rate is often limited primarily by solute diffusion. By an argument similar to that given above for heat transport we conclude that solute diffusion in the liquid is destabilizing and can lead to dendritic growth.

Let us consider solidification of a pure component in detail. For the planar interface situation of Figure VI-15a, conservation of energy in the liquid (phase B) requires that the following equation be satisfied in a coordinate system moving with interfacial velocity V:

$$- V \frac{dT_i}{dz} = D_T \frac{d^2T_i}{dz^2} \qquad [VI\text{-}47]$$

This equation has the solution

$$T_i = T_I + \frac{G\ D_T}{V} \left[1 - \exp\left(-\frac{Vz}{D_T}\right)\right] \qquad [VI\text{-}48]$$

where T_I and G are the temperature and temperature gradient at the interface. These quantities and the velocity V are related by the boundary conditions which are discussed next.

Conservation of energy at the interface implies that

$$k\ G + V\rho\lambda = 0 \qquad [VI\text{-}49]$$

where k and λ are the liquid thermal conductivity and the latent heat of fusion. In this simplified derivation the solid and liquid densities are assumed to have the same value ρ.

For the remaining boundary condition we use an empirical expression which allows for the possibility that the kinetics of crystal structure formation at the interface may significantly influence the solidification rate:

$$V = b\ (T_o - T_I) \qquad [VI\text{-}50]$$

Here T_o is the equilibrium melting temperature of a flat interface and b an empirical coefficient. In the limiting case of large b, we have $T_I \simeq T_o$. In this special case crystal formation kinetics are unimportant and the solidification rate is limited solely by heat transport in the liquid.

We now proceed as in the stability analyses of Chapter V and suppose that perturbation shifts the interface from the plane z = 0 in the moving coordinate system to the wavy surface z* given by $B\ e^{i\alpha x}\ e^{\beta t}$. We further suppose that the temperature distribution after perturbation is given by

$$T\,(x,z,t) \;=\; T_i(z) + T_p(z)\, e^{i\alpha x}\, e^{\beta t} \qquad \text{[VI-51]}$$

In this case the differential energy equation becomes

$$\beta\, T_p - V\,\frac{dT_p}{dz} = D_T \left(\frac{d^2T_p}{dz^2} - \alpha^2\, T_p \right) \qquad \text{[VI-52]}$$

The solution of this equation which vanishes far from the interface is

$$T_p \;=\; a_1\, e^{-\omega z} \qquad \text{[VI-53]}$$

$$\omega \;=\; \frac{V}{2D_T} + \left(\frac{V^2}{4D_T^{\,2}} + \alpha^2 + \frac{\beta}{D_T} \right)^{1/2} \qquad \text{[VI-54]}$$

This solution is valid only when $(\alpha^2 + \beta/D_T) > 0$, i.e., for unstable situations and for stable situations where the decay rate is not too rapid. Because it turns out that the unstable wave numbers are typically fairly large (about 1000 cm^{-1}), we can usually ignore (β/D_T) in comparison with α^2, a simplification which is employed in the remainder of the derivation.

Local energy conservation along the wavy interface requires that

$$k\left(-\,\omega\, a_1 - \frac{VGB}{D_T} \right) + \beta\,\lambda\,\rho\, B \;=\; 0 \qquad \text{[VI-55]}$$

Similarly, local application of Eq. [VI-50] leads to

$$\beta\, B + b\,\left(a_1 + GB + \frac{\gamma\alpha^2 T_0}{\lambda\rho}\, B\right) \;=\; 0 \qquad \text{[VI-56]}$$

The last term in parentheses in Eq. [VI-56] represents the effect of interfacial tension and curvature on the equilibrium freezing temperature. The basic idea is that curvature produces a difference in pressure between solid and melt. Since the chemical potentials in the two phases are functions of pressure and since they must be equal at equilibrium, the equilibrium freezing temperature depends on curvature (see Problem VI-7).

Eqs. [VI-55] and [VI-56] are linear, homogeneous equations in a_1 and B. For a nontrivial solution to exist, their determinant of coefficients must vanish, a condition which leads to

$$\beta = \frac{k\ G(-\omega + \frac{V}{D_T}) - (\omega\ k\ \gamma\ T_0 \alpha^2)/(\lambda\rho)}{\lambda\rho(1 + \frac{\omega\ k}{b\lambda\rho})} \qquad [VI-57]$$

The first term in the numerator of this equation represents the basic effect of heat transport on interfacial stability. Since $G < 0$ and $\omega > (V/D_T)$, this term makes a positive contribution to β, i.e., it is destabilizing. The second term represents a stabilizing effect produced by interfacial tension. It is most important for short wavelengths (large α) and indeed outweighs the destabilizing effect of transport for sufficiently large α.

The term $(\omega\, k/b\rho\lambda)$ in the denominator of Eq. [VI-57] represents the ratio of the rates of heat transport and interface crystallization processes. When this ratio is small owing to, for instance, a large value of b, crystallization is rapid and the solidification rate is limited by heat transport. But when the ratio is large, the kinetics of interfacial crystallization control the solidification. Note that, when b has a finite value, kinetic effects are always rate controlling for perturbations with sufficiently short wavelengths since the time required for heat transport over this length scale becomes very small. When kinetic effects dominate, Eq. [VI-57] shows that β is small in magnitude. Thus, interfacial kinetic effects cannot remove the basic instability produced by transport, but they can slow drastically the rate of growth of the unstable perturbations.

When an initially plane fluid interface is deformed, interfacial tension initiates flow which acts to restore the planar configuration, as we saw in the discussion of wave motion in Chapter V. This stabilizing effect is normally strong enough during ordinary condensation or vaporization to overwhelm transport effects on interfacial stability such as those in the above analysis of solidification (44). However, transport effects on stability are important for certain moving "fronts" in fluid systems which are not true interfaces but instead regions of rapid composition changes. Examples are moving fronts where fog forms or vaporizes (45)(see Section 10) and moving condensation or combustion fronts in porous media (46-48).

An intriguing phenomenon important in vaporization of organic liquids under high vaccum conditions is instability due to the "vapor recoil" mechanism (49,50, see Figure VI-16). In this case the difference in momentum between liquid entering and vapor leaving the interface becomes

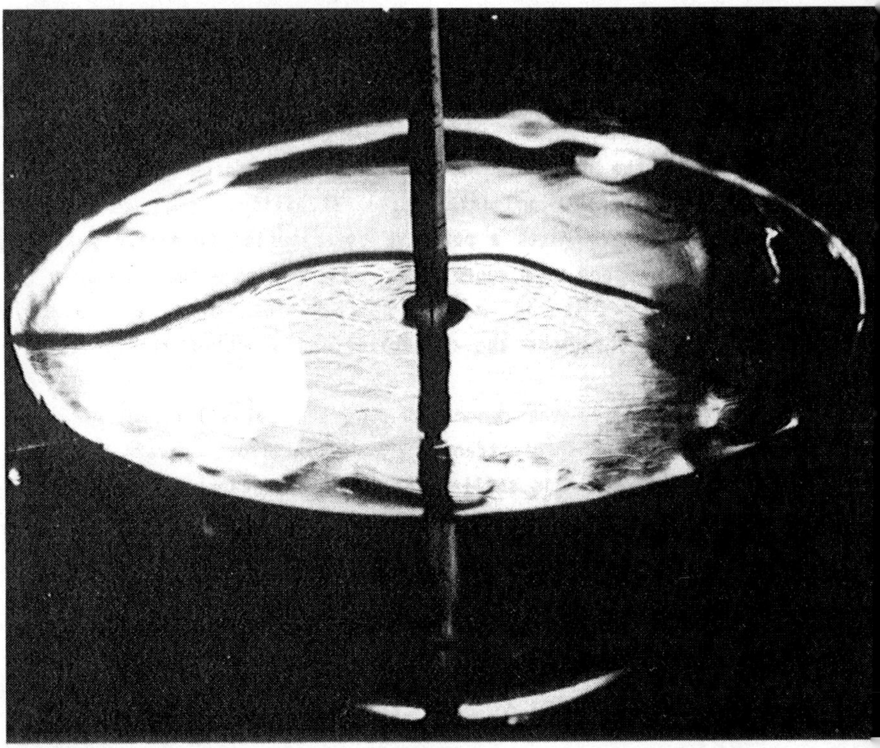

Figure VI-16. The surface of a rapidly evaporating mineral oil: evaporation is much more rapid in the foreground owing to vapor recoil instability. From ref. (50) with permission.

important, i.e., the last two terms of Eq. [V-35] must be included in the interfacial momentum balance. Since vapor density is very low, the interfacial mass balance Eq. [V-32] indicates that vapor velocity normal to the interface must be high. Accordingly, vapor momentum must be high as well and liquid pressure must exceed vapor pressure in order for the normal component of the interfacial momentum balance Eq. [V-35] to be satisfied. Hydrostatic effects thus dictate that surface elevation be lowest in locations where the vaporization rate is highest.

Suppose that some bulk liquid reaches the surface at point P of Figure VI-17, producing a local increase in temperature. The result is a local increase in vaporization rate and, according to the above argument, a slight additional depression of the surface. Palmer's analysis (50) shows that the shear stress produced by vapor flow along the wavy surface near P

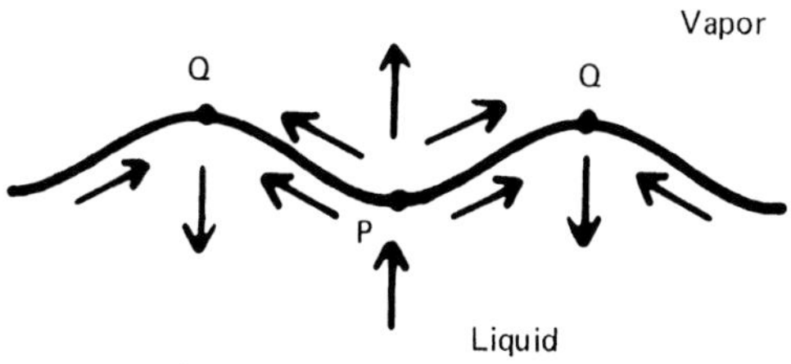

Figure VI-17. Schematic of convection in liquid produced by vapor recoil mechanism. The local evaporation rate increases at point P and decreases at points Q.

drags liquid outward from P, which, in turn, brings more warm liquid from the bulk liquid to P. The initial perturbation is thus enhanced, and an instability involving appreciable convection near the interface develops. As a result, the vaporization rate increases by severalfold over that for a stagnant fluid surface.

Example VI-2. Characteristics of Interfacial Instability During Solidification

Find the wavelength of the fastest growing disturbance and its rate of growth for solidification of water when (a) the rate of crystal formation is very fast ($b \to \infty$) and (b) $b = 10^{-5}$ cm/sec °K. Assume that the temperature gradient (-G) in the water at the interface is 10°C/cm and use the following values

ρ	$= 1$ gm/cm^3	λ	$=$ 80 cal/gm
k	$= 1.4 \times 10^{-3}$ cal/sec·cm K	γ	$=$ 100 mN/m
D_T	$= 1.4 \times 10^{-3}$ cm^2/sec	T_0	$=$ 273°K

Solution

(a) For $b \to \infty$ we have, according to Eq. [VI-50], $T_I = T_0$ at the interface. From Eq. [VI-49] we can calculate that interfacial velocity V is 1.75×10^{-4} cm/sec. If $(V/D_T\alpha) \ll 1$, $\omega \simeq \alpha$ and Eq. [VI-57] becomes

$$\beta = \frac{-k\ G\ \alpha - k\ \gamma\ T_0 \alpha^3/\lambda\rho}{\lambda\rho} \qquad \text{[VI-E2-1]}$$

By differentiating this equation with respect of α and setting the derivative equal to zero, we find

$$\alpha_m = \left(\frac{-G\lambda\rho}{3\gamma T_o} \right)^{1/2} \qquad \text{[VI-E2-2]}$$

With the above values, α_m is 639 cm^{-1}, i.e., the wavelength ($2\pi/\alpha_m$) of the fastest growing perturbation is 98.3 μm. Note that $(V/D_T\alpha_m) = 1.9\times10^{-4}$, so that the assumption made above is justified. The time factor β_m for growth of this disturbance is .069 sec^{-1}, i.e., the constant β_m^{-1} is 14.7 sec.

(b) When b is very small, the rate of crystal formation controls the solidification rate. If $(\alpha k/b\lambda\rho) >> 1$ and $(V/D_T\alpha) >> 1$, Eq. [VI-57] simplifies to

$$\beta = b\left(- G - \frac{\gamma T_o \alpha^2}{\lambda\, \rho}\right) \qquad \text{[VI-E2-3]}$$

Comparing this equation with Eq. [VI-E2-1], we see that the wave number α_c at marginal stability, i.e., where $\beta = 0$, is the same as in (a) for a given temperature gradient G. But the time factor β now increases with decreasing α and reaches its maximum value $\beta_m = -b\,G$ in the limit $\alpha \to 0$. Thus, β_m in the present case is 10^{-4} sec^{-1}, corresponding to a time constant of 10^4 sec, even though G and V have the same values as in (a). We conclude that development of the instability is greatly slowed when solidification kinetics are important.

Note that for $(\alpha k/b\lambda\rho)$ to have a magnitude of order unity, α must be about 0.57 cm^{-1} or the wavelength ($2\pi/\alpha$) about 11 cm. As seen in (a) the wavelengths of interest are much shorter than this value, so that the assumption made in deriving Eq. [VI-E2-3] is valid. Note further, that with $V = 1.75 \times 10^{-4}$ cm/sec, the temperature drop at the interface is 17.5°C, according to Eq. [VI-50].

8. STABILITY OF MOVING INTERFACES WITH CHEMICAL REACTION

Moving interfaces where one or more chemical reactions occur are subject to transport-related instabilities similar to those discussed in Section 7 which lead to dendritic growth during solidification. However, the wide variety of possible reaction schemes and kinetic relationships makes possible different types of behavior.

Let us consider a solid S that is being consumed by reaction with a species A which diffuses from the bulk fluid to the interface (Figure VI-18). Based on the qualitative argument given in Section 7 we would expect

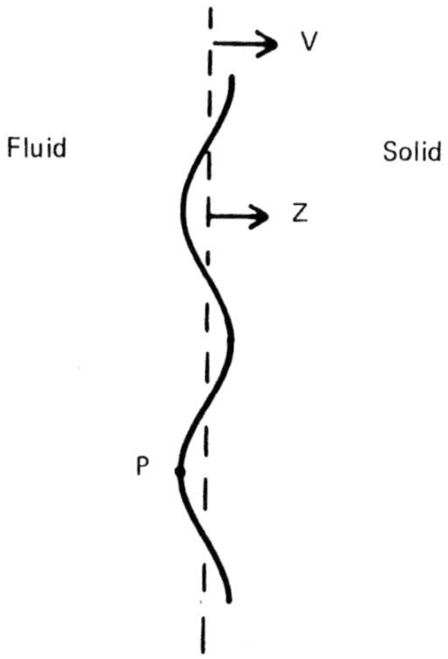

Figure VI-18. Schematic perturbation of reacting solid surface.

enhanced diffusion of A to points such as P when the interface is deformed with a resulting increase in local reaction rate and interfacial velocity. This effect is clearly a stabilizing one.

But suppose now that the reaction is exothermic, so that heat must be transported from the interface into the bulk liquid. Now the perturbation also produces a local increase in heat flux at P, which favors a reduction in local temperature and hence in reaction rate, a destabilizing effect. If it exceeds the stabilizing effect of reactant diffusion, instability can result (51-53).

The stability analysis is very similar to that of Section 7 except that there are equations analogous to Eqs. [VI-52] and [VI-53] for both heat and mass transport in the liquid. Moreover, interfacial conservation equations both for energy and for the reactant A must be invoked.

The key equation in the analysis, however, is the boundary condition giving the local reaction rate r. For simplicity we assume that the reaction proceeds by formation of an activated complex C* in equilibrium with the reactants A and S. The equilibrium constant for this reaction is taken as K_e. The complex then decays irreversibly with a rate constant k_1 to form the product B in the fluid phase: thus, we have

$$A + S \overset{K_e}{\rightleftarrows} C^* \overset{k_1}{\rightarrow} B \qquad [VI-58]$$

With this scheme the reaction rate r can be written in terms of K_e and k_1 as follows:

$$r = k_1 c_{C*} \exp\left(\frac{-E_a}{RT}\right)$$
$$= k_1 K_e c_A (f_s/f_s^o) \exp\left(\frac{-E_a}{RT}\right) \qquad \text{[VI-59]}$$

where c_{C*} and c_A are concentrations of C* and A at the interface, E_a is the activation energy for the reaction, and (f_s/f_s^o) is the ratio of the fugacity f_s of the reacting solid to its standard state value f_s^o.

Now f_s^o is evaluated at the pressure p_I exerted by the fluid at the initial flat interface, which is also the pressure in the solid under these conditions if the Young-Laplace equation is assumed to apply. But once the interface is deformed as in Figure VI-18, the pressure in the solid near the interface becomes a function of position. Indeed, if pressure in the fluid phase is presumed to be uniform, we can use the Young-Laplace equation and the well known dependence of fugacity upon pressure to write

$$\frac{f_s}{f_s^o} = \exp\left(-\frac{v_s\gamma}{RT}\alpha^2 z^*\right) \qquad \text{[VI-60]}$$

Substituting this expression into Eq. [VI-57], we obtain

$$r = k_1 K_e C_A \exp\left(\frac{-\gamma v_s}{RT}\alpha^2 z^*\right)\exp\left(\frac{-E_a}{RT}\right) \qquad \text{[VI-61]}$$

According to Eq. [VI-60], solid fugacity increases over the standard state value at points such as P of Figure VI-18, where $z^* < 0$. An increase in the local reaction rate at P results. We see, therefore, that, as before, interfacial tension has a stabilizing effect which enters the analysis through the equation describing interface kinetics. The orgin of the effect is made plain here, however, because the kinetic equation is based on a physical model of reaction and is not simply an empirical expression like Eq. [VI-50].

When the stability analysis is carried out in the usual way, the time factor β for short wavelengths where $(V/\alpha D)$ and $(V/\alpha D_T)$ are small is found to be given by

$$\beta = \frac{V\left(\frac{G_A}{c_{Ao}} + \frac{E_a G k}{RT_o^2(k+k_s)}\right) - \frac{\gamma M_s \alpha^2}{\rho_s RT_o}}{1 + \frac{V\rho_s}{c_{Ao} D M_s \alpha} + \frac{E_a V \Delta H_R \rho_s}{RT_o^2 \alpha(k+k_s)}} \qquad \text{[VI-62]}$$

Here c_{A0} and T_o are reactant concentration and temperature at the interface before perturbation, G_A and G are the initial concentration and temperature gradients, ΔH_R is the heat of reaction per unit mass of solid S consumed, ρ_s and M_s are the density and molecular weight of S, and k and k_s are the thermal conductivities of the fluid and solid respectively. The initial temperature in the solid phase has been presumed uniform, and $(\beta/D_T\alpha^2)$ and $(\beta/D\alpha^2)$ are taken as small quantities (cf. discussion following Eq.VI-54]).

Since $G_A < 0$, it is clear from the numerator of Eq. [VI-62] that reactant diffusion has a stabilizing effect, as expected. If the reaction is exothermic, heat transport in the fluid is away from the interface, $G>0$, and heat transport is destabilizing. Finally, the effect of interfacial tension is stabilizing and of greatest importance for short wavelengths (small α) , the same as found in Section 7 for solidification. Thus instability is possible if the heat transport effect outweighs the combined effect of reactant diffusion and interfacial tension. Of course, $G < 0$ for an endothermic reaction, all three effects are stabilizing, and no instability is possible.

An exothermic reaction has autocatalytic properties in the sense that heat generated by the reaction increases temperature and hence speeds up the reaction. It might be expected that other reaction schemes having autocatalytic features could cause interfacial instability. Suppose, for instance that the following two reactions occur:

$$S + aA + b_1B \rightarrow cC + \text{other products} \qquad [VI-63]$$

$$cC + dD + eE \rightarrow b_2B + \text{other products} \qquad [VI-64]$$

Here species A, B, C, D, and E exist only in the fluid phase. Further suppose that $b_2 > b_1$ so that some B is produced by the overall reaction obtained by adding Eqs. [VI-63] and [VI-64]. Finally, suppose that the rate of Eq. [VI-63] is much slower than that of Eq. [VI-64]. Then the rate of the overall reaction is, for all practical purposes, equal to the rate of Eq. [VI-63] and hence proportional to $c_B^{\,b_1}$, where c_B is the concentration of B at the solid-fluid interface. In other words, the rate of the overall reaction is proportional to the concentration of one of its products B. The reaction is thus of an autocatalytic nature and instability is possible (see Problem VI-8).

A possible example of a reaction scheme of this type involves leaching of the metal ore chalcopyrite ($CuFeS_2$). In this case Eq. [VI-63] is the basic dissolution reaction with the reactants A and B being dissolved oxygen and hydrogen ion respectively, while C is ferric ion Fe^{+3}. The second reaction involves precipitation of Fe^{+3} by sulfate ions D and water E to form, for example, hydrogen jarosite ($Fe_3(SO_4)_2(OH)_5 \cdot 2H_2O$). Hydrogen ions are also produced. A reaction scheme of this type is reported by Braun *et al* (54) who also describe certain observations in which ore particles are leached nonuniformily, i.e., the amount of leaching is different at different positions along the particle surface. Such behavior was associated with an increased overall leaching rate for the entire particle, a desirable effect. A possible explanation of the nonuniform leaching is instability of the reacting surface as a result of the mechanism just described in accordance with the analysis outlined in Problem VI-8.

So far we have dealt with reactions in which a solid has been consumed. When a solid surface is built up by a reaction, transport of reactants in the fluid phase is destabilizing, just as is the case during solidification (55). An example of considerable practical importance is electrodeposition. In this case instability is normally undesirable because it leads to an irregular surface which does not have the preferred bright appearance. It has been found that instability can sometimes be prevented by including small quantities of certain surface-active organic additives in the electrodeposition bath (56). The additives diffuse to the solid-liquid interface and are either incorporated into the developing deposit or consumed by an electrochemical reaction. Since they are surface active, they adsorb at the interface and produce a decrease in the rate of the electrodeposition reaction, possibly because they block some sites where metal ions would otherwise deposit.

Although details of the action of organic additives remain poorly understood, their qualitative effect can be described empirically by an equation of the following type for the electrodeposition rate r:

$$r = k \quad c_M/c_A^m \qquad [VI-65]$$

where c_M and c_A are local concentrations of metal ions and additive at the interface and m an empirical constant. If the additive has a large stabilizing effect, i.e., if small amounts of additive cause large changes in electrodeposition rates, m must be a large positive number. When an initially flat interface is deformed, arguments similar to those of the initial paragraph of Section 7 indicate that both metal ion and additive concentrations increase at points where the solid projects into the liquid. In view of Eq. [VI-65], transport of the metal ion must be destabilizing, while transport of the additive is stabilizing. Whether instability develops depends on the relative magnitudes of ion and additive effects (see Problem VI-9). As a minimum, the additive reduces the growth rate of unstable perturbations.

These examples provide some idea of the variety of phenomena which can influence the stability of reacting solid surfaces. No doubt numerous other reaction schemes showing interesting behavior could be devised.

9. TRANSPORT-RELATED SPONTANEOUS EMULSIFICATION

Another intriguing phenomenon involving phase transformation and transport processes is spontaneous emulsification. When toluene containing some ethyl alcohol is carefully contacted with water, for instance, cloudiness immediately appears in the aqueous phase near the interface (see Figure VI-19). It has long been recognized that spontaneous emulsification in such cases is associated with diffusion of the various species (57), but theoretical analysis of the phenomenon is relatively recent (58). The analysis is based on the calculation of "diffusion paths", as discussed below, and is similar in approach to that used to predict the occurrence of isolated precipitation in metallurgical systems (59).

Consider a ternary system having the phase behavior shown in Figure VI-20. Many systems containing an oil, water, and a short-chain, slightly polar organic compound such as an alcohol or a fatty acid are of this type, including the toluene-water-ethanol system mentioned above. Now suppose that two semi-infinite phases having mass fractions (ω_{10}, ω_{20}) and (ω'_{10}, ω'_{20}) of species 1 and 2 are placed carefully in contact at position $x = 0$ and time $t = 0$. We assume that no convection develops, that transport of each species depends only on its own concentration gradient, that the diffusion coefficient of each species in each phase is independent of composition, that equilibrium between phases is maintained at the interface, and that the mass density of the system is uniform throughout. With these assumptions the concentration profiles can be found by solving

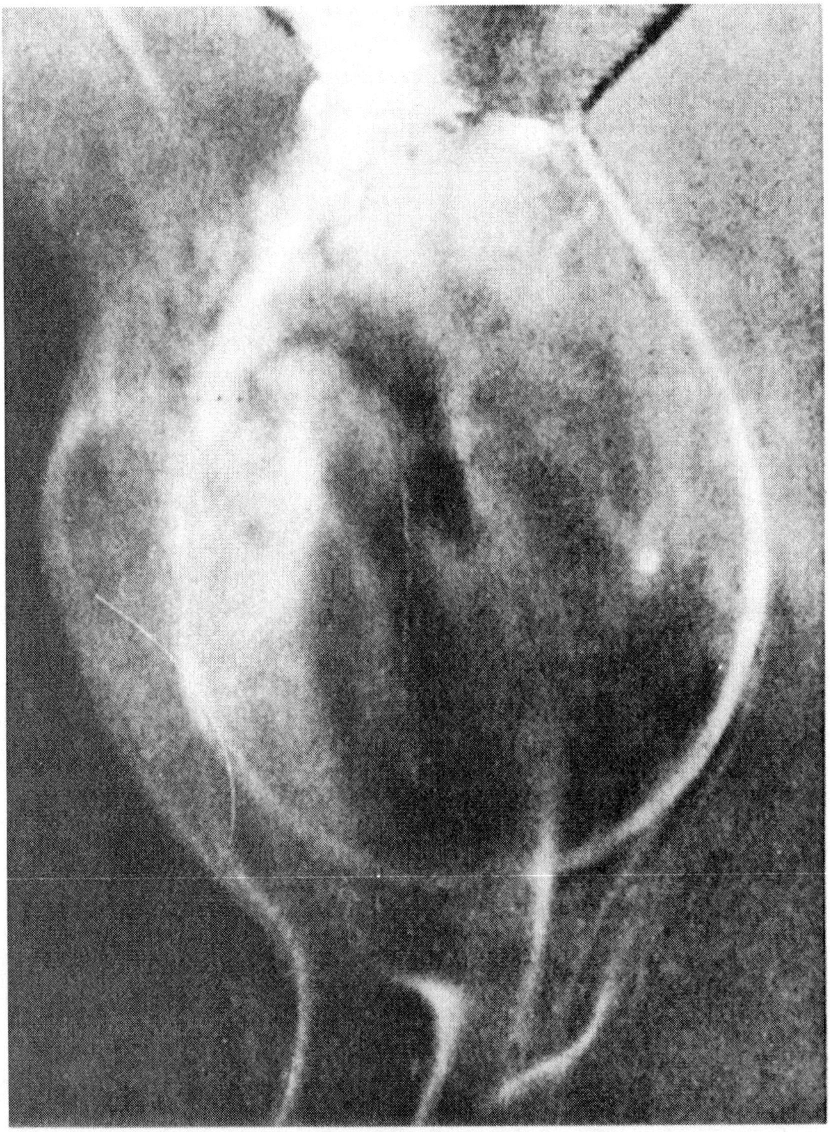

Figure VI-19. Spontaneous emulsification when a drop of an ethyl alcohol-toluene mixture is placed in water. Reproduced from ref. 57 with the permission of Professor J.T. Davies.

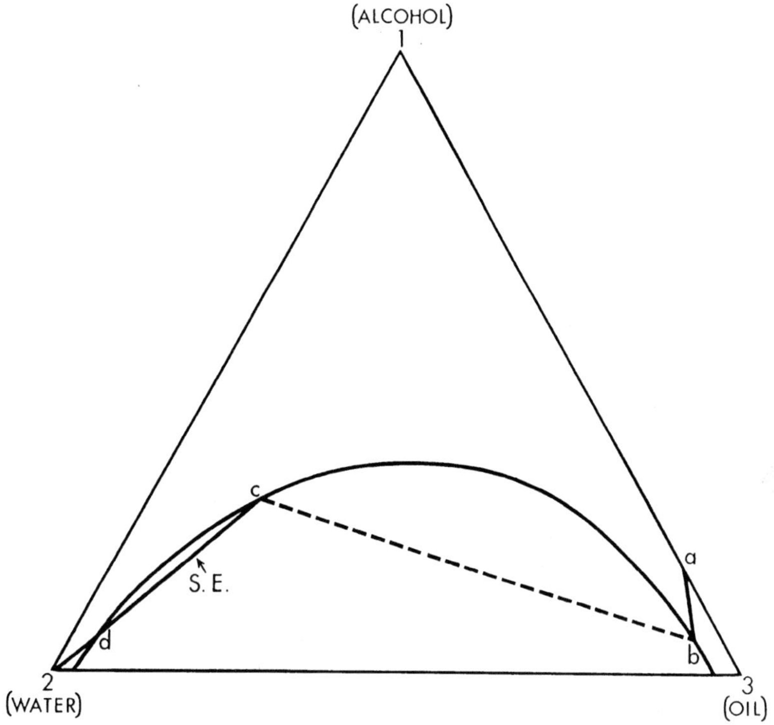

Figure VI-20. Sample diffusion path exhibiting spontaneous emulsification (S.E.). The initial compositions are point a (ω_{10}, ω_{20}) and pure 2 (ω_{10}', ω_{20}').

the following four differential equations, two in each phase (58):

$$\frac{\partial \omega_i}{\partial t} = D_i \frac{\partial^2 \omega_i}{\partial x^2}$$
$$\frac{\partial \omega_i'}{\partial t} = D_i' \frac{\partial^2 \omega_i'}{\partial x^2} \qquad (i = 1, 2) \qquad \text{[VI-66]}$$

Here ω_i and ω_i' are the mass fractions of species i in the two phases and D_i and D_i' are the corresponding diffusion coefficients. The condition that the mass fractions in each phase must sum to unity can be used to determine ω_3 and ω_3' for the third component from the solution to Eq. [VI-66].

Since both the initial condition and the boundary conditions for $x \to \pm\infty$ are that the initial compositions must exist, a similarity solution is possible (60). It has the form

$$\omega_i = \omega_{io} + B_i \left(1 - \text{erf} \frac{x}{(4D_i t)^{1/2}}\right)$$

[VI-67]

$$\omega_i' = \omega_{io}' + B_i' \left(1 + \text{erf} \frac{x}{(4D_i' t)^{1/2}}\right)$$

The constants B_i and B_i' must be determined from boundary conditions at the moving interface, the position of which we denote by $\varepsilon(t)$. Two of these conditions are the following component mass balances:

$$D_i \left.\frac{\partial \omega_i}{\partial x}\right|_\varepsilon - D_i' \left.\frac{\partial \omega_i'}{\partial x}\right|_\varepsilon = [\omega_i'(\varepsilon) - \omega_i(\varepsilon)] \frac{d\varepsilon}{dt} \quad ; \quad (i = 1, 2) \qquad \text{[VI-68]}$$

where $\omega_i(\varepsilon)$ and $\omega_i'(\varepsilon)$ are the mass fractions of species i in the two phases at the interface. These equations are consistent with the similarity solution Eq. [VI-67] if the interfacial compositions are constant and if $\varepsilon(t)$ has the following form

$$\varepsilon(t) = bt^{1/2} \qquad \text{[VI-69]}$$

with b a constant.

The remaining boundary conditions are the phase equilibrium relationships at the interface given by the applicable phase diagram. It may be convenient in calculations to describe the coexistence curve and tie-lines by empirical equations such as those proposed by Hand (61). Whatever the details of the method used, the problem is well posed once initial compositions, diffusion coefficients, and equilibrium data are specified. Solution yields the concentration profiles in both phases.

According to the similarity solution Eq. [VI-67] the mass fractions ω_i are functions of the single variable $(x/t^{1/2})$, which ranges from $-\infty$ to b in one phase and from b to $+\infty$ in the other. Hence, the *set* of compositions in the system is time independent, and this "diffusion path" can be plotted on the phase diagram as shown in Figure VI-20. The diffusion paths in the two phases (solid lines) are straight lines as shown if the diffusion coefficients of the various species are equal in each phase. Otherwise they have some curvature.

Note that in Figure VI-20 part of the diffusion path for the primed phase, viz., the segment cd, lies within the two-phase region of the phase diagram. Ruschak and Miller (58) proposed that when such behavior is found, spontaneous emulsification should be expected in the phase or phases having supersaturated compositions. Their experiments for three toluene-water-solute systems having tie-lines of different slope confirmed the ability of this scheme to predict both when emulsion would form and in which phase.

In oil-water-surfactant systems phase diagrams are typically much more complicated than that of Figure VI-20, as discussed in Chapter IV. Employing both contacting experiments in test tubes and optical microscopy, Benton *et al* (62) observed intermediate microemulsion or brine phases forming between oil and surfactant-brine mixtures containing liquid crystal. They also found that the occurrence of spontaneous emulsification under some circumstances in these systems could be explained in terms of diffusion paths passing through two-phase regions as above. Recently, Raney *et al* (63) developed a novel optical microscopy technique which enabled them to see more clearly the development of intermediate phases and to measure their rates of growth. They observed one and sometimes two intermediate phases forming. Also interfacial distances from the initial position of contact were found to be proportional to $t^{1/2}$ as predicted by diffusion path theory except in a few cases where densities of the new phases were such that extensive gravitationally-produced convection ensued. These results may be perinent to certain mechanisms of detergency where soil removal involves formation of intermediate phases (64, 65).

10. OTHER INTERFACIAL PHENOMENA INVOLVING DISPERSE PHASE FORMATION

Spontaneous emulsification is but one example of disperse phase formation where diffusion effects are important, albeit one of interest to many workers in interfacial phenomena. Numerous other situations exist where a disperse solid or fluid phase forms as a result of phase transformation or chemical reaction. Moving fronts can develop, for instance, between regions containing and free of the disperse phase. Such fronts involving the growth or shrinkage of regions of fog were considered by Toor (66), and their stability was analyzed by Miller and Jain (45) using a method similar to that described in Section 7. It was found that a front where fog is consumed by evaporation of drops is stable. But a front where fog forms by invading a region of supersaturated vapor is unstable (45).

Another interesting phenomenon is formation of a disperse phase in narrow, individual bands separated by regions containing no disperse phase. The bands are often called "Liesegang rings". Figure VI-21 illustrates the formation of discrete bands of $Mg(OH)_2$ precipitate in test tubes where aqueous solutions of NH_4OH (top) and $MgSO_4$ (bottom have been contacted and diffusion allowed to proceed. A small amount of a gelling agent has been added to prevent convection.

Theoretical descriptions of such behavior have been given in terms of a mechanism first put forward by Ostwald around the turn of the century, namely that a critical local supersaturation is required to initiate local precipitation (see 67-69). But recent experimental studies have raised serious questions concerning the applicability of this mechanism (70). In particular, the solid has been observed to form as a colloidal dispersion before the appearance of bands. Also it has been found that bands sometimes develop, in contrast with predictions of the above theory, in systems of uniform initial composition (71).

We focus for simplicity on this last situation and on the instability mechanism put forward by Lovett *et al* (72), which, according to the recent experiments, may also be pertinent in Liesegang ring formation where initial concentration gradients exist. First we note that solution of the diffusion equation for a single spherical particle of a pure species A in an infinite expanse of a solvent having a bulk concentration c_∞ of A leads to the following expression for particle growth rate:

$$\frac{da}{dt} = \frac{D\, v_{As} c_e}{a} \left(\frac{c_\infty}{c_e} - 1 - \frac{2\gamma v_{As}}{a\, RT} \right) \qquad [VI\text{-}70]$$

where a is the particle radius, D the diffusivity of A in the solvent, v_{As} the molar volume of the particle, γ the interfacial tension, and c_e the concentration of A in the solvent when it has been equilibrated with a large bulk phase of A. We note from Eq. [VI-70] that the concentration $c_{A\infty}$ for which there is no particle growth, i.e., for which (da/dt) vanishes, is larger for particles of smaller radius as a result of the term involving interfacial tension. In other words small particles are more soluble than large particles, so that the latter should be expected to grow at the expense of the former, a process which reduces interfacial area and free energy. It is this basic behavior which leads in the analysis below to the development of bands.

We consider a uniform initial dispersion where the values of c_∞* and a* of solute concentration and radius are such that (da/dt), as given by

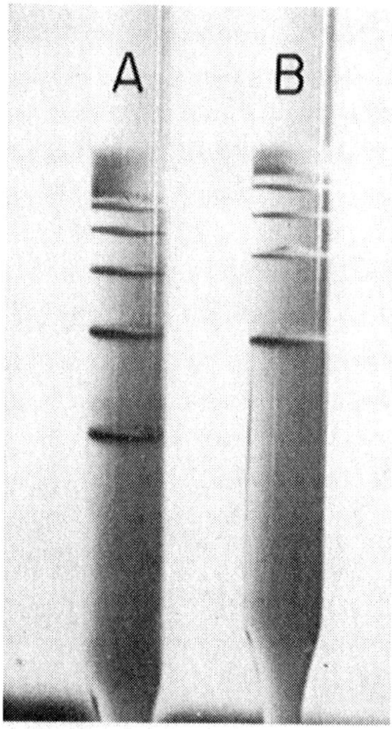

Figure VI-21. Liesegang Ring Formation. From ref. (70) with permission.

Eq. [VI-70], is zero. Now suppose that $c_{A\infty}$ and a are perturbed in such a way that

$$c_{\infty} = c_{\infty}^{*} + c_{\infty p}\, e^{i\alpha x}\, e^{\beta t} \qquad [VI-71]$$

$$a = a^{*} + a_{p}\, e^{i\alpha x}\, e^{\beta t} \qquad [VI-72]$$

where $c_{\infty p}$ and a_p are the perturbation amplitudes and α and β are the wave number and time factor as in previous stability analyses. The diffusion equation in the bulk solvent is, in the absence of convection, given by

$$\frac{\partial c_{\infty}}{\partial t} = D\,\frac{\partial^2 c_{\infty}}{\partial x^2} + r \qquad [VI-73]$$

Here r is the net rate of solute addition to the solvent by particle dissolution and is given by

$$r = -\frac{da}{dt} \bullet \frac{4\pi a^2}{v_{As}} \bullet n \qquad \text{[VI-74]}$$

with n the number of particles per unit volume.

We can use Eq. [VI-70] to calculate (da/dt) following the perturbation:

$$\frac{da}{dt} = \beta\, a_p\, e^{i\alpha x+\beta t} = \frac{D\, v_{As} c_e}{a^*} \left[\frac{c_{\infty p}}{c_e} - \left(\frac{c_\infty}{c_e} - 1\right) \frac{a_p}{a^*} + \frac{4\gamma\, v_{As}}{RT} \frac{a_p}{a^{*2}} \right] e^{i\alpha x+\beta t} \qquad \text{[VI-75]}$$

From this equation we find

$$c_{\infty p} = a_p \left[\frac{\beta a^*}{D v_{As}} - \frac{2\gamma v_{As} c_e}{a^{*2}\, RT} \right] \qquad \text{[VI-76]}$$

Finally, substituting Eqs. [VI-71], [VI-72], and [VI-74]-[VI-76] into Eq. [VI-73], we obtain a quadratic equation which yields the following solutions for the time factor β:

$$\beta = -\frac{b}{2} \pm \left[\frac{b^2}{4} + \frac{2\gamma\alpha^2 D^2\, v_{As}^{\,2}\, c_e}{a^{*3}\, RT} \right]^{1/2} \qquad \text{[VI-77]}$$

$$b = 4\pi\, Dn\, a^* + D\, \alpha^2 - \frac{2\gamma v_{As}^{\,2} c_e D}{a^{*3}\, RT} \qquad \text{[VI-78]}$$

Inspection of this equation reveals that with $b > 0$ there is always one positive root for β , i.e., the system is always unstable. However, the positive root becomes very small in the limit of long wavelengths ($\alpha \rightarrow 0$) , the expected result since solute must diffuse over relatively long distances in order for the perturbation to grow. Lovett *et al* (72) showed that when diffusion of the particles themselves is considered, perturbations of very short wavelengths are stable. That is, particle diffusion virtually eliminates particle concentration gradients and precludes discrete bands from forming. An intermediate wavelength λ_m exists where perturbation growth rate reaches a maximum value. For reasonable values of the physical properties, λ_m was calculated to be of the order of 1 cm and the time constant for growth of the order of 1 day. Both these values are generally consistent with experimental observations.

Although this analysis yields insight on how bands of precipitate may form, understanding of the origin of Liesegang rings is not yet complete.

For instance, several studies including some quite old (73) and some very recent (74) suggest that flocculation of a colloidal dispersion can play an important role in the development of visible precipitate bands.

REFERENCES

General References on Phenomena Involving Transport Near Interfaces

Berg, J.C. (1972) "Interfacial Phenomena in Fluid Phase Separation Processes," in *Recent Developments in Separation Science*, N.N. Li (ed), **vol. 2**, CRC Press, Cleveland, p.1.

Berg, J.C. (1982) *Canad. Met. Quart.* **21**, 121.

Levich, V.G. and Krylov, V.S. (1969) "Surface-Tension-Driven Phenomena," *Ann. Rev. Fluid Mech.* **1**, 293.

Lucassen-Reynders, E.H. (1981) "Surface Elasticity and Viscosity in Compression/Dilation," in *Anionic Surfactants, Physical Chemistry of Surfactant Action*, E.H. Lucassen-Reynders (ed), Marcel Dekker, New York, p. 173.

Miller, C.A. (1978) "Stability of Interfaces," in *Surface and Colloid Science*, E. Matijevic (ed), **vol. 10**, Plenum, New York, p. 227.

Sawistowski, H. (1971) "Interfacial Phenomena" in *Recent Advances in Liquid-Liquid Extracion*, C. Hanson (ed), Pergamon Press, Oxford, p.293.

Sekerka, R.F. (1968) *J. Cryst. Growth* **3, 4**, 71.

van den Tempel, M. and Lucassen-Reynders, E.H. (1983) *Adv. Colloid Interface Sci.* **18**, 281.

Textual References

1. Benard, H. (1900) *Revue Gen. Sci. Pur. Appl.* **11**, 1261, 1309.
2. Pearson, J.R.A. (1958) *J. Fluid Mech.* **4**, 489.
3. Hansen, C.M. and Pierce, P.E. (1973) *Ind. Eng. Chem. Prod. Res. Dev.* **12**, 67.
4. Scriven, L.E. and Sternling, C.V. (1964) *J. Fluid Mech.* **19**, 321.
5. Smith, K.A. (1966) *J. Fluid Mech.* **24**, 401.
6. Berg, J.C., Acrivos, A., and Boudart, M. (1966) *Adv. Chem. Eng.* **6**, 61.
7. Berg, J.C. and Acrivos, A. (1965) *Chem. Eng. Sci.* **20**, 737.
8. Palmer, H.J. and Berg, J.C. (1972) *J. Fluid Mech.* **51**, 385.

9. Vidal, A., and Acrivos, A. (1968) *Ind. Eng. Chem. Fundam.* **7**, 53.

10. Nield, D.A. (1964) *J. Fluid Mech.* **19**, 341.

11. Palmer, H.J. and Berg, J.C. (1971) *J. Fluid Mech.* **47**, 779.

12. Wollkind, D.J. and Segel, L.A. (1970) *Philos. Trans. Roy. Sec. London* **A268**, 351.

13. Hinkbein, T.E. and Berg, J.C. (1970) *Int. J. Ht. Mass Transf.* **21**, 1241.

14. Brian, P.L.T. (1971) *AIChE J.* **17**, 765.

15. Sternling, C.V. and Scriven, L.E. (1959) *AIChE J.* **5**, 514.

16. Berg, J.C. (1972) "Interfacial Phenomena in Fluid Phase Separation Processes," in *Recent Developments in Separation Science,* N.N. Li (ed), **vol. 2**, CRC Press, Cleveland, p. 1.

17. Gouda, J.H. and Joos, P. (1975) *Chem. Eng. Sci.* **30**, 521.

18. Burkholder, H.C. and Berg, J.C. (1974) *AIChE J.* **20**, 863, 872.

19. Bainbridge, G.S. and Sawistowski, H. (1964) *Chem. Eng. Sci.* **19**, 992.

20. Coyle, R.W., Berg, J.C., and Nina, J.C. (1981) *Chem. Eng. Sci.* **36**, 19.

21. Berg, J.C. (1982) *Canad. Met. Quart.* **21**, 121.

22. Ruckenstein, E. and Berbente, C. (1964) *Chem. Eng. Sci.* **19**, 329.

23. Nelson, Jr., N.K. and Berg, J.C. (1982) *Chem. Eng. Sci.* **37**, 1067.

24. Vedove, W.D. and Sanfeld, A. (1981) *J. Colloid Interface Sci.* **84**, 318, 328.

25. Sanfeld, A., Steinchen, A. Hennenberg, M., Bisch, P.M., van Lansweerde Gallez, D. and Vedove, W.D. (1979) "Mechanical, Chemical, and Electrical Constrants and Hydrodynamic Interfacial Stability," in *Dynamics and Instability of Fluid Interfaces*, T.S. Sorensen (ed), Springer-Verlag, Berlin, p. 168.

26. Sorensen, T.S. (1979) "Instabilities Induced by Mass Transfer, Low Surface Tension in and Gravity at Isothermal and Deformabale Fluid Interfaces," in *Dynamics and Instability of Fluid Interfaces*, T.S. Sorensen (ed), Springer-Verlag, Berlin, p. 1.

27. Perez de Ortiz, E.S. and Sawistowski, H. (1973) *Chem. Eng. Sci.* **28**, 2051, 2063.

28. Sorensen, T.S. and Hennenberg, M. (1979) "Instability of a Spherical Drop with Surface Chemical Reactions and Transfer of Surfactants," in *Dynamics and Instability of Fluid Interfaces,* T.S. Sorensen (ed), Springer-Verlag, Berlin, p.276.

29. Sawistowski, H. (1971) "Interfacial Phenomena" in *Recent Advances in Liquid-Liquid Extraction,* C. Hanson (ed), Pergamon Press, Oxford, p.293.

30. Linde, H., Schwartz, P., and Wilke, H. (1979) "Dissipative Structures and Nonlinear Kinetics of the Marangoni Instability," in *Dynamics and Instability of Fluid Interfaces,* T.S. Sorensen (ed), Springer-Verlag, Berlin, p. 75.

31. Kenning, D.B.R. (1968) *Appl. Mech. Rev.* **21**, 1101.

32. Levich, V.G. and Krylov, V.S. (1969) "Surface-Tension-Driven Phenomena," *Ann. Rev. Fluid Mech.* **1**, 293.

33. Zuiderweg, F.J. and Harmens, A. (1958) *Chem. Eng. Sci.* **9**, 89.

34. Wang, K.H., Ludviksson, V., and Lightfoot, E.N. (1971) *AIChE J.* **17**, 1402.

35. Johnson, D.K. and Berg, J.C. (1983) "The Effect of Surfactants on Thin Film Behavior in Distillation Equipment," Preprint 62c for AIChE National Meeting, Houston.

36. Bascom, W.D., Cottington, R.L., and Singleterry, C.R. (1964) "Dynamic Surface Phenomena in the Spontaneous Spreading of Oils on Solids," in *Adv. Chemistry Series,* **vol. 43**, Am. Chem. Soc., Washington, p. 355.

37. Hansen, R.S. (1961) *J. Colloid Sci.* **16**, 549.

38. Ward, A.F. and Tordai, L. (1946) *J. Chem. Phys.* **14**, 453.

39. Maru, H.C. and Wasan, D.T. (1979) *Chem. Eng. Sci.* **34**, 1283, 1295.

40. Lucassen, J. and Barnes, G.T. (1972) *J. Chem. Soc. Faraday I* **68**, 2129.

41. Lucassen, J. and van den Tempel, M. (1972) *J. Colloid Interface Sci.* **41**, 491.

42. England, D.C. and Berg, J.C. (1971) *AIChE J.* **17**, 313.

43. Rubin, E. and Radke, C.J. (1980) *Chem. Eng. Sci.* **35**, 1129.

44. Miller, C.A. (1973) *AIChE J.* **19**, 909.

45. Miller, C.A. and Jain, K. (1973) *Chem. Eng. Sci.* **28**, 157.

46. Miller, C.A. (1975) *AIChE J.* **21**, 474.

47. Armento, M.E. and Miller, C.A. (1977) *Soc. Pet. Eng. J.* **17**, 423.

48. Gunn, R.D. and Krantz, W.B. (1982) *Soc. Pet. Eng. J.* **20**, 267.

49. Hickman, K. (1952) *Ind. Eng. Chem.* **44**, 1892.

50. Palmer, H.J. (1976) *J. Fluid Mech.* **75**, 487.

51. Cannon, K.J. and Denbigh, K.G. (1957) *Chem. Eng. Sci.* **6**, 145, 155.

52. Knapp, R. and Aris, R. (1972) *Arch. Rat. Mech. Anal.* **44**, 165.

53. Miller, C.A. (1978) "Stability of Interfaces," in *Surface and Colloid Science*, E. Matijevic (ed), **vol. 10**, Plenum, New York, p. 227.

54. Braun, R.L., Lewis, A.E. and Wadsworth, M.E. (1974) "In-Place Leaching of Primary Sulfide Ores", in *Solution Mining Symposium*, F.F. Aplan (ed), AIME, New York.

55. Seshan, P.K. (1975) "Electrodeposition Cells: A Theoretical Investigation into Their Performance and Deposit Growth Stability." Ph.D. Thesis, Carnegie-Mellon University.

56. Edwards, J. (1964) *Trans. Inst. Met. Finishing* **41**, 169.

57. Davies, J.T. and Haydon, D.A. (1957) *Proc. Int. Congr. Surf. Act. 2nd* 1, 417.

58. Ruschak, K.J. and Miller, C.A. (1972) *Ind. Eng. Chem. Fundam.* **11**, 534.

59. Kirkaldy, J.S. and Brown, L.F. (1963) *Canad. Met. Quart.* **2**, 89.

60. Bird, R.B., Stewart, W.E., and Lightfoot, E.N. (1960) *Transport Phenomena*, Wiley, New York.

61. Hand, D.B. (1930) *J. Phys. Chem.* **34**, 1961.

62. Benton, W.J., Miller, C.A., and Fort, Jr. T. (1982) *J. Disp. Sci. Tech.* **3**,1.

63. Raney, K.H., Benton, W.J. and Miller, C.A. (1983) "Diffusion Phenomena and Spontaneous Emulsification in Oil-Water-Surfactant Systems," ACS Symposium Series volume on emulsions and microemulsions, in press.

64. Stevenson, D.G. (1961) "Ancillary Effects in Detergent Action," in *Surface Activity and Detergency*, K. Durham (ed), Macmillan, London, p. 146.

65. Lawrence, A.S.C. (1961) "Polar Interaction in Detergency," in *Surface Activity and Detergency*, K. Durham (ed), Macmillan, London, p.158.

66. Toor, H.L. (1971) *AIChE J.* **17**, 5.

67. Wagner, C. (1950) *J. Colloid Sci.* **5**, 85.

68. Prager, S. (1956) *J. Chem. Phys.* **25**, 279.

69. Klueh, R.L. and Mullins, W.W. (1969) *Acta Met.* **17**, 69.

70. Kai, S., Muller, S.C., and Ross, J. (1982) *J. Chem. Phys.* **76**, 1392.

71. Flicker, M. and Ross, J. (1974) *J. Chem. Phys.* **60**, 3458.

72. Lovett, R., Ortoleva, P. and Ross, J. (1978) *J. Chem. Phys.* **69**, 947.

73. Dhar, N.R. and Chatterji, N.G. (1925) *Kolloid-Z.* **37**, 2, 89.

74. Kanniah, N., Gnanam, F.D., and Ramasamy, P. (1983) *J. Colloid Interface Sci.* **94**, 412.

75. Defay, R. and Petre, G. (1971) "Dynamic Surface Tension" in *Surface and Colloid Science*, E. Matijevic (ed), **vol. 3**, Wiley, New York p. 27.

76. van Voorst Vader, F., Erkens, Th.F., and van den Tempel, M. (1964) *Trans. Faraday Soc.* **60**, 1170.

77. Jackson, R. (1977) *Ind. Eng. Chem. Fundam.* **16**, 304.

78. Avsec, D. (1939) *Publ. Sci. Tech. Minist. Air France* **155**.

79. Lucassen-Reynders, E.H. (1981) "Surface Elasticity and Viscosity", in *Anionic Surfactants*, E.H. Lucassen-Reynders (ed), Marcel Dekker, New York, p. 173.

PROBLEMS

VI-1. Adapt the derivation of Section 3 to apply to the case of a solute with some surface activity which desorbs from a thin layer of liquid into a gas phase. Derive the condition for marginal stability for the case of a flat interface (N_{Cr} = 0). Neglect surface diffusion effects.

VI-2. Consider a thin layer of liquid where the temperature is uniform at each lateral position but where a uniform lateral temperature gradient (dT/dx) is imposed. Assuming that film thickness is everywhere equal to h and that, at steady state, a pressure gradient develops so that there is no net lateral flow of liquid, calculate the pressure gradient in the film and the velocity distribution. Comment on whether the uniform film thickness and the existence of a pressure gradient in the film are consistent features of this problem.

VI-3. Starting with Eq. [VI-1], derive the differential interfacial mass balance Eq. [VI-2].

VI-4. In the falling meniscus method for measuring dynamic interfacial tensions, a vertical tube of nonuniform diameter is fabricated having at

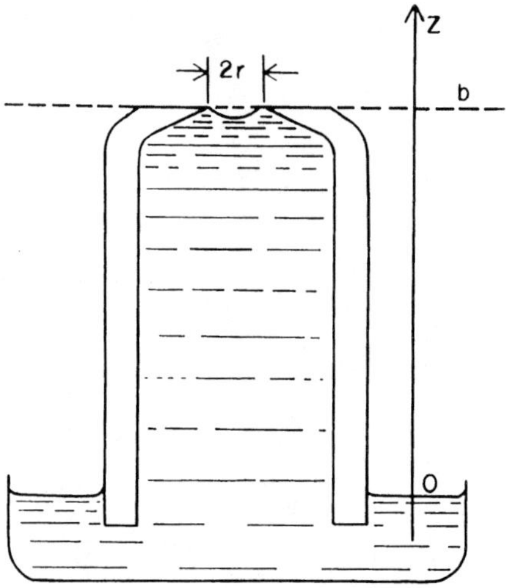

Figure VI-22. Principle of the falling meniscus method. From ref. (73) with permission.

its top a small circular opening of radius r with a sharp edge (see Figure VI-22). The tube is placed in a large pool of liquid with the small hole a height h above the surface of the pool. A fresh surface is created at the hole at time t = 0. Assuming that surface tension decreases with time as a result of diffusion of a surface-active material to the interface, find the value of the surface tension for which the liquid level in the tube will fall precipitously. Ignore deviation of the meniscus shape from a spherical conformation although this factor should be considered in accurate work (see (75)).

VI-5. Suppose that in the arrangement of Figure VI-11 the barriers are moved apart in such a way that tangential velocity along the surface between them is given by

$$v_x = ax$$

where a is a constant. If a steady state is reached where interfacial tension is time-independent, derive an expression for surfactant concentration c_0 in the solution just beneath the surface. Assume that c_0 and the surface concentration Γ are related by a linear isotherm

$$\Gamma = A\, c_0$$

Also assume that concentration gradients are negligible except in the immediate vicinity of the interface. This technique for studying dynamic interfacial tension was proposed by van Voorst Vader *et al* (76).

Note that since the adsorption isotherm and the corresponding surface equation of state (see Section 5 of Chapter II) can be used to calculate c_0 for an experimental value of interfacial tension γ, the expression you have derived for c_0 could be used to obtain the diffusivity D.

VI-6. Prove the statement made in the text that for a soluble surfactant the quantity Γ_{os} $(- d\gamma/d\Gamma_s)$ in Eq [V-71] must be replaced by the first expression for E in Eq. [VI-42]. Then show that in the low viscosity limit E is given by the second expression, i.e., that containing the parameter B.

VI-7. Show that the equilibrium melting temperature T_M of a pure material varies with interfacial curvature in accordance with the following equation (cf. last term of Eq. [VI-56] and accompanying discussion)

$$T_M = T_o + \frac{2H\gamma T_o}{\lambda\ \rho}$$

where T_o is the equilibrium melting temperature of a flat interface, and the sign convention is such that curvature is negative when the center of curvature is in the solid phase.

VI-8. Analyze the stability of a reacting solid surface when the reaction scheme is that described by Eqs. [VI-63] and [VI-64]. Assume that the liquid is stirred so that the concentrations of A and B have their bulk values c_{A0} and c_{B0} outside a stagnant film of thickness d and that the concentration profiles in the film are linear before perturbation. Neglect interfacial tension effects.

(a) Show that the concentration perturbation c_{Ap} for A in the stagnant film has the form $[g_A(z)\, e^{i\alpha x+\beta t}]$ with

$$g_A = A_1(\cosh \alpha q_A z + \coth \alpha q_A d \sinh \alpha q_A z)$$

$$\text{with } q_A^2 = 1 + (\beta/D_A\alpha^2)$$

Write the correponding expression for c_{Bp}.

(b) Apply suitable boundary conditions and derive the following expression for the time factor β in the limit q_A, $q_B \to 1$

$$\beta = \frac{\frac{k}{c_s} c_{As}^a c_{Bs}^{b_1} \left[\frac{b_1}{c_{Bs}} \frac{(c_{Bs} - c_{Bo})}{d} - \frac{a}{c_{As}} \frac{(c_{Ao} - c_{As})}{d} \right]}{1 + k c_{As}^a c_{Bs}^{b_1} \left[\frac{a^2 \tanh \alpha d}{c_{As} D_A \alpha} - \frac{b_1 (b_2 - b_1) \tanh \alpha d}{c_{Bs} D_B \alpha} \right]}$$

where c_{As} and c_{Bs} are the concentrations of A and B at the initial plane interface.

VI-9. Analyze the stability of an initially plane interface where a metallic phase is growing as a result of electrodeposition and where the electrodeposition rate is given by Eq. [VI-65]. Assume that the effect of the applied electrical potential is included in the constant k and that the organic additive is slightly soluble in the metal, its distribution coefficient between metal and the electrodeposition bath being b. For simplicity, neglect the stabilizing effect of interfacial tension. Show that the time factor β is given by

$$\beta = \frac{\frac{G_M}{c_{MI}} \left(1 - \frac{V}{\omega_M D_M}\right) - \frac{m G_A}{c_{AI}} \left(1 - \frac{V(1+b)}{\omega_A D_A + bV}\right)}{\frac{c_{Ms} c_{AI}^m}{k c_{MI}} + \frac{c_{Ms}}{c_{MI} \omega_M D_M} - \frac{mb}{\omega_A D_A + bV}}$$

VI-10. Section 9 dealt with prediction of diffusion paths and spontaneous emulsification for a particular unsteady state situation. We consider here determination of the diffusion path at steady state (77).

Consider one-dimensional, steady-state, ternary diffusion along a capillary tube as shown in Figure VI-23a. Two immiscible liquids occupy the two halves of the tube with the position of the interface between them taken as the origin of the coordinate system (z=0). Compositions A and D at the ends of the tube are known. Compositions B and C at the interface are not known initially. But if local equilibrium is assumed at the interface, B and C must be at the ends of a tie line on the (known) ternary phase diagram as shown in Figure VI-23b. The question is which tie line? Once the tie line is determined, the concentration profiles are known and the question of whether spontaneous emulsification occurs can be settled.

Let $\omega_i(z)$ be the mass fraction of component i at position z for i = 1, 2, 3. If the density ρ within each liquid phase is uniform, the mass

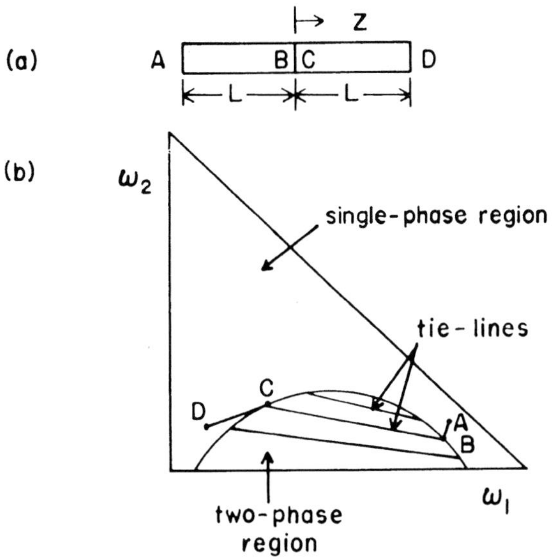

Figure VI-23. Steady state diffusion path in ternary system (Problem VI-10).

flux j_i with respect to the mass average velocity has the form

$$j_1 = -\rho D_i \frac{d\omega_i}{dz}$$

where D_i is the diffusivity of component i. Note that we have neglected the dependence of j_i on the concentration gradients $(d\omega_j/dz)$ of the other two components.

(a) Show that $\omega_i(z)$ in the left half of the capillary tube (between A and B) is given by

$$\omega_i(z) = \frac{n_i}{n} + (\omega_{iA} - \frac{n_i}{n}) \exp\left[\frac{n(z+L)}{\rho D_i}\right] , \quad i=1,2,3$$

where n_i = mass flux of i with respect to fixed coordinates,

$$n = n_1 + n_2 + n_3 ,$$

and D_i is independent of composition

(b) If $D_1 = D_2 = D_3 = D$, show that $\omega_1(z)$ is a linear function of $\omega_2(z)$ in the left half of the capillary tube. Hence, all compositions in this region must be on a straight line joining A and B

(c) Derive an equation similar to that of (a) for the right half of the capillary tube, assuming that the density is ρ' and that the diffusivity is D'_i. As in (b), we shall assume that $D'_1 = D'_2 = D'_3 = D'$.

(d) What boundary conditions apply at the interface? Show that if L, ρ, ρ', D, D', ω_{iA}, ω_{iD} and the complete phase diagram are known, these conditions are sufficient to solve for all unknown constants in your equations and hence to determine the concentration profiles and fluxes.

(e) By manipulating the equations of (a) and (c), one can show that when $\rho D = \rho' D'$

$$\frac{\omega_{1B} - \omega_{1D}}{\omega_{2B} - \omega_{2D}} = \frac{\omega_{1C} - \omega_{1A}}{\omega_{2C} - \omega_{2A}}$$

In view of this result, can you suggest a simple geometric construction to locate points B and C on the phase diagram given the locations of points A and D?

VII

Dynamic Interfaces

1. INTRODUCTION

An important feature of all mass transfer operations and of a significant number of reaction systems in chemical engineering is the critical role played by interfacial phenomena. Liquid-liquid and gas-liquid systems are characterized by convective-diffusive transfer at interfaces which keep distorting, e.g., in distillation, gas absorption and liquid-liquid extraction. One fluid phase in such systems is often dispersed in another. Consequently, the nature of mass transfer at dynamic interfaces in disperse systems is important. The dynamic behavior of drops and bubbles, e.g., their shapes under various flow conditions and their breakage and coalescence, has been studied for many years although much remains to be learned. Information about such behavior is required to predict the mass transfer rates in disperse systems (1).

In some cases only the shapes of fluid-fluid interfaces need be studied, for instance in coating problems. Here, the material to be coated is in the form of a flat plate or wire which is withdrawn continuously from a pool of liquid. The liquid adheres to the surface as a thin film of constant thickness and is dried to form the coating. Considering both that the coat thicknesses are very small and that the uniformity of the coat is often essential to the product, there exists a great need for precision (2). The knowledge of the thickness of the liquid layer, i.e., the shape of the liquid interface (see Figure VII-3) as it deposits on the solid surface, is needed. This is well-known as the dip coating problem.

Similar problems where the final products or processes depend on the shapes of moving fluid-fluid interfaces are widely known: in extrusion of polymer melts, in fiber spinning, in formation of droplets from jets, in spreading of ink drops, oil slicks, etc.

In spite of numerous attempts no general strategy has emerged for solving the equations of fluid mechanics to obtain the shapes of fluid interfaces (3). A solid-fluid interface is rigid and can be made to have regular shapes, e.g., planar, spherical, cylindrical. However, liquid-fluid interfaces are rarely simple in shape. Moreover, their locations and shapes are determined by force balances (i.e., the solutions of the momentum equations) and are not known beforehand. This last difficulty has proven to be so enormous that even numerical solution to a complete problem can rarely be obtained.

Indeed the difficulty is even greater. Not only is interfacial position influenced by the flow, but the presence of the interface alters the flow. In the hydrostatics discussed in Chapter I, it was seen that the pressure is discontinuous across the interface separating phases A and B, by

$$p_A - p_B = -2\gamma H \qquad [VII\text{-}1]$$

where H is the mean curvature of the interface and p is the pressure evaluated at the interface. It may be anticipated that since pressure differences give rise to a flow, the shape of an interface will affect fluid flow.

2. SURFACES

As indicated above, dynamic interfaces frequently have complex shapes. In this section we provide additional information on the geometric properties of surfaces needed to apply some of the basic principles discussed previously to interfaces of arbitrary shape. For simplicity only certain pertinent results are given; the reader may consult Slattery (4) for a detailed account.

The knowledge of how to describe a surface is essential. Consider a surface given by

$$z = f(x,y) \qquad [VII\text{-}2]$$

in rectangular coordinates. Eq. [VII-2] can be rewritten as

$$g(x,y,z) = z - f = 0 \qquad [VII-3]$$

On differentiating a curve in a plane we obtain the tangent line. The question arises as to what constitutes a derivative of a surface and how such a differentiation may be performed. The differentiation of a variable in three dimensions is done vectorially through the gradient operator

$$\nabla = \mathbf{e_x} \frac{\partial}{\partial x} + \mathbf{e_y} \frac{\partial}{\partial y} + \mathbf{e_z} \frac{\partial}{\partial z} \qquad [VII-4]$$

Here, $\mathbf{e_x}$, $\mathbf{e_y}$ and $\mathbf{e_z}$ are the unit vectors in the x, y and z-directions respectively. On a surface, a tangent *plane* (and not just a single tangent) can be drawn. The quantity ∇g is perpendicular to the plane at that point. The unit vector

$$\mathbf{n} = \frac{\nabla g}{|\nabla g|} \qquad [VII-5]$$

is called the unit normal to the surface. Note that $|\nabla g| = (\nabla g \cdot \nabla g)^{1/2}$ and $\mathbf{n} \cdot \mathbf{n} = 1$. If two mutually perpendicular unit vectors $\mathbf{t_1}$ and $\mathbf{t_2}$ are chosen on the tangent plane, then the unit vector set ($\mathbf{n}$, $\mathbf{t_1}$, $\mathbf{t_2}$) describes the orientation of the surface and the vector or tensor quantities which describe the interfacial properties.

As a special case, the surface

$$z = h(x) \qquad [VII-6]$$

can be considered. It is sketched in Figure VII-1. There is no variation in the y-direction. The unit normal is found from Eq. [VII-5] to be

$$\mathbf{n} = [- \frac{dh}{dx} \mathbf{e_x} + \mathbf{e_z}][1 + (\frac{dh}{dx})^2]^{-1/2} \qquad [VII-7]$$

One tangent vector, $\mathbf{e_y}$, is evident. The other is obtained from the condition that it is perpendicular to $\mathbf{n}$ and also perpendicular to $\mathbf{e_y}$, i.e.,

$$\mathbf{t} = [\mathbf{e_x} + (\frac{dh}{dx}) \mathbf{e_z}][1 + (\frac{dh}{dx})^2]^{-1/2} \qquad [VII-8]$$

It is to be noted that $\mathbf{n} \cdot \mathbf{n} = \mathbf{e_y} \cdot \mathbf{e_y} = \mathbf{t} \cdot \mathbf{t} = 1$, as well as $\mathbf{n} \cdot \mathbf{t} = \mathbf{t} \cdot \mathbf{e_y} = \mathbf{e_y} \cdot \mathbf{n} = 0$. Further, any tangent obtained as a linear combination of $\mathbf{e_y}$ and $\mathbf{t}$ is also a tangent, i.e., perpendicular to $\mathbf{n}$.

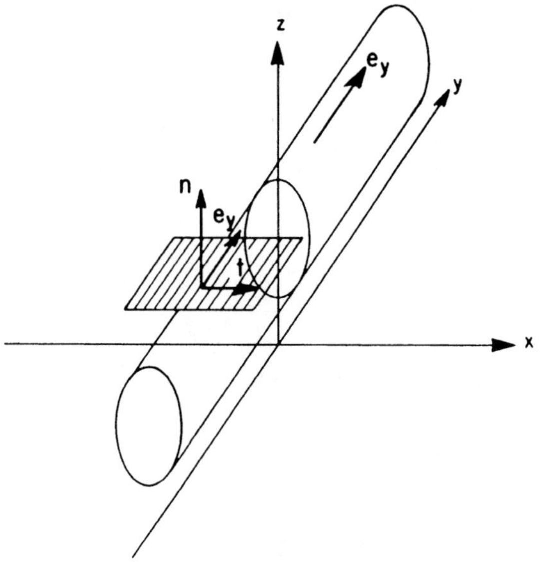

Figure VII-1. The equation of the surface is given by Eq. [VII-6]. There are no variations in the y-direction. The two tangents **t** and $\mathbf{e}_y$ are shown on the tangent plane (shaded). The unit normal is **n**.

However the choice of $\mathbf{e}_y$ and **t** as tangents is very convenient, since there is no variation in the y-direction and $\mathbf{e}_y$ will not enter into calculations. As a matter of fact, on any surface it is possible to draw two sets of curves which are everywhere perpendicular to one another. It can be seen in Figure VII-1 that lines parallel to y-axis on the surface (direction $\mathbf{e}_y$) and curves formed by the intersection of the x-z planes with the surface (direction **t**), form such sets. Two such directions at a point on the surface are the principal directions at that point, and the two curves are said to form the lines of curvature. The curvatures c_1 and c_2 of the lines of curvature at that point are the maximum and minimum values of c there and are called the principal curvatures of the surface (5). Their inverses, that is, $r_1 = c_1^{-1}$ and $r_2 = c_2^{-1}$ are the two principal radii of curvature, and $H = \frac{1}{2}(c_1+c_2)$ is the mean curvature. The quantities r_1, r_2 and H have been introduced in Chapter I, and the Gaussian curvature $K = c_1c_2$ in Chapter V. The importance of H and K lies in the fact that they are invariants, that is, they do not depend on the coordinate system.

It is not only necessary to know about the variation in the shape of a surface but also to know the variations in the properties on the surface.

If ϕ (x,y,z) is a property of a system in the bulk (say the temperature), then its derivative is given as $\nabla\phi$. If ϕ is evaluated on the surface as $\phi_s = \phi[x,y,f(x,y)]$, then its variation is given through the gradient operator ∇_s as $\nabla_s \phi_s$.

The operator $\boldsymbol{\nabla}_s$ is the component of the operator $\boldsymbol{\nabla}$ parallel to the surface. Like the unit tensor $\mathbf{I} = \mathbf{e}_x\mathbf{e}_x + \mathbf{e}_y\mathbf{e}_y + \mathbf{e}_z\mathbf{e}_z$ in three dimensions, one may write an unit tensor $\mathbf{I}_s$ for a surface, where $\mathbf{I}_s = \mathbf{t}_1\mathbf{t}_1 + \mathbf{t}_2\mathbf{t}_2$ or $\mathbf{I}_s = \mathbf{I} - \mathbf{nn}$. The component $\boldsymbol{\nabla}_s$ is obtained from $\boldsymbol{\nabla}$ with the projection $\boldsymbol{\nabla}_s = \mathbf{I}_s \cdot \boldsymbol{\nabla}$. Similarly, one obtains the component of the three dimensional velocity $\mathbf{v}$ on the surface by projection. The result is $\mathbf{v}_t = [\mathbf{I}_s \cdot \mathbf{v}]_s$. The brackets signify that the quantity needs to be evaluated on the surface. Two important results for curvatures can be obtained with this operator (5)

$$\boldsymbol{\nabla}_s \cdot \mathbf{n} = -2H \qquad [VII-9]$$

and

$$2K = \mathbf{n} \cdot \nabla_s^2\mathbf{n} + (\boldsymbol{\nabla}_s \cdot \mathbf{n})^2 \qquad [VII-10]$$

where $\nabla_s^2 = \boldsymbol{\nabla}_s \cdot \boldsymbol{\nabla}_s$ is the surface Laplacian operator analogous to that in three dimensions $\nabla^2 = \boldsymbol{\nabla} \cdot \boldsymbol{\nabla}$

If the equation of a surface is known, $\mathbf{n}$ and hence H can be calculated from Eqs. [VII-5] and [VII-9]. In practice curvatures of very few surfaces can be determined explicity because of the complexity. The curvature of even the simple surface given by Eq. [VII-6] turns out to be complex, and is given by

$$2H = \frac{\dfrac{d^2h}{dx^2}}{[1 + (\frac{dh}{dx})^2]^{3/2}} \qquad [VII-11]$$

Some relations for surface operations on simple surfaces are given in Table VII-I.

3. BASIC EQUATIONS OF FLUID MECHANICS

In terms of the formalism presented in the previous section, the boundary conditions derived in Chapter V, Section 2 are rewritten for the particular situation of interest here, which is the case of two immiscible fluid media, labeled A and B, separated by an interface. The shape of the

TABLE VII-I. Expressions Involving the Surface Gradient Operator in Simple Coordinate Systems

Planar surface: $z = A$

$$\nabla_s g = \frac{\partial g}{\partial x}\mathbf{e_x} + \frac{\partial g}{\partial y}\mathbf{e_y}$$

$$\nabla_s \bullet \mathbf{p_s} = \frac{\partial p_x}{\partial x} + \frac{\partial p_y}{\partial y}$$

$$\nabla_s^2 g = \frac{\partial^2 g}{\partial x^2} + \frac{\partial^2 g}{\partial y^2}$$

Cylindrical surface: $r = A$

$$\nabla_s g = \frac{1}{A}\frac{\partial g}{\partial \theta}\mathbf{e}_\theta + \frac{\partial g}{\partial z}\mathbf{e}_z$$

$$\nabla_s \bullet \mathbf{p_s} = \frac{1}{A}\frac{\partial p_\theta}{\partial \theta} + \frac{\partial p_z}{\partial z}$$

$$\nabla_s^2 g = \frac{1}{A^2}\frac{\partial^2 g}{\partial \theta^2} + \frac{\partial^2 g}{\partial z^2}$$

Spherical surface: $r = R$

$$\nabla_s g = \frac{1}{R}\frac{\partial g}{\partial \theta}\mathbf{e_\theta} + \frac{1}{R\sin\theta}\frac{\partial g}{\partial \theta}\mathbf{e_\phi}$$

$$\nabla_s \bullet \mathbf{p_s} = \frac{1}{R\sin\theta}\frac{\partial}{\partial \theta}(p_\theta \sin\theta) + \frac{1}{R\sin\theta}\frac{\partial}{\partial \phi}p_\phi$$

$$\nabla_s^2 g = \frac{1}{R^2\sin\theta}\frac{\partial}{\partial \theta}\left(\sin\theta\frac{\partial g}{\partial \theta}\right) + \frac{1}{R^2\sin^2\theta}\frac{\partial^2 g}{\partial \phi^2}$$

interface is given vectorially as $\mathbf{a}(\mathbf{r},t)$ and the rate at which it moves is given by $\dot{\mathbf{a}} = (d\mathbf{a}/dt)$. For simplicity, the fluids are chosen to be incompressible and Newtonian, whence the equations of motion and continuity become

$$\rho_i\left[\frac{\partial \mathbf{v}_i}{\partial t} + \mathbf{v}_i \cdot \nabla \mathbf{v}_i\right] = \mu_i \nabla^2 \mathbf{v}_i - \nabla p_i + \rho_i \hat{\mathbf{F}}_i \qquad \text{[VII-12]}$$

and

$$\nabla \cdot \mathbf{v}_i = 0 \qquad \text{[VII-13]}$$

where i = A in phase A and B in phase B. $\hat{\mathbf{F}}_i$ is the body force. ρ_i and μ_i are the densities and viscosities. $\mathbf{v}_i$ is the velocity and p_i the pressure.

Eqs. [VII-12] and [VII-13] are subject to appropriate boundary conditions away from the interface, while at the interface we have, first of all,

$$\mathbf{n} \cdot \rho_A(\mathbf{v}_A - \dot{\mathbf{a}}) = \mathbf{n} \cdot \rho_B(\mathbf{v}_B - \dot{\mathbf{a}}) \qquad \text{[VII-14]}$$

This equation is obtained on neglecting terms in Γ in the interfacial mass balance Eq. [V-32]. Here $\mathbf{n}$ is the unit normal to the interface and $\dot{\mathbf{a}}$ is the velocity of the interface. The two terms are equal to the flux (mass transfer) through the interface, in absence of which Eq. [VII-14] reduces to

$$\mathbf{n} \cdot \mathbf{v}_A = \mathbf{n} \cdot \dot{\mathbf{a}}\,, \qquad \text{[VII-15]}$$

$$\mathbf{n} \cdot \mathbf{v}_B = \mathbf{n} \cdot \dot{\mathbf{a}}\,. \qquad \text{[VII-16]}$$

The tangential velocities are continuous at the interface, the generalization of Eq. [V-40]

$$\mathbf{t} \cdot \mathbf{v}_A = \mathbf{t} \cdot \mathbf{v}_B = \mathbf{t} \cdot \dot{\mathbf{a}} \qquad \text{[VII-17]}$$

The interfacial momentum balance Eq. [V-35] will now be written for the case of negligible surface excess mass and momentum and two Newtonian fluids. The stress tensor in a Newtonian fluid is written as $p\mathbf{I} - \mu[\nabla \mathbf{v} + (\nabla \mathbf{v})^T]$, where $\mathbf{I}$ is the identity tensor and T represents the transpose of a tensor. When inertial forces are considered as well, the total force on an interface exerted by the ith phase is $\mathbf{n} \cdot \{\rho_i(\mathbf{v_i} - \dot{\mathbf{a}})\mathbf{v_i} + p_i\, \mathbf{I} - \mu_i[\nabla\mathbf{v_i} + (\nabla\mathbf{v_i})^T]\}$. The dot product with $\mathbf{n}$ denotes that the forces act on a surface characterized by $\mathbf{n}$. It is in making a force balance on the interface that the effects of interfacial tension make themselves felt. The balance is known to be (see Chapter V)

$$\mathbf{n} \cdot \{\rho_A(\mathbf{v_A} - \dot{\mathbf{a}})\mathbf{v_A} + p_A\, \mathbf{I} - \mu_A[\nabla\mathbf{v_A} + (\nabla\mathbf{v_A})^T]\}$$

$$= \mathbf{n} \cdot \{\rho_B\,(\mathbf{v_B} - \dot{\mathbf{a}})\mathbf{v_B} + p_B\, \mathbf{I} - \mu_B[\nabla\mathbf{v_B} + (\nabla\mathbf{v_B}\,)^T]\}$$

$$- \nabla_{\mathbf{s}}\, \gamma - \mathbf{n}2H\gamma \;. \qquad \text{[VII-18]}$$

The last two terms on the right hand side in Eq. [VII-18] show that the forces are discontinuous across the interface due to the interfacial tension, which resists deformation of the interface. One may resolve Eq. [VII-18] into a tangential component by taking the dot product with $\mathbf{t}$. The result, on using Eqs. [VII-14] and [VII-17] and the relations $\mathbf{n} \cdot \mathbf{I} = \mathbf{n}$, $\mathbf{n} \cdot \mathbf{t} = 0$, is

$$- \mu_A\, \mathbf{n} \cdot [\nabla\mathbf{v_A} + (\nabla\mathbf{v_A})^T] \cdot \mathbf{t} \;=\; - \mu_B\, \mathbf{n} \cdot [\nabla\mathbf{v_B} + (\nabla\mathbf{v_B})^T] \cdot \mathbf{t}$$

$$- \nabla_{\mathbf{s}}\, \gamma \cdot \mathbf{t}, \text{ or}$$

$$\tau_A \;=\; \tau_B - \nabla_{\mathbf{s}}\gamma \cdot \mathbf{t} \qquad \text{[VII-19]}$$

where τ demptes the viscous *shear* part of the stress $-\mu[\nabla\mathbf{v} + (\nabla\mathbf{v})^T]$ at the interface, and $\mathbf{t}$ is a unit tangent on the tangent plane. In the absence of mass transfer, the balance in the normal direction yields

$$p_A - \mu_A\, \mathbf{n} \cdot [\nabla\mathbf{v_A} + (\nabla\mathbf{v_A})^T] \cdot \mathbf{n} = p_B - \mu_B\mathbf{n} \cdot [\nabla\mathbf{v_B} + (\nabla\mathbf{v_B})^T] \cdot \mathbf{n}$$

$$- 2H\gamma, \text{ or}$$

$$N_A \;=\; N_B - 2H\gamma \qquad \text{[VII-20]}$$

where N deonotes the normal stress at the interface which includes the pressure.

As mentioned previously, the fluid mechanical problems must be solved in order to determine the shapes of the interfaces. However, the solution to the problems demands that the boundary conditions for the fluid flow problem be satisfied *first*, with the equations of the boundaries still unknown. Combined with this coupling of the interfacial shape and the fluid mechanical problem is yet another difficulty, which is that the shapes of the interfaces are not "regular" like those of spheres, cylinders or planes. A cursory look at introductory or advanced texts in fluid mechanics (6,7) shows that symmetry in the geometry is critical to solving a problem in a straightforward (although at times lengthy) manner. For example, flow in the annulus between infinitely long co-axial cylinders may be considered simple because of the symmetry in the problem. However, flow around a sphere settling inside a cylinder, even when it is falling along the axis of the cylinder, is a difficult problem to solve.

Consequently analytical methods are mostly confined to creeping flows. Roughly, there are two types of problems which can be solved. The first of these deals with interfaces which show small deviations from simple geometric forms, as for instance the case of a "slightly deformed sphere" settling in an infinite fluid. The second type constitutes cases where interfacial position changes, but only very slowly. Then, its variation can be neglected to the first approximation and the lubrication theory approximation or the slender body approximation applied. It is noteworthy that both the above methods yield approximate solutions.

The cases of a sphere and slightly deformed sphere in a uniform flow field are considered first in Sections 4 and 5. The mathematical method used conventionally in these problems is the regular asymptotic expansion. The reader is introduced to this method. In Section 6 the dip coating problem under the lubrication theory approximation is examined. (The closely related slender body approximation is outlined in Problem VII-4). A more sophisticated method of matched asymptotic expansions is used to solve this problem and its main features are explained there. More realism is brought to the study of interfacial dynamics by incorporating the non-ideal nature of surfaces. This is done through a physico-chemical approach in Section 7 and through a surface rheological approach in Section 8. Finally, dynamic contact angles are analyzed in Section 9 through a problem which needs a complete analysis in matched asymptotic expansions. Too few problems have been solved numerically to attempt a general discussion here on that subject. Much less is known about solutions in cases where the interfaces are affected by inertia or turbulence.

4. FLOW PAST A DROPLET

The fluid mechanical problem of solving for the flow around a droplet is a standard one and has formed the starting point of many of the methods and theories on how interfacial effects influence fluid flow. A droplet of material A is immersed in an infinite medium B. The latter flows upward past the droplet and exerts a drag force on the droplet. This drag force is balanced by gravity, so that the droplet stays in place and steady state is reached. Far from the droplet, the velocity in phase B assumes a constant (upward) value U.

The problem is solved at very low Reynolds numbers, when the left hand side in the momentum equation, Eq. [VII-12], may be neglected. As shown in Figure VII-2, a z-axis passing through the center of the drop is chosen in the same direction as the velocity U. Consequently, far away from the drop $\mathbf{v}_B \to U\mathbf{e}_z$. Although this condition suggests the use of a cylindrical coordinate system (z,r,θ), a spherical coordinate system (r,θ,ϕ) with the origin at the center of the drop is used instead, in order to make it easier to satisfy the boundary conditions on the drop surface. The drop is assumed to have a spherical shape of radius a. From the symmetry of the figure, $v_\phi = 0$ (no swirl) and v_θ, v_r are functions of r and θ only. Together with the previous conditions, these simplifications allow one to write the r and θ components of the momentum equation, Eq. [VII-12] and the continuity equation, Eq. [VII-13]. On eliminating the pressure by cross differentiation of the components of the momentum equation and subtracting much as in Section 2a of Chapter V, and on defining a stream function ψ to assure that continuity is satisfied, one obtains the biharmonic stream function equation [VII-21]. The velocity components are given in terms of the stream function in Eqs. [VII-22] and [VII-23].

$$E^4\psi_i = 0 \qquad \text{[VII-21]}$$

$$v_{ri} = -\frac{1}{r^2\sin\theta}\frac{\partial\psi_i}{\partial\theta} \qquad \text{[VII-22]}$$

$$v_{\theta i} = \frac{1}{r\sin\theta}\frac{\partial\psi_i}{\partial r} \qquad \text{[VII-23]}$$

In general biharmonic equations specify the fluid flow at low Reynolds numbers when the velocity component in the third direction is zero and all the fluid mechanical quantities are independent of the coordinate in that

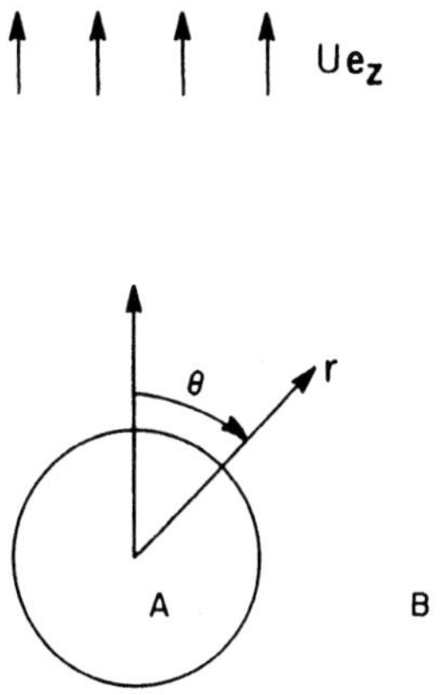

Figure VII-2. The spherical coordinate system is employed with the origin at the center of the drop of phase A. Far away from the drop the velocity field is $U\mathbf{e}_z$ in phase B.

direction. Bird, Stewart and Lightfoot (8) discuss the biharmonic operator E^4 and show how Eq. [VII-21] can be solved for flow around a solid sphere and how quantities like the drag force on the sphere can be evaluated.

For the case of a fluid sphere Eqs. [VII-21]-[VII-23] apply in both the inner and outer fluids and the boundary conditions are different. The solution has been provided by Levich (9). We first list the boundary conditions

(1) $\psi_B \to -\frac{1}{2}Ur^2\sin^2\theta$ as $r \to \infty$. This condition is that of $\mathbf{v}_B \to U\mathbf{e}_z$ as $|\mathbf{r}| \to \infty$ mentioned earlier.

(2) $v_{\theta A} = v_{\theta B}$ on $r = a$, from Eq. [VII-17],

(3) $v_{rA} = v_{rB} = 0$ on $r = a$, from Eqs. [VII-15,16], and

(4) $\tau_{r\theta A} = \tau_{r\theta B}$ on $r = a$, from Eq. [VII-19], where it has been assumed that γ is a constant on the surface.

An additional condition that all fluid mechanical quantities in A are finite at $r = 0$ is imposed as well. The boundary conditions are rewritten in terms of the stream functions using Eqs. [VII-22] and [VII-23]. The expressions for τ in terms of the velocities (see 8) are also invoked. When the results are substituted into the general solutions of Eq. [VII-21] for fluids A and B, one obtains

$$\psi_A = \frac{Ua^2\sin^2\theta}{4} \frac{\mu_B}{\mu_A+\mu_B} [(\frac{r}{a})^2 - (\frac{r}{a})^4] \qquad [VII-24]$$

$$\psi_B = - \frac{Ua^2\sin^2\theta}{4} [2(\frac{r}{a})^2 - \frac{3\mu_A+2\mu_B}{\mu_A+\mu_B} (\frac{r}{a}) + \frac{\mu_B}{\mu_A+\mu_B} - (\frac{a}{r})] \qquad [VII-25]$$

The drag force **F** is the quantity of interest and is equal to the net gravitational force

$$\mathbf{F} = 2\pi Ua\, \mu_B [\frac{3\mu_A+2\mu_B}{\mu_A+\mu_B}]\mathbf{e}_z = \frac{4\pi a^3}{3} (\rho_A - \rho_B) g\, \mathbf{e}_z \qquad [VII-26]$$

This equation can be solved for the velocity U.

The analysis usually stops here without the normal stress balance given by Eq. [VII-26]. Here $N_i = p_i - 2\mu_i\, (\partial v_{ri} / \partial r)$ and the balance yields the relation

$$p_A^o - p_B^o = \frac{2\gamma}{a} \qquad [VII-27]$$

where p_A^o and p_B^o are the two constants of integration that appear on evaluating the pressures. The other terms in the normal stress balance cancel out on using Eq. [VII-26]. The value of $H = -a^{-1}$ has been used in Eq. [VII-27]. Thus Eq. [VII-27] can be used to *evaluate* one of the pressures if the other is known or held to be the datum pressure. It is to be noted that if $U \to 0$, $\mathbf{v_A}$ and $\mathbf{v_B} \to 0$ and p_A and p_B reduce to the constants that appear in Eq. [VII-27]. One may say that p_A^o and p_B^o are the static pressures.

5. ASYMPTOTIC ANALYSIS

It is fortunate that in the previous problem the normal stress balance is satisfied by the spherical drop profile $r = a$. However no satisfactory proof that this is the only solution has been given. Indeed, one may anticipate some deviation in shape when inertial effects are considered. If one assumes that this deviation is small so that the profile is given by $r = a(1 + \xi)$ with $|\xi| \ll 1$, then from the symmetry ξ is a function of θ alone. If the drop is still to have the same volume, then

$$\frac{4}{3}\pi a^3 - \int_0^{\pi} \{ \int_0^{2\pi} [\int_0^{a(1+\xi)} r^2 dr] d\phi\} \sin\theta d\theta = 0$$

This expression reduces to

$$1 - \frac{1}{2} \int_{-1}^{1} (1+\xi)^3 \, dq = 0$$

where $q = \cos\theta$, and further to

$$\int_{-1}^{1} \xi dq = 0 \quad . \tag{VII-28}$$

Similarly, if the drop is still to have its center of mass at the origin, then

$$\int_0^{\pi} \{ \int_0^{2\pi} [\int_0^{a(1+\xi)} r\cos\theta \, r^2 \, dr] d\phi \} \sin\theta \, d\theta = 0$$

which reduces to

$$\int_{-1}^{1} \xi q dq = 0 \tag{VII-29}$$

Finally for $|\xi| << 1$, one has (9,10)

$$2H \simeq - \frac{1}{a} [2 - 2\xi - \frac{d}{dq} \{(1-q^2) \frac{d\xi}{dq} \}] \tag{VII-30}$$

Taylor and Acrivos (10) found an approximate expression for ξ applicable for small values of the Reynolds number N_{Re} (= $Ua\rho_B / \mu_B$) and capillary number N_{Ca} (= $U\mu_B / \gamma$). They first obtained the creeping flow solution $\psi_i^{(0)}$ which satisfied all boundary conditions, those at the drop-fluid interface being satisfied at $r = a$. The normal stress balance, Eq. [VII-20], which was not used in this initial procedure, was then applied with N_i being evaluated at $r = a$ to obtain a first approximation $\xi^{(0)}$ satisfying the conditions given by Eqs. [VII-28] and [VII-29]. Eq. [VII-30] was used for the curvature. Since the normal stress condition is satisfied exactly by the creeping flow solution given by Eqs. [VII-24] and [VII-25] (see Eq. [VII-27]), it was found that $\xi^{(0)} = 0$.

Hence, in order to determine the deviation from spherical shape, it was necessary to include higher order terms proportional to U^2 in the stream function. Two expressions having this form are possible: $\psi_i \sim \psi_i^{(0)} + N_{Re} \, \psi_i^{(1)}$ and $\psi_i \sim \psi_i^{(0)} + N_{Ca} \, \psi_i^{(1)}$. Taylor and Acrivos (10) pointed out that only for the former of these is the correction $\psi_i^{(1)}$ itself affected by the inertial terms (11), i.e., those on the left hand side of the momentum equation Eq. [VII-12]. With this form for ψ_i and with the boundary at $r = a\,(1+\xi^{(0)}) = a$, they obtained from the normal stress balance

$$\xi^{(1)} = -N_{Re}N_{Ca}\lambda P_2(q)$$

where P_2 is the Legendre polynomial of order two, and

$$\lambda = \frac{1}{4(\kappa+1)^3}\left\{\left(\frac{81}{80}\kappa^3 + \frac{57}{20}\kappa^2 + \frac{103}{40}\kappa + \frac{3}{4}\right) - \frac{(\sigma-1)}{12}(\kappa+1)\right\} \qquad \text{[VII-31]}$$

Also $\kappa = (\mu_A/\mu_B)$ and $\sigma = (\rho_A/\rho_B)$.

To obtain the next term, they took

$$\psi_i \sim \psi_i^{(0)} + N_{Re}\,\psi_i^{(1)} + N_{Re}N_{Ca}\psi_i^{(2)}$$

and imposed the boundary conditions (with the exception of the normal stress balance) at the new boundary $r = a\,(1-\lambda N_{Re}N_{Ca}P_2)$. A Taylor series was used to express any function $f(r)$ at the interface as $f(a) + (\partial f/\partial r)_a\,(-\lambda a N_{Re}N_{Ca}P_2)$ plus small terms which can be neglected. Hence at this level of approximation ψ_i must have terms of the order of $N_{Re}N_{Ca}$ completely represented when the boundary conditions are satisfied on the surface of the slightly deformed sphere. From the normal stress balance one obtains the second iteration for ξ:

$$\xi^{(2)} = -N_{Re}N_{Ca}\,\lambda P_2 + \frac{3\lambda(11\kappa+10)}{70\,(\kappa+1)}\,N_{Re}N_{Ca}^2\,P_3 \qquad \text{[VII-32]}$$

where $P_3(q)$ is the Legendre polynomial of order three. Taylor and Acrivos stopped here due to complexity of equations involved. Obtaining higher order approximations to ψ_i is a very difficult problem in fluid mechanics.

From Eq. [VII-32] one finds that a drop is spherical for creeping flow but becomes spheroidal (pill shaped) at somewhat higher Reynolds numbers (P_2 term) and a spherical cap at still higher velocities (P_3 term). If only $\xi^{(1)}$ were considered, then as $\kappa \to \infty$, $\xi^{(1)} = -0.25\ N_{Re}N_{Ca}P_2$, while for $\kappa = 0$ and $\sigma = 0$ (gas bubble) $\xi^{(1)} = -0.21\ N_{Re}N_{Ca}\,P_2$. That is, the shape is relatively independent of the physical properties. It is also noteworthy that in related problems the series solution obtained is rarely as complex as in this case. The $\xi^{(0)}$ term which is obtained from $\psi_i^{(0)}$ usually suffices. In the case of drops $\xi^{(0)}$ turns out to zero and higher order approximations are needed. See Problem VII-1 for further details.

In this low Reynolds number, or small U, problem the successive correction terms are proportional to higher powers of U. Hence one may be tempted to assume that as the last term $\to 0$, the solution is a series solution. However, the construction of the solution is *contingent* on the small magnitude of U (or N_{Ca}, N_{Re} and $N_{Re}N_{Ca}$). This gives rise to only an asymptotic expansion of the solution. The expansion has the following features:

(1) The solution is *approximate* and its construction is based on a small parameter which exists in the problem.

(2) The practical aspect of a series solution is its speed of convergence, because, after all, in order to get numbers from the solution the series has to be truncated. It is only the asymptotic series which addresses itself to estimating the magnitude of the remainder directly. The magnitude of the remainder in an asymptotic expansion is the magnitude of the first term dropped. This is not true in an exact infinite series in general. From Eq. [VII-32], the remainder here is of the order of U^4 (or $N_{Re}{}^2N_{Ca}{}^2$ or $N_{Re}N_{Ca}{}^3$). However, there is little to suggest that by calculating more terms and decreasing the remainder in this fashion the series obtained is convergent. In fact, divergent series are sometimes obtained.

(3) In conclusion, progressively adding terms to an asymptotic series is of little value and even dangerous if the series is not convergent. Besides being prudent, it is at least very desirable from the view of mathematical simplicity to stop at one or two terms. Indeed a very satisfactory approximation is obtained when the assumed smallness of U (or N_{Ca}, N_{Re} and $N_{Re}N_{Ca}$) is justified.

Based on the Taylor-Acrivos solution just described one can suggest a general solution scheme. When an interface deviates slightly from a regular shape where one of the coordinates ζ_k has a constant value A, the solution proceeds as follows:

(a) Solve for $\psi = \psi^{(0)}$ as if the interface were indeed at $\zeta_k = A$, using all boundary conditions except the normal stress balance.

(b) Write the curvature as $2H \sim 2(H^{(0)} + H^{(1)})$, where $H^{(0)}$ corresponds to the shape given by $\zeta_k = A$ and $H^{(1)}$ is the correction when the interface deviates slightly from the regular shape. Eq. [VII-20] is now non-dimensionalized by dividing by $(\mu_B U/A)$ to give

$$\overline{N}_A^{(0)} - \overline{N}_B^{(0)} = -\frac{2}{N_{Ca}}[\overline{H}^{(0)} + \overline{H}^{(1)}] \qquad \text{[VII-33]}$$

where the overbars denote dimensionless quantities. Rearranging, one has

$$-\overline{H}^{(1)} = \frac{N_{Ca}}{2}[\overline{N}_A^{(0)} - \overline{N}_B^{(0)} + \frac{2}{N_{Ca}}\overline{H}^{(0)}] \qquad \text{[VII-34]}$$

The left hand side is usually in a differential form. The right hand side is N_{Ca} times a function. This happens because the constant of integration which appears in ($\overline{N}_A^{(0)} - \overline{N}_B^{(0)}$) cancels with ($2\overline{H}^{(0)}/N_{Ca}$) which usually has a constant value. If this cancellation does not occur, the asymptotic scheme will not apply since $\overline{H}^{(1)}$, the correction, will be comparable to $\overline{H}^{(0)}$. Table VII-II provides some expressions for $\overline{H}^{(0)}$ and $\overline{H}^{(1)}$.

constant. This procedure also allows the use of the assumption that, to a good approximation, the flow is fully developed with $v_x \sim 0$ and v_y a function of x alone. Then the equations of motion become

$$0 = -\frac{\partial p}{\partial y} + \mu \frac{\partial^2 v_y}{\partial x^2} - \rho g \qquad \text{[VII-35]}$$

TABLE VII-II Expressions for Curvature of Surfaces Slightly Deformed from Regular Shapes.

Surface	$\overline{H}^{(0)}$	$\overline{H}^{(1)}$
Sphere of radius $A[1+\xi(q,\phi)]$	-1	$[\xi + \frac{1}{2}\frac{\partial}{\partial q}\{(1-q^2)\frac{\partial \xi}{\partial q}\} + \frac{1}{2(1-q^2)}\frac{\partial^2 \xi}{\partial \phi^2}]$
Cylinder of radius $A[1+\xi(\theta,z)]$	$-\frac{1}{2}$	$\frac{1}{2}[\xi + A^2\frac{\partial^2 \xi}{\partial z^2} + \frac{\partial^2 \xi}{\partial \theta^2}]$
Almost planar surface at $z = A[1 + \xi(x,y)]$	0	$A^2[\frac{\partial^2 \xi}{\partial x^2} + \frac{\partial^2 \xi}{\partial y^2}]$

$$0 = -\frac{\partial p}{\partial x} . \qquad [VII-36]$$

Eq. [VII-36] says that p is a function of y alone. Consequently, Eq. [VII-35] can be rewritten as

$$\frac{\partial^2 v_y}{\partial x^2} = \frac{1}{\mu}\left(\frac{\partial p}{\partial y} + \rho g\right) \qquad [VII-37]$$

where the right hand side is not a function of x.

Eq. [VII-37] is to be solved subject to the boundary conditions

$$v_y \Big|_{x=0} = V \quad \text{(no slip)} \qquad [VII-38]$$

$$\frac{\partial v_y}{\partial x}\Big|_{x=h} = 0 \quad \text{(zero shear at the liquid-air interface).} \qquad [VII-39]$$

In this section only minimum attention has been paid to the methods for solving the stream function equations, since the basic purpose has been to illustrate the method for obtaining the correction ξ. It is noteworthy that the solutions in the form of stream functions for many regular geometric shapes are available and can be used to calculate the first correction $\xi^{(0)}$ (7). In cylindrical and spherical coordinates the general solutions have been provided by Haberman and Sayre (13) and their method can be applied to many of the separable coordinate systems (7). Where the coordinates are not separable the solution and the method used by Lee and Leal (14) are useful. The higher order corrections in ξ require the knowledge of general solutions of the biharmonic equation in those coordinate systems. Although such solutions exist in a few coordinate systems, it is worth noting that $\xi^{(0)}$ is a sufficient approximation to ξ in most cases.

6. DIP COATING

In this problem a flat plate is withdrawn with a steady velocity V from a pool of liquid. When V is sufficiently large, entrainment will occur as shown in Figure VII-3. Sufficiently far away from the meniscus the film reaches a constant thickness. The problem differs from previous ones by the important fact that it is not easy to visualize the profile as a small deviation from a simple geometric shape.

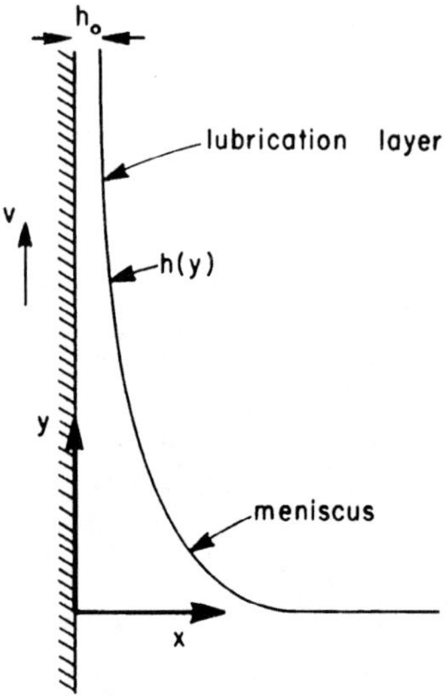

Figure VII-3. The arrangement of the dip coating experiment is shown. The plate is being withdrawn at a constant velocity V. The film profile h reaches a constant value h_o in the region far from the meniscus.

The lubrication theory approximation is used to formulate this problem. In this scheme it is assumed that the film is almost flat. Consequently if x = h(y) is the profile shape, then $|dh/dy| << 1$. Thus the boundary conditions applicable on h(y) are used, treating h to be almost a The solution is

$$v_y = V + \frac{1}{2\mu}\left(\frac{\partial p}{\partial y} + \rho g\right)(x^2 - 2xh). \qquad \text{[VII-40]}$$

The pressure p is evaluated from the normal stress balance:

$$p\Big|_{x=h} = -\gamma \frac{d^2h}{dy^2} \qquad \text{[VII-41]}$$

which, in view of Eq. [VI-36], is the pressure for any x at a given y. Note that the curvature has been obtained from Eq. [VII-11] with $|dh/dy|<<1$. Further,

$$\frac{\partial p}{\partial y} = -\gamma \frac{d^3h}{dy^3} \qquad \text{[VII-42]}$$

Substituting Eq. [VII-42] in [VII-40], one has

$$v_y = V + \frac{1}{2\mu}\left(\rho g - \gamma \frac{d^3h}{dy^3}\right)(x^2 - 2xh) \qquad \text{[VII-43]}$$

One writes now the equivalent form of the continuity equation. If Q is the volumetric flow rate in the film, then under steady state it is a constant for all values of y. Hence

$$Q = \int_0^h v_y \, dx = Vh - \frac{h^3}{3\mu}\left(\rho g - \gamma \frac{d^3h}{dy^3}\right) \qquad \text{[VII-44]}$$

This third order ordinary differential equation needs three boundary conditions for solution and a fourth one to evaluate the unknown constant Q.

Far away from the meniscus the thickness reaches a steady value, i.e., the film has a flat profile. There the curvature is zero and the second term within parentheses in Eq. [VII-44] is hence zero. If the steady value h_0 is very small, the term $Vh_0 \gg (\rho g h_0^3/3\mu)$, and consequently

$$Q = Vh_0 \qquad \text{[VII-45]}$$

Further, as noted previously, as the slope and the curvature go to zero at large y, the boundary conditions become

$$h \to h_0 \quad \text{for} \quad y \to \infty \qquad \text{[VII-46]}$$

$$\frac{dh}{dy} \to 0 \quad \text{for} \quad y \to \infty \qquad \text{[VII-47]}$$

$$\frac{d^2h}{dy^2} \to 0 \quad \text{for} \quad y \to \infty \, . \qquad \text{[VII-48]}$$

Eqs. [VII-44] and [VII-46] - [VII-48] may be non-dimensionalized to

$$\frac{d^3L}{d\lambda^3} = \frac{1-L}{L^3} \qquad \text{[VII-49]}$$

$$L \to 1 \quad \text{as} \quad \lambda \to \infty \qquad \text{[VII-50]}$$

$$\frac{dL}{d\lambda} \to 0 \quad \text{as} \quad \lambda \to \infty \qquad \text{[VII-51]}$$

$$\frac{d^2L}{d\lambda^2} \to 0 \quad \text{as} \quad \lambda \to \infty \qquad \text{[VII-52]}$$

where $L = (Vh/Q)$ and $\lambda = (3\mu V^4/\gamma Q^3)^{1/3} y$. Eq. [VII-49] can be solved numerically to yield L as function of λ. More important is the quantity

$$\left(\frac{d^2L}{d\lambda^2}\right)_{L\to\infty} = \alpha \qquad \text{[VII-53]}$$

where α is evaluated from this solution to be 0.63.

If the meniscus region is examined it is seen that $h \to \infty$ as $y \to 0$. Consequently, the average value of the velocity there is ~ (Q/h), which is very small as Q is a constant and $h \to \infty$. Thus dynamic effects can be neglected and the profile is very close to that of a static meniscus. The method of determining static meniscus profiles has been given in Chapter I. Starting from the solution of L* as a function of λ, one obtains

$$\left(\frac{d^2L^*}{d\lambda^2}\right)_{L^* \to 0} = \left(\frac{2\rho g}{\gamma}\right)^{1/2} \frac{Q}{V^{5/3}} \left(\frac{\gamma}{3\mu}\right)^{2/3} \qquad \text{[VII-54]}$$

The symbols have the same meaning as before except that the star on L* is used to denote the fact that L* represents the static meniscus. It is assumed now that the right hand side in Eq. [VII-54] is equal to α given by Eq. [VII-53] leading to

$$N_B = 0.86\, N_{Ca}^{4/3} \qquad \text{[VII-55]}$$

where the Bond number $N_B = \rho g h_0^2/\gamma$. Eq. [VII-55] is known to be valid for $N_{Ca} < 0.01$.

The principle behind equating the right hand side of Eq. [VII-53] to the right hand side of Eq. [VII-54] can be explained as follows. The lubrication theory approximation is valid away from the meniscus and consequently cannot satisfy the boundary conditions in the meniscus region. Consequently the solution contains a number of unknowns equal to the number of boundary conditions in the meniscus region. Similarly the solution in the meniscus region cannot satisfy the boundary conditions in the lubrication layer. If the solution in the lubrication layer is extrapolated into the meniscus region ($L \to \infty$), and if the solution in the meniscus region is extrapolated into the lubrication layer ($L^* \to 0$), they should agree with each other. The conditions under which they "agree" with each other leads to the evaluation of the unknowns. This process is called matching. The matched solution is the sum of the two, less the parts common to both solutions. It is noteworthy that the two solutions are the first terms of asymptotic series and the solution is referred to as the matched asymptotic solution.

This solution, which was obtained by Landau and Levich (15), matches only the second derivatives, $(d^2L/d\lambda^2)$ to $(d^2L^*/d\lambda^2)$. Since the solutions $L(\lambda)$ and $L^*(\lambda)$ are the first terms in the asymptotic solutions in these two regions, Ruschak and Scriven (16), proceeded further to obtain the small quantity on which the two asymptotic series were based. As seen in Section 5, the construction of higher order approximations is based on such a small quantity, which was N_{Re} or N_{Ca}, i.e., U, in this problem. Ruschak and Scriven, however, found that analytic solutions could not be obtained for the higher order terms. Hence Eq. [VII-65] remains the only known analytic result, although better insight into the physics of the problem is obtained as a result of the Ruschak-Scriven analysis. More details on the method of formal matched asymptotic expansions are given in Section 9.

Besides the dip coating problem, the lubrication theory approximation and the closely related slender body approximation (see Problem VII-4) have been applied successfully in a number of cases. The critical assumption is that the variation of the velocity profile in the direction of flow is ignored in the first approximation. The flow under this approximation becomes one-dimensional in most cases, or at least retains a simple form.

7. SPHERICAL DROP REVISTED

Previously the drag on a spherical droplet falling through an infinite fluid was obtained at Eq. [VII-26]. On balancing this force with gravity the terminal velocity $U^* = -U$ is obtained as

$$U^* = \frac{2a^2 g\rho_B}{3\mu_B}(\sigma-1)\frac{(\kappa+1)}{(3\kappa+2)} \qquad \text{[VII-56]}$$

where $\sigma = \rho_A/\rho_B$ and $\kappa = \mu_A/\mu_B$. When $\kappa \to \infty$, $U^* \to U^*_\infty = [2\rho_B g a^2(\sigma-1)/9\mu_B]$, which is the Stokes sedimentation velocity for a *solid* sphere. This limit is expected since the inner fluid becomes undeformable when the viscosity is very large. Similarly if $\kappa \to 0$, $U^* \to U^*_0 = [\rho_B g a^2(\sigma-1)/3\mu_B]$. The limit $\kappa \to 0$ is closely approximated by gas bubbles rising in a liquid. The important feature is that $U^*_0 > U^*_\infty$. However, from experiments the terminal velocity of a gas bubble in water was often seen to be given better by the Stokes velocity U^*_∞ than by U^*_0 as predicted by theory. It was postulated that this behavior was due to the fact that under non-equilibrium conditions the equilibrium surface tension γ is supplemented by a dynamic contribution $\hat{\gamma}$. The dynamic surface tension vanishes when equilibrium is attained. Since the velocities on the interface vary from point to point, so will $\hat{\gamma}$, leading to a surface stress given by $\nabla_s \hat{\gamma}$. The origin of this stress has been discussed earlier.

There are two different ways of accounting for $\hat{\gamma}$ or $\nabla_s \hat{\gamma}$. The first consists of postulating a physical mechanism, which in this case is the effect of the presence of minute quantities of surface-active impurities. The second is to define an effective surface viscosity, a matter considered further in Section 8.

With the first of these models $\hat{\gamma}$ is the change in γ due to adsorption of the impurities. The transport of these materials in bulk is governed at steady state by the conservation equation

$$\mathbf{v} \cdot \nabla c = \nabla \cdot (D \nabla c) \qquad \text{[VII-57]}$$

where c is the concentration in the bulk phase and D is the diffusion coefficient. The mass transfer to the droplet surface is given by $j^* = D\,(\partial c/\partial r)_a$. This term j^* can also be expressed in terms of the adsorption-desorption process as follows:

$$j^* = Q\,(\Gamma,\, c(a)) - P(\Gamma) \qquad \text{[VII-58]}$$

where Q is the rate of adsorption, dependent on the surface concentration Γ and the concentration $c(a)$ at the interface, and P is the rate of desorption.

Finally one has the equation of conservation of the adsorbed species, which at steady state simplifies from Eq. [VI-2] to

$$\nabla_s \cdot (\dot{\mathbf{a}}_t\, \Gamma) = \nabla_s \cdot (D_s \nabla_s \Gamma) + j^* \qquad \text{[VII-59]}$$

where j^* is the "rate of production" of the species on the interface due to the transport from the bulk to the interface. D_s is the surface diffusion coefficient and $\dot{\mathbf{a}}_t$ is the tangential velocity. At a clean interface it can be obtained from Eqs. [VII-23] and [VII-24] as $\dot{\mathbf{a}}_t = v_\theta^* \mathbf{e}_\theta$, where

$$v_\theta^* = -\frac{U \sin\theta}{2\,(1+\kappa)} \, . \qquad \text{[VII-60]}$$

Of course, this expression is modified when surfactant is present although v_θ^* remains proportional to sin θ.

The general procedure for solution with a contaminated interface is as follows. Eq. [VII-21] is solved to obtain the general form of the stream function in liquid B (see Problem VII-5). The resulting velocity distribution is then used to solve Eq. [VII-57] for the concentration distribution. Unknown constants apprearing in these general solutions are found using suitable boundary conditions. Among these are conditions (1) - (4) given following Eq. [VII-23] with the tangential stress condition modified to include the surface tension gradient. Also the interfacial mass balance condition for the surfactant is needed along with the condition that surfactant concentration is some known value c_0 far from the drop. This method of solution based on a physico-chemical mechanism is due to Frumkin and Levich (17) and has been discussed in detail by Levich (9). One simplified version of the solution is given below.

In this special case, it is considered that the adsorption-desorption is the rate limiting step: Consequently $c(a) = c_0$, the bulk concentration. It is further assumed that Γ deviates only a little from the equilibrium value Γ_0 given by

$$Q\ (\Gamma_0, c_0) - P(\Gamma_0) = 0 \qquad \text{[VII-61]}$$

to $\Gamma = \Gamma_0 + \Gamma'$. Eq. [VII-58] can be linearized to give

$$j^* = -\alpha\Gamma' \qquad \text{[VII-62]}$$

where

$$\alpha = \left(\frac{\partial P}{\partial \Gamma} - \frac{\partial Q}{\partial \Gamma}\right)_{\Gamma = \Gamma_0\ ,\ c = c_0} \qquad \text{[VII-63]}$$

Substituting Eqs. [VII-62] and [VII-60] into Eq. [VII-59] and approximating Γ on the left hand side of Eq. [VII-59] with Γ_0, one has, on using the relations from Table VII-I,

$$-\frac{\Gamma_0}{a\ \sin\theta}\frac{d}{d\theta}\left[bU\ \sin\theta\ \ \sin\theta\right] = \frac{D_s}{a^2\sin\theta}\frac{d}{d\theta}\left(\sin\theta\ \frac{d\Gamma'}{d\theta}\right) - \alpha\Gamma' \qquad \text{[VII-64]}$$

where b is an unknown constant and D_s has been assumed constant. If D_s is negligible,

$$\Gamma' = \frac{2\Gamma_0 Ub}{\alpha\, a} \cos\theta \qquad \text{[VII-65]}$$

In this case $\nabla_s \tilde{\gamma}$ becomes

$$\nabla_s \tilde{\gamma} = \left(\frac{\partial \gamma}{\partial \Gamma}\right)_{\Gamma_0} \frac{1}{a} \frac{d}{d\theta} \left[\frac{2\Gamma_0 Ub \cos\theta}{\alpha\, a}\right] \mathbf{e}_\theta , \qquad \text{[VII-66]}$$

where the term $(\partial\gamma/\partial\Gamma)_{\Gamma_0}$ can be determined explicitly if the adsorption isotherm is known. Eq. [VII-66] is now substituted into Eq. [VII-19] to obtain the boundary condition on the shear stresses. When this equation and the other fluid mechanical conditions are used to solve for U and b, the result is

$$U = 3\, U_\infty \frac{1 + \kappa + e}{2 + 3\kappa + 3e} \qquad \text{[VII-67]}$$

where U_∞ is the Stokes velocity and

$$e = \frac{2\Gamma_0}{3\mu_B \alpha\, a} \left(\frac{\partial \gamma}{\partial \Gamma}\right)_{\Gamma_0} \qquad \text{[VII-68]}$$

A simple interpretation of Eq. [VII-67] is obtained on noting that $U \to U_\infty$ as $e \to \infty$. This limit is attained when the contaminant is surface active and surface elasticity $[\Gamma_0(\partial\gamma/\partial\Gamma)]$ is very large. More specific information on the action of surfactants is revealed on noting that the velocity at the interface v^*_θ is

$$v^*_\theta = -\frac{3U_\infty}{2} \frac{\sin\theta}{2 + 3\kappa + 3e} \qquad \text{[VII-69]}$$

which obviously goes to zero as $e \to \infty$. The physical picture that emerges is that adsorbed surfactant is swept upward due to tangential flow along the interface. Desorption occurs near the top of the drop. Thus, the surfacant concentration is high at the top and low at the bottom of the drop. This sets up a force proportional to $\nabla_s \gamma = (\partial\gamma/\partial\Gamma)\, \nabla_s \Gamma$. It is directed from the top, the region of higher surface concentration and lower surface tension, towards the bottom. Obviously the direction of this force being against the direction of v^*_θ, it tends to decrease this velocity. When v^*_θ, is zero, i.e., $e \to \infty$, the surface is immobile and $U \to U_\infty$. This situation is analogous to the case of an inextensible surface considered in the damping of capillary waves in Chapter V, Section 3.

A general conclusion is also reached that the presence of surfactants makes the interface more rigid and provides resistance to lateral interfacial deformation, the measure of the latter being the interfacial velocities. It is noteworthy that the normal stress balance is satisfied with this solution, the same result as found in Section 4 in the absence of surfactants.

The effect of the variation of surface tension, $\nabla_s \gamma$ has been investigated here. On noting that the equations of motion and the boundary conditions, particularly Eq. [VII-19] where the term $\nabla_s \gamma$ appears, are all linear, it is possible to say that a simple proportionality exists between $\nabla_s \gamma$ and the quantity e that appears in Eq. [VII-67]. Here $\nabla_s \gamma = a^{-1}(\partial\gamma/\partial\theta)\, \mathbf{e}_\theta$ and e incorporate the nature of variation of γ.

8. SURFACE RHEOLOGY

Although the model of the preceding section reveals the basic mechanism by which surfactants retard drop motion, it appears that it has little predictive abilities. To predict U for instance, it is necessary to know a host of physical properties for components which are essentially contaminants. Nevertheless there is a great need to know the "stiffness" of fluid-liquid interfaces in real systems which are almost always contaminated. Unless enough information is available to formulate suitable boundary conditions, the correct fluid mechanical problem cannot be solved, nor can useful quantities like the drag force, pressure drop, etc., be extracted. Frequent lack of availability of the requisite data has led investigators to a phenomenological form of describing the "stiffness" of an interface.

If an interface is stiff, then it has a resistance to deformation. In three dimensions the compressibility is linked to $\nabla \cdot \mathbf{v}$, which is zero for incompressible fluids. Further, the rate of strain $\frac{1}{2}[\nabla\mathbf{v} + (\nabla\mathbf{v})^T]$ also describes the deformation. The resulting stress for a Newtonian fluid is $[p + (\frac{2}{3}\mu - \kappa)\, \nabla \cdot \mathbf{v}]\, I - \mu\, [\nabla\mathbf{v} + (\nabla\mathbf{v})^T]$, where κ is referred to as the bulk viscosity. For incompressible fluids, the stress reduces to $p\mathbf{I} - \mu[(\nabla\mathbf{v}) + (\nabla\mathbf{v})^T]$. The surface stress, by analogy, would be $[-\gamma + (\frac{2}{3}\hat{\varepsilon} - \hat{\kappa})\nabla_s \cdot \acute{\mathbf{a}}\,]\mathbf{I}_s - \hat{\varepsilon}[\nabla_s \mathring{\mathbf{a}} + (\nabla_s \mathring{\mathbf{a}})^T]$. The negative sign in front of γ denotes tension as opposed to p which is usually compressive. Here $\mathbf{I}_s$ is the surface identity tensor defined previously, while $\hat{\varepsilon}$ and $\hat{\kappa}$ are two appropriate constants.

Conventionally, the surface viscous stress $\mathbf{T_s}$ is written in a slightly different form (4,18)

$$\mathbf{T_s} = -(\eta-\varepsilon)(\nabla_s \cdot \dot{\mathbf{a}})\, \mathbf{I_s} - \varepsilon\, [\mathbf{I_s} \cdot \nabla_s \dot{\mathbf{a}} + (\nabla_s \dot{\mathbf{a}})^T \cdot \mathbf{I_s}] \qquad \text{[VII-70]}$$

where η is the surface dilational viscosity and ε is the surface shear viscosity. The term in γ is dropped in Eq. [VII-70] since its effects on the force balances have already been included in Eqs. [VII-19] and [VII-20]. Use of a stress *tensor* $\mathbf{T_s}$ provides a more general treatment than the use of a dynamic surface tension in the form of a scalar $\tilde{\gamma}$, i.e., the surface stress has the form $-\gamma\mathbf{I_s} + \mathbf{T_s}$ instead of $-(\gamma+\tilde{\gamma})\mathbf{I_s}$. However their actions as surface stresses in the forms of $\nabla_s \cdot \mathbf{T_s}$ and $-\nabla_s\tilde{\gamma}$ are similar. The first, however, is more general and also contains terms in the direction perpendicular to the surface. In particular we have

$$\tau_A - \tau_B = [-\nabla_s\gamma - \eta\, \nabla_s\, (\nabla_s \cdot \dot{\mathbf{a}}_t) - \varepsilon\, \nabla_s^2\, \dot{\mathbf{a}}_t] \cdot \mathbf{t} \qquad \text{[VII-71]}$$

The last two terms on the right hand side provide the effects of the surface viscosities. Here the effects of the component $\dot{a}_n\mathbf{n}$ have been ignored for simplicity. If there is no local expansion or contraction of the interface, one has

$$\nabla_s \cdot \dot{\mathbf{a}}_t = 0 \qquad \text{[VII-72]}$$

which means that the effects of the surface dialational viscosity η are not felt.

With $\varepsilon = 0$, Boussinesq (19) obtained for a fluid sphere

$$U = 3U_\infty \frac{1 + \kappa + e}{2 + 3\kappa + 3e} \qquad \text{[VII-73]}$$

where $e = (2\eta/a\mu_B)$. As $e \to \infty$, $U \to U_\infty$ and it can be shown from Eq. [VII-71], that $v_\theta^* \to 0$ everywhere on the surface.

With Eqs. [VII-71] and [VII-72] the system of equations relevant to a given experimental set-up can be solved to give the surface rheological coefficients. The deep channel viscometer of Mannheimer and Schechter (20) is popular and also provides a very nice instance where the complete solution of the equations of motion is available. A sketch of the apparatus is shown in Figure VII-4. A cylindrical annulus is filled with

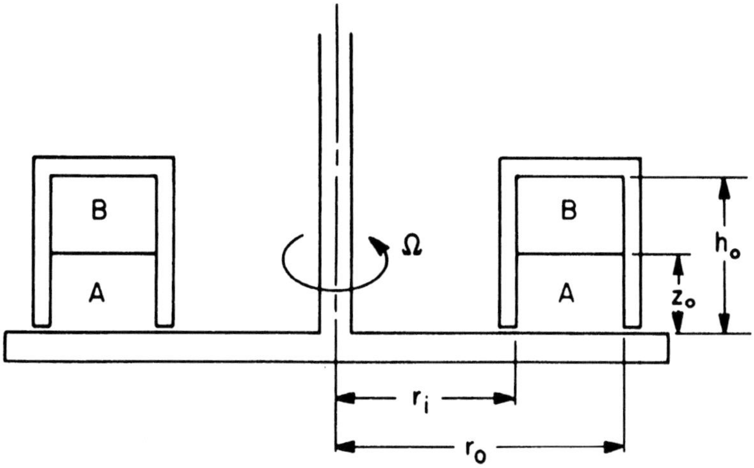

Figure VII-4. The deep channel viscometer. The longitudinal section of a radially symmetric figure is shown.

two liquids A and B. The annulus is held stationary and the floor rotated with a constant angular velocity Ω. r_i and r_o are the inner and the outer radii of the annulus in the cylindrical coordinate system and z_o designates the position of the interface. It is seen here that $v_{\theta A}$ and $v_{\theta B}$ are the only nonzero velocity components in the two phases, that they are functions of r and z only, and that the pressures satisfy the hydrostatic equilibrium equations. The solution to the equations of motion subject to the conventional boundary conditions are

$$V_{\theta B} = \sum_{i=o}^{\infty} (A_i \sinh E_i D + B \cosh E_i D)(\sinh E_i Z - \tanh E_i H \cosh E_i Z)$$

$$\psi_1(RE_i)/(\sinh E_i D - \tanh E_i H \cosh E_i D) \qquad [VII-74]$$

$$V_{\theta A} = \sum_{i=o}^{\infty} (A_i \sinh E_i Z + B_i \cosh E_i Z)\ \psi_1(RE_i) \qquad [VII-75]$$

where

$$V_\theta = \frac{v_\theta}{\Omega r_o}, \quad Z = \frac{z}{(r_o - r_i)}, \quad R = \frac{r}{(r_o - r_i)}, \quad H = \frac{h_o}{(r_o - r_i)},$$

$$D = \frac{z_o}{(r_o - r_i)}, \quad A = \frac{r_o}{(r_o - r_i)}, \quad B = \frac{r_i}{(r_o - r_i)} \text{ and}$$

$$B_i = \frac{2[B^2\psi_0(BE_i) - A^2\psi_0(AE_i)]}{A\, E_i[A^2\psi_0{}^2(AE_i) - B^2\psi_0{}^2(BE_i)]} ,$$

$$\psi_0(RE_i) = J_0(RE_i)Y_1((AE_i) - J_1(AE_i)Y_0(AE_i) \text{ , and}$$

$$\psi_1(RE_i) = J_1(RE_i)Y_1(AE_i) - J_1(AE_i)Y_1(AE_i).$$

J_ν and Y_ν are Bessel functions of the first and second kind and the E_i's are the roots of $\psi_1(RE_i) = 0$. Eq. [VII-75] has been derived using all boundary conditions except the tangential stress balance at $z = z_0$.

In view of the facts that $V_{\theta B} = V_{\theta A} = V^*_\theta$ at $z = z_0$ where $\dot{V}^*_\theta = (v^*_\theta / \Omega r_o)$ and that v^*_θ is not a function of θ, the term $\nabla_s \cdot \dot{\mathbf{a}}$, where $\mathbf{a} = v^*_\theta \mathbf{e}_\theta$, can be shown to be zero and Eq. [VII-72] becomes

$$\varepsilon[\frac{d^2V^*_\theta}{dR^2} + \frac{1}{R}\frac{dV^*_\theta}{dR} - \frac{V^*_\theta}{R^2}] = -\mu_B \frac{\partial V_{\theta B}}{\partial Z}\Big|_{Z=D} + \mu_A \frac{\partial V_{\theta A}}{\partial Z}\Big|_{Z=D} \qquad \text{[VII-76]}$$

When $(\mu_B / \mu_A) \simeq 0$, i.e., when the upper fluid is a gas, the interfacial velocity is given by

$$V^*_\theta = \frac{4}{\pi} \sum_{n=1}^{\infty} \frac{\sin\,(2n-1)\pi Y}{(2n-1)[(2n-1)\pi S\, \sinh\,(2n-1)\pi D + \cosh\,(2n-1)\pi D]} \qquad \text{[VII-77]}$$

where $Y = (r-r_i)/(r_o-r_i)$ and $S = [\,\varepsilon/\mu_A(r_o-r_i)]$. For deep canals only the first term in the series need be retained.

Combining the resulting equations for surfactant covered and surfactant free surfaces, one obtains the equation

$$\varepsilon \simeq \frac{(r_o-r_i)\mu_A}{E_1} \left(\frac{t_c}{t^*_c} -1\right) \coth E_1D \qquad \text{[VII-78]}$$

Here t_c is the time needed for a point on the interface at the centerline $r = \frac{1}{2}(r_i + r_o)$ to travel an arc of a given length and t^*_c is the time needed for a corresponding surface with $\varepsilon = 0$. These times are measured by following small particles placed on the interface. Obviously, a calibration with a pure liquid needs to be performed. For all practical purposes E_1 in Eq. [VII-78] can be approximated as π, and for deep channels $\coth \pi D \simeq 1$. The surface viscosity ε is expressed as surface poise (s.p.) = 1 (dyne . sec/cm). An interface can be considered rigid at $\varepsilon \simeq 5$ s.p. Mannheimer and Schechter give details of the equipment in their paper. More discussion on the subject can be found in Slattery (4).

It becomes apparent now that the effects of surface viscosities can be included in the boundary conditions for all problems involving dynamic interfaces, including those considered in Chapters V and VI.

9. DYNAMIC CONTACT LINES

The study of non-equilibrium contact lines is the logical extension of the study of dynamic interfaces. The obvious question to ask here first is if Young's equation describes the contact angle at the contact line under dynamic conditions. For study of this question, dynamic contact lines may be grouped in two categories. The first is that of spontaneous spreading, where there are no external forces and the contact line moves by the action of the surface tension forces. The rates of spreading are very small--in the range of 10^{-2} mm.s^{-1}. Most of the shapes of spreading drops under these conditions appear to be spherical caps (21,22). Where sufficient accuracies have been used for measurements (ellipsometry, scanning electron microscopy, etc.)(23,24), the results show that the dynamic contact angles do indeed appear to have their equilibrium values. It is seen that the profiles of the menisci change sharply very close to the contact lines, indeed by so much that optical microscopy with an error of ~ ± 1 μm in the linear measurements can give incorrect values of the contact angles. Consequently in such cases investigators refer to their results as the "apparent contact angles", suggesting that these are the contact angles formed by extrapolating the macroscopic profiles to the horizontal surface.

The second type of spreading is forced spreading, where forces from outside are imposed on the system and give rise to the movement of the contact line. The special case of a slug of one fluid displacing an immiscible fluid in a tube is shown in Figure VII-5. The displacing fluid wets the solid surface and when the slug velocity is zero, the apparent contact angle is zero (a). As the slug velocity is increased, (b) and (c), the apparent contact angle increases, ultimately reaching 180° (25-27). In this case it is difficult to imagine how the apparent contact angle could be so large and the profile could still maintain 0° at the contact line which the previous case of spontaneous spreading suggests. Indeed it is more plausible to suggest that in this case the velocities are so high that the contact angle is determined from the balance of hydrodynamic forces at the interface, Eqs. [VII-19] and [VII-20], rather than by the surface forces which lead to Young's equation.

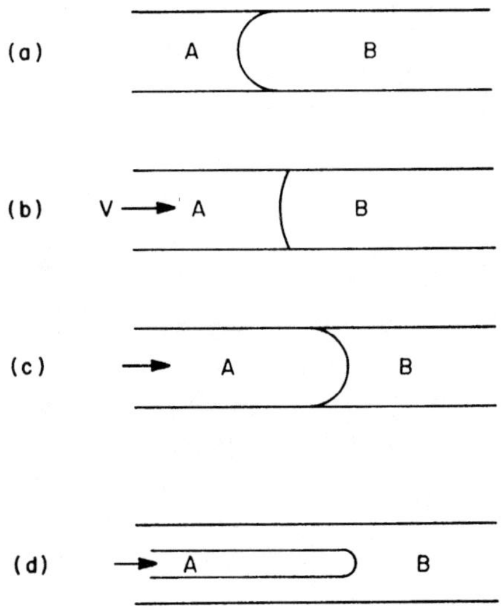

Figure VII-5. The nature of displacement of the fluid B by the fluid A is shown. Case (a) is the static case. In case (c) the velocity of displacement is higher than in case (b). At very high displacement velocities, a lubrication layer of phase B is left behind as shown in (d).

In summary, one notes that dynamic contact angles tend to their equilibrium values λ as contact line velocities decrease to zero. They also become independent of λ as contact line velocities increase. It is noteworthy that in Figure VII-5 it is supposed that the contact line always exists. That is not true in general. At sufficiently large velocities (d) entrainment of one of the phases occurs and the contact line disappears (26). It is to be noted that in dipcoating a static meniscus profile yields to an entrained film as shown in Figure VII-3 only at some sufficiently large velocity of the plate. In view of the large variety of cases, one solves the equations of motion for these problems only under an a priori assumption on the nature of the profile.

The problems involving dynamic contact lines contain additional difficulties that arise from the fact that these lines describe the intersection of *two* surfaces. The physical problem can require two different types of boundary contitions to be satisfied, one condition on each of the surfaces. However, *both* need to be satisfied at the contact line, leading to an overspecification and physically unrealistic results. Consider the case of a drop spreading on a solid surface. At the air-liquid interface the zero shear stress boundary condition applies, whereas at the solid-liquid interface the no-slip boundary condition is applicable. In Figure VII-6 approximate velocity profiles in a drop near the contact line region are shown. Since the contact line 0 is moving outward, the average velocity over the drop thickness must be nonzero. As the thickness decreases, the velocity gradient must increase as the contact line is approached to preserve the nonzero value of the average velocity. In fact, the solution to this problem shows that the velocity gradient becomes infinite at the contact line (26,28). This means that the shear stress, pressure gradient, viscous dissipation, etc., at the contact line are infinite! Obviously this is absurd, since if the shear stress at the dynamic contact line were infinite then it would take infinite force to dip a piece of wire into a pool of liquid. For the simple case considered in Figure VII-6, it suffices to say that this absurdity arises due to the fact that a material point on the dynamic contact line is required both to move on the solid surface and to obey the no-slip boundary condition. In a more general context Dussan and Davis (29) show that the two velocities on the two interfaces in the limit when the contact line is approached are different, leading to a discontinuity in the velocity field and an infinite velocity gradient at the contact line.

It is suggested that to eliminate this problem a slip boundary condition be used in the vicinity of the contact line (29,30). This slip velocity is a tangential velocity v_t on the liquid-solid interface and eliminates the need to have an infinite velocity gradient at the contact line in order to reach a finite average velocity. An empirical form for slip may be chosen as

$$\beta \frac{\partial v_t}{\partial x_n} = v_t \qquad \text{[VII-79]}$$

Here β is a constant. The left hand side of Eq. [VII-79] is proportional to the shear stress on the solid surface; the equation thus represents a physical argument that if the forces on the surface are sufficiently large, they can move the liquid molecules next to the surface. Since as $\beta \to 0$, the familiar no-slip boundary condition is obtained, β has to be very small in reality for the slip to be undetectable in conventional fluid mechanics.

The fact that β has the dimensions of length has an important implication. Wherever the representative length scale of the fluid mechanical problem greatly exceeds β, the effects of slip are not felt.

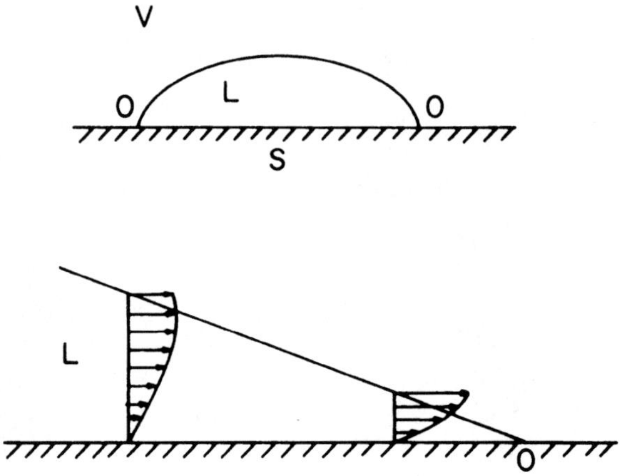

Figure VII-6. On the top, a liquid (L) drop is lying on a solid surface (S). The third phase is the vapor (V). The contact line is shown as O. In the figure below, the velocity profiles are shown under dynamic conditions, i.e., the point O is moving to the right. The velocity gradient is seen to get sharper as the point O is approached.

Thus for a spreading drop of volume V, the length scale in the bulk of the drop is $V^{1/3} >> \beta$, and consequently the effects of the slip in the major part of the drop are not felt. This disparity between two length scales which appear in a single problem is seen often in fluid mechanics. In the contact line region the thickness of the drop h approaches β. Hence in this region the slip *dominates* the fluid mechanics. From the previous discussion it becomes clear that the method of matched asymptotic solutions is needed to solve this problem. One asymptotic solution applies to the major part of the drop and is characterized by the length scale $V^{1/3}$. The other asymptotic solution is characterized by β and applies to the contact line region. Here the ratio $\beta/V^{1/3}$ is the required small quantity on which the asymptotic solutions are to be built.

In search of a basis for slip, one observes that in the immediate neighborhood of the contact line region the continuum approach breaks down. Consequently the molecular activity there, like adsorption, relaxation, reorientation, etc., could be important. However, if these molecular effects are to be used to predict the slip velocity they should not yield a length scale comparable to molecular dimensions because such dimensions are not admissible under the continuum treatment. Ruckenstein and Dunn (31) obtained a slip velocity having the form

$$v_t = -\frac{D_s}{n_L kT}\frac{\partial\phi}{\partial x_t} \qquad [VII-80]$$

on the solid surface. Here kT is the product of the Boltzman constant and the absolute temperature. n_L is the number of molecules in the liquid per unit volume. x_t is the coordinate in the direction tangential to the interface. (D_s/kT) is the mobility of the liquid on the solid surface and $\phi = p + \phi$, where p is the pressure and ϕ is the molecular potential of the type discussed in Chapter II, Section 3 and Chapter V, Section 6 except that it is evaluated on the solid surface. D_s is viewed as the surface diffusivity. Eq. [VII-80] says that the velocity at the interface is equal to the product of the mobility and a thermodynamic force expressed as the gradient of a potential in the tangential direction.

Later Neogi and Miller (32) derived Eq. [VI-80] using a different method. They also solved the equations of motion for a drop sufficiently small that the effects of gravity could be neglected. At large times of spreading the drop is flat and thin and the lubrication theory approximation can be used. For this problem their method of matched asymptotic expansions yielded drop shapes and the rates of spreading. The latter was found to be proportional to $[\ln|1/\varepsilon|]^{-1}$ with $\varepsilon = (3\mu D_s/n_L kT)^{1/2}/\hat{X}_0$ where $(3\mu D_s/n_L kT)^{1/2}$ is a form of the slip length and $\hat{X}_0$ is the macroscopic length scale proportional to $V^{1/3}$. Even though the slip length can be small, such that $\varepsilon \sim 10^{-6}$, its effect on hydrodynamics is more pronounced as $[\ln|1/\varepsilon|]^{-1} \sim 0.06$.

The above method eliminates the infinite stresses and pressure gradients at the contact lines. It also provides an expression for the rate of spreading, which in the case of spontaneous spreading discussed here can be obtained only by defining a slip condition. In spite of these favorable properties the slip condition of Eq. [VII-80] has limitations that arise from the fact that most solid surfaces have surface irregularities larger than the scale of molecular effects. Even machine polished metal surfaces have irregularities of some 1 to 10 μm (33), compared to the range of the molecular effects which are confined below 0.1 μm, as discussed in Chapter III. Thus a slip over the length scale dictated by roughness is necessary to handle the more realistic problems.

Neogi and Miller (34) modelled surface roughness by assuming the rough surface to be the surface of a porous medium. The slip velocity is given by Darcy's law for flow through a porous medium,

$$v_t = -\frac{k}{\mu}\frac{\partial p}{\partial x_t} \qquad \text{[VII-81]}$$

on the surface of the solid. The slip length in this case is found to be $(3k)^{1/2}$ where k is the permeability of the porous medium. It was also found to the first approximation that $3k \sim \sigma^2$ where σ^2 is the variance of the surface irregularities. The spreading velocity was found to be proportional to $[\ln|1/\varepsilon|]^{-1}$ where $\varepsilon = [(3k)^{1/2}/\hat{X}_0]$. The lubrication theory approximation and matched asymptotic expansions were again used. The available experimental data are limited to short times of spreading where lubrication theory does not apply. However under crude adjustments for this difference they fitted their result to some data on the rate of spreading to obtain an estimate of surface irregularities as $1 \sim 5\mu m$, whereas $\sim 3\mu m$ was reported in the original paper (21,22).

Hocking (35) has also obtained an expression for slip on a rough surface in the form of Eq. [VII-79]. With this slip the equations of motion of a drop spreading on the surface have been solved (36) for drops that have large curvatures such that the lubrication theory approximation cannot be used. The drop shape and the rate of spreading have been obtained using the method of matched asymptotic expansions. The spreading rate was found to be proportional to $[\ln|1/\varepsilon|]^{-1}$ where $\varepsilon = (\beta/a_0)$ and a_0 is the initial radius of the basal circle of the drop and thus proportional to $V^{1/3}$.

It becomes apparent that the method of matched asymptotic expansions is indispensable in obtaining the solution to the equations of motion in problems involving dynamic contact angles. It is worth examining the standard and extremely illuminating texts on this method (37,38), which is of great importance in fluid mechanics. A variety of problems containing dynamic contact lines have been solved by these methods, some of which have been reviewed by Dussan (39). A simple illustration is given below.

A thin liquid film lies on the solid surface which forms the floor of a narrow horizontal slit. Through the slit, air is blown at a steady rate. The air is seen to exert a constant shear stress on the liquid surface due to which the film thickness varies linearly with the distance from the leading edge which is also the contact line (9,40). Very close to the contact line the profile changes to retain the equilibrium contact angle at the contact line. The equations of motion and continuity under the lubrication theory approximation reduce to (41)(see Problem VII-11).

$$\frac{\partial H}{\partial \tau} = - \frac{\partial}{\partial X} [H\{(H^2 + \varepsilon^2) \frac{\partial^3 H}{\partial X^3} + N_{Ca} H\}] \qquad [VII-82]$$

where $H = h/b$, $X = x/b$, $\varepsilon^2 = (3\mu D_s/n_L kT)/b^2$, $\tau = (\gamma t/2\mu b)$ and the capillary number $N_{Ca} = 9\mu_0 V/2\gamma$. The arrangement showing the thickness h and the coordinate x in the horizontal direction may be seen in Figure VII-7. The term ε is due to the slip velocity and is a small quantity. initial and boundary conditions are written as

(1) $(\partial H/\partial X) \rightarrow \theta(\tau)$ as $X \rightarrow \infty$. This condition arises out of the experimental observations and will not be imposed.
(2) $\theta(0) = \theta_0$, the initial slope
(3) $(\partial H/\partial X) = \lambda$ for $X = X_0$, i.e., the equilibrium contact angle is retained at the contact line. The approximation that $\tan \lambda \simeq \lambda$ for the small contact angles assumed here, has been made.
(4) $H = 0$ for $X = X_0$, at the contact line.
(5) $X_0(0) = 0$. The initial position of the contact line forms the origin.

In solving this problem with the matched asymptotic expansions it is seen that there are two driving forces, viz., the air drag and the curvature. The air drag acts everywhere and leads to a slope of θ. However, near the contact line region the curvature effects overwhelm this force to change the slope from θ to λ. Thus a bulk or an outer region can be identified where the air drag dominates and only conditions (1) and (2) apply. To obtain an asymptotic expansion, one rescales the variable X at distances far from the contact line X_0 to $p = [(X-X_0)/u(\varepsilon)]$. Here u is large, in fact it becomes infinite as $\varepsilon \rightarrow 0$. Thus, to examine the profile away from the contact line it is sufficient to analyze the region where p is comparable to 1. One chooses H as $H \sim uH_0 + u_1 H_1 + \ldots$, such that as

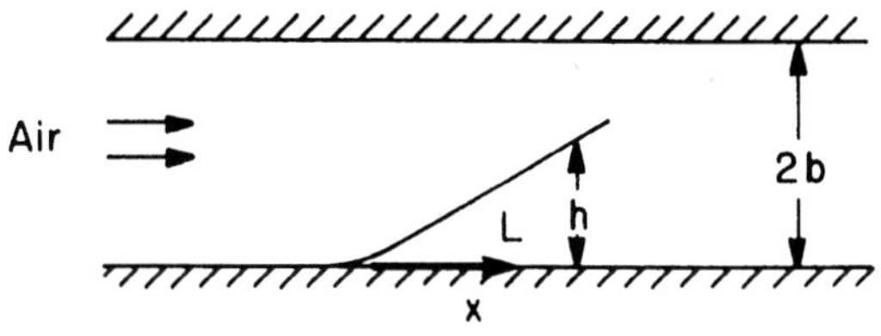

Figure VII-7. The arrangement of the blown-off liquid (L) film. The slit width is 2b and h is the local film thickness. The thickness h is shown on an exaggerated scale.

$\varepsilon \to 0$, $(u_1(\varepsilon)/u) \to 0$, i.e., $u_1 H_1 << uH_0$. Substituting this series into Eq. [VII-82] and using the rescaled variable p for small ε, one has as $\varepsilon \to 0$

$$\frac{\partial H_0}{\partial \tau} = - N_{Ca} \frac{\partial}{\partial p} [H_0^2]. \qquad \text{[VII-83]}$$

The coefficient of H_0 in the series is found to be u if a non-trivial equation such as Eq. [VII-83] is to be obtained. Eq. [VII-83] can be solved by the separation of variables method, to obtain as one solution

$$H_0 = \theta\, p \qquad \text{[VII-84]}$$

$$\theta = \frac{1}{\frac{1}{\theta_0} + 2N_{Ca}\tau} \,. \qquad \text{[VII-85]}$$

uH_0 constitutes the first approximation to the outer solution. Note that a similar substitution has been made in the boundary conditions and the limit $\varepsilon \to 0$ has been taken to get the boundary conditions for H_0.

Just as the outer solution stresses the role of the air drag, the inner solution, i.e., the solution in the vicinity of the contact line, has to be dominated by the slip and the curvature effects, and boundary conditions (3)-(5) apply. The series chosen is

$$H \sim \varepsilon \hat{H}_0 + \varepsilon\left(\frac{dX_0}{d\tau}\right) \hat{H}_1 + \ldots \qquad \text{[VII-86]}$$

where $(dX_0/d\tau)$, the rate of movement of the contact line, is assumed to $\to 0$ as $\varepsilon \to 0$. $\hat{H}_0$ and $\hat{H}_1$ are functions of $s = (X-X_0)/\varepsilon$, a rescaled variable that allows one to examine the region close to the contact line. Substituting the rescaled variables into Eq. [VI-82] and taking the limit $\varepsilon \to 0$, one has

$$\frac{\partial}{\partial s} [\hat{H}_0 (\hat{H}_0^2 + 1) \frac{\partial^3}{\partial s^3} \hat{H}_0] = 0. \qquad \text{[VII-87]}$$

Similarly the boundary conditions reduce to $(\partial \hat{H}_0 / \partial s)_0 = \lambda$ and $\hat{H}_0(o) = 0$. The solution to Eq. [VII-87] becomes

$$\hat{H}_0 = \lambda s \qquad \text{[VII-88]}$$

when subject to a boundedness condition. Now, substituting Eq. [VII-86] into Eq. [VII-82], using Eq. [VII-87] to eliminate the leading term, and dividing the entire expression by $(dX_0/d\tau)$, one has

$$-\frac{\partial}{\partial s}\hat{H}_0 = -\frac{\partial}{\partial s}[\hat{H}_0(\hat{H}_0^2+1)\frac{\partial^3}{\partial s^3}\hat{H}_1] \qquad [VII-89]$$

on taking the limit $\varepsilon \to 0$. Similarly, suitable manipulations lead to the conditions that $(\partial \hat{H}_1/\partial s)_0 = 0$ and $\hat{H}_1(o) = 0$. The solution subject to Eq. [VII-88] and the condition of boundedness is

$$\tilde{H}_1 = \frac{1}{2\lambda^3}(\lambda^2 s^2-1)\tan^{-1}(\lambda s) + \frac{s}{2\lambda^2} - \frac{s}{2\lambda}\ \ell n\ (1+\theta^2 s^2) + Bs^2 \qquad [VII-90]$$

where B is a constant of integration.

Now if the inner solution $\varepsilon\,\hat{H}_0 + \varepsilon\,(dX_0/d\tau)\,\hat{H}_1$ and the outer solution uH_0 have been correctly constructed they should match. That is, when extrapolated into an intermediate region they should agree. A suitable coordinate for the intermediate region is chosen here as z, where $p = \eta z/u$ and $s = \eta z/\varepsilon$. η is a small quantity such that as $\varepsilon \to 0$, $\eta/u \to 0$ and $\eta/\varepsilon \to \infty$. Consequently $p \to 0$ and $s \to \infty$ when z is of the order of 1, and the two solutions will be forced to go into the domains of one another. The difference between the two solutions is examined, and

$$u\,H_0 - [\varepsilon\tilde{H}_0 + \varepsilon\,(\frac{dX_0}{d\tau})\,\tilde{H}_1] = \frac{\eta z}{\frac{1}{\theta_0} + 2\,n_{Ca}\tau} - [\lambda\eta z + \frac{dX_0}{d\tau}\{-\frac{\eta z}{\lambda^2}\,\ell n|\frac{1}{\varepsilon}| +$$

$$+ (\frac{\pi}{4\lambda} + B)\frac{\eta^2 z^2}{\varepsilon} + \text{high order terms}\}]$$

The largest of the terms omitted are proportional to $\eta\ \ell n\ \eta$. Obviously the leading terms cancel if

$$\frac{1}{\frac{1}{\theta_0} + 2N_{Ca}\tau} - \lambda + \frac{1}{\lambda^2}\frac{dX_0}{d\tau}\,\ell n|\frac{1}{\varepsilon}| = 0$$

$$B = -\frac{\pi}{4\lambda}$$

The first equality provides the rate of spreading as

$$\frac{dX_0}{d\tau} \sim \lambda^2[\ell n|\frac{1}{\varepsilon}|]^{-1}\ (\lambda-\theta)$$

which may be integrated subject to the initial condition. The remainder is ignored as it is now small, of the order of $[\ell n|1/\varepsilon|]^{-1}(\eta\ \ell n\ \eta)$, and provides the small errors in the estimated profiles and the rates of spreading. Note that both $[\ell n|1/\varepsilon|]^{-1}$ and $\eta\ \ell n\ \eta \to 0$ as $\varepsilon \to 0$. In the matching process, it is observed that the outer solution is contained in the inner solution, hence the latter provides the "overall" solution. It is noteworthy that the justification for the choice of asymptotic expansions, scaling variables, etc., lies in the fact that the two asymptotic expansions match.

REFERENCES

General References

Levich, V.G. (1962) *Physicochemical Hydrodynamics*, Prentice-Hall, Englewood Cliffs, NJ.

Slattery, J.C., *Interfacial Transport Phenomena*, in preparation.

Textual References

1. Clift, R., Grace, J.R., and Weber, M.E. (1978) *Bubbles, Drops, and Particles*, Acad. Press, NY; Azbel, D., (1981) *Two-phase Flows in Chemical Engineering*, Cambridge University Press.

2. Deryagin, B.V., and Levi, S.M. (1964) *Film Coating Theory*, Focal Press, London.

3. Higgins, B.G., Silliman, W.J., Brown, R.A., and Scriven, L.E. (1977) *Ind. Eng. Chem. Fund.* **16**, 393.

4. Slattery, J.C., *Interfacial Transport Phenomena*, in preparation.

5. Weatherburn, C.A. (1961) *Differential Geometry of Three Dimensions*, Cambridge Univ. Press, Cambridge.

6. Slattery, J.C., (1981) *Momentum, Energy and Mass Transfer in Continua*, 2nd. ed., Robert E. Krieger Pub. Co., Huntington, N.Y.

7. Happel, J., and Brenner, H. (1965) *Low Reynolds Number Hydrodynamics*, Prentice-Hall, Inc., Englewood Cliffs, N.J.

8. Bird, R.B., Stewart, W.E., and Lightfoot, E.N. (1960) *Transport Phenomena*, John Wiley, N.Y.

9. Levich, V.G. (1962) *Physicochemical Hydrodynamics*, Prentice-Hall, Englewood Cliffs, N.J.

10. Taylor, T.D. and Acrivos, A. (1964) *J. Fluid Mech*. **18**, 466.

11. Proudman, I., and Pearson, J.R.A. (1957) *J. Fluid Mech*. **2**, 237.

12. Landau, L.D., and Lifshitz, E.M. (1959) *Fluid Mechanics*, Addison-Welsey Pub. Co., Inc., Reading, MA.

13. Haberman, W.L., and Sayre, R.M. (1958) *Motion of Rigid and Fluid Spheres in Stationary and Moving Liquids Inside Cylindrical Tubes*, U.S. Dept. of Navy, David Taylor Model Basin, Report No. 1143.

14. Lee, S.H., and Leal, L.G. (1980) *J. Fluid Mech*. **98**, 193.

15. Landau, L.D., and Levich, V.G. (1942) *Acta Physicochemica URSS* **17**, 42. See also Ref. (9) and Deryajin, B.V. (1943) *Dokl. Akad. Nauk SSSR* **39**, 11.

16. Ruschak, K.J., and Scriven, L.E. (1977) *J. Fluid Mech*. **81**, 305.

17. Frumkin, A.N., and Levich, V.G. (1947) *Zhur. Fiz. Khim*. **21**, 1183.

18. Scriven, L.E. (1960) *Chem. Eng. Sci*. **12**, 98.

19. Boussinesq, J. (1913) *Comp. Rend*. **156**, 983, 1035, 1124, **157**, 89, *Ann. Chim. et Phys*. (1913) **29**, 349, 357, 364.

20. Mannheimer, R.J., and Schechter, R.S. (1970) *J. Colloid Interface Sci*. **32**, 195. See also Burton, R.A., and Mannheimer, R.J. (1967) in *Advances in Chemistry Series* No. **63**, American Chemical Society, Washington, D.C., p. 315, Osborne, M.F.M. (1968) *Kolloid-Z. Polym*. **224**, 150; Mannheimer, R.J., and Schechter, R.S. (1968) *J. Colloid Interface Sci*. **27**, 324.

21. Schonhorn, H., Frisch, H.L., and Kwei, T.K. (1966) *J. Appl. Phys.* **37**, 4967.

22. Van Oene, H., Chang, Y.F., and Newman, S. (1969) *J. Adhesion* **1**, 54.

23. Bascom, W.D., Cottington, R.L., and Singleterry, C.R. (1964) in *Advances in Chemistry Series*, **v. 43**, American Chemical Society, Washington, D.C., p. 355.

24. Radigan, W., Ghiradella, H., Frisch, H.L., Schonhorn, H., and Kwei, T.K. (1974) *J. Colloid Interface Sci.* **49**, 241.

25. Rose, W., and Heins, R.W. (1962) *J. Colloid Interface Sci.* **17**, 39.

26. Hansen, R.S., and Toong, T.V. (1971) *J. Colloid Interface Sci.* **36**, 410.

27. Hoffman, R.L. (1975) *J. Colloid Interface Sci.* **50**, 228.

28. Lopez, J., Miller, C.A., and Ruckenstein, E. (1976) *J. Colloid Interface Sci.* **56**, 460.

29. Dussan, E.B. and Davis, S.H. (1974) *J. Fluid Mech.* **65**, 71.

30. Huh, C. and Scriven, L.E. (1971) *J. Colloid Interface Sci.* **35**, 85.

31. Ruckenstein, E. and Dunn, C.S. (1977) *J. Colloid Interface Sci.* **59**, 135.

32. Neogi, P. and Miller, C.A. (1982) *J. Colloid Interface Sci.* **86**, 525.

33. Mussel, L.I. and Glang, R., ed. (1970) *Handbook of Thin Film Technology*, McGraw-Hill.

34. Neogi, P. and Miller, C.A. (1982) *J. Colloid Interface Sci.* **92**, 338.

35. Hocking, L.M. (1982) *J. Fluid Mech.* **76**, 801.

36. Hocking, L.M. and Rivers, A.D. (1982) *J. Fluid Mech.* **121**, 425.

37. Van Dyke, M. (1975) *Perturbation Methods in Fluid Mechanics*, Annotated Ed., Parabolic Press.

38. Cole, J.D. (1968) *Perturbation Methods in Applied Mechanics*, Blaisdell Pub. Co.

39. Dussan, V̲, E.B. (1979) *Ann. Rev. Fluid Mech.* **11**, 371.

40. Derjaguin, B., Strakhovsky, G., and Malysheva, D. (1944) *Acta Phys. Chem. URSS* **19**, 541.

41. Neogi, P. (1982) *J. Colloid Interface Sci.* **89**, 358; **90**, 554.

PROBLEMS

VII-1. (a) Instead of the series $\psi_i \sim \psi_i^{(0)} + N_{Re}\psi_i^{(1)} + N_{Re}N_{Ca}\psi_i^{(2)}$, in the problem in Section 6, consider using a series

$$\psi_i \sim \psi_i^{(0)} + N_{Re}\psi_i^{(1)} + u\psi_i^{(2)} + \ldots \qquad \text{(i)}$$

where u is small but unknown. For (i) to be an asymptotic series, it is also required that as N_{Re}, $u \to 0$, $u/N_{Re} \to 0$. The momentum equation is

$$N_{Re}D^4\psi_i = E^4\psi_i \qquad \text{(ii)}$$

where

$$D^4\psi_i = \frac{1}{r^2\sin^2\theta}\frac{\partial(\psi_i, E^2\psi_i)}{\partial(r,\theta)} - \frac{2E^2\psi_i}{r^2\sin^2\theta}\left(\frac{\partial\psi_i}{\partial r}\cos\theta - \frac{1}{r}\frac{\partial\psi_i}{\partial r}\sin\theta\right)$$

and the Jacobian is given by

$$\frac{\partial(\psi_i, E^2\psi_i)}{\partial(r,\theta)} = \frac{\partial\psi_i}{\partial r}\frac{\partial E^2\psi_i}{\partial\theta} - \frac{\partial\psi_i}{\partial\theta}\frac{\partial E^2\psi_i}{\partial r}$$

When the series (i) is substituted into (ii), the result is

$$N_{Re}D^4\psi_i^{(0)} + N_{Re}^2D^4\psi_i^{(1)} + N_{Re}uD^4\psi_i^{(2)} + \ldots$$

$$= E^4\psi_i^{(0)} + N_{Re}E^4\psi_i^{(1)} + uE^4\psi_i^{(2)} + \ldots \qquad \text{(iii)}$$

On taking the limit N_{Re}, $u \to 0$,

$$E^4\psi_i^{(0)} = 0 \tag{iv}$$

is obtained. Substituting (iv) in (iii) and dividing with N_{Re}, one has

$$D^4\psi_i^{(0)} + N_{Re}D^4\psi_i^{(1)} + uD^4\psi_i^{(2)} + \ldots$$

$$= E^4\psi_i^{(1)} + \left(\frac{u}{N_{Re}}\right) E^4\psi_i^{(2)} + \ldots \tag{v}$$

On taking the limit N_{Re}, $u \to 0$ and using the property that $u/N_{Re} \to 0$ in that limit, (v) becomes

$$D^4\psi_i^{(0)} = E^4\psi_i^{(1)} \tag{vi}$$

Continue this process to obtain next

$$E^4\psi_i^{(2)} = 0 \tag{vii}$$

under the assumption that $N_{Re}^2/u \to 0$ as N_{Re}, $u \to 0$.

(b) Examine now the normal stress balance

$$(\overline{N}_A^{(0)} + N_{Re}\,\overline{N}_A^{(1)} + u\overline{N}_A^{(2)}) - (\overline{N}_B^{(0)} + N_{Re}\,\overline{N}_B^{(1)} + u\,\overline{N}_B^{(2)})$$

$$= -\frac{2}{N_{Ca}}[\overline{H}^{(0)} + v_1\,\overline{H}^{(1)} + v_2\,\overline{H}^{(2)}] \text{ on } r = 1 + v_1\,\xi^{(1)} + v_2\,\xi^{(2)} \tag{viii}$$

where $\overline{N}_A^{(j)}$ are the normal stresses obtained from $\psi_i^{(j)}$, v_1 and v_2 are small quantities with $v_2 \ll v_1$ and $\overline{H}^{(0)}$ is the curvature of the shape given by $r = 1$, $v_1\overline{H}^{(1)}$ of $v_1\xi^{(1)}$ and $v_2\overline{H}^{(2)}$ of $v_2\xi^{(2)}$. (It can be verified from Table VII-2 that the order of ξ will be the same as H.) Taylor's series expansion can be used to simplify (viii), for example

$$\overline{N}_A^{(0)}\Big|_{r = 1 + v_1\xi^{(1)} + v_2\xi^{(2)}} \sim \overline{N}_A^{(0)}\Big|_{r = 1} + \frac{\partial\overline{N}_A^{(0)}}{\partial r}\Big|_{r = 1}(1 - 1 - v_1\xi^{(1)} - v_2\xi^{(2)})$$

$$+ \frac{1}{2}\,\frac{\partial^2\overline{N}_A^{(0)}}{\partial r^2}\Big|_{r = 1}(1 - 1 - v_1\xi^{(1)} - v_2\xi^{(2)})^2$$

Substituting such expansions into (viii) and using the method shown in part (a), where in addition to the limits N_{Re}, $u \to 0$ it is also necessary to take the limit $N_{Ca} \to 0$, show that

$$\bar{N}_A^{(0)} - \bar{N}_B^{(0)} = -\frac{2}{N_{Ca}} \bar{H}^{(0)} \qquad \text{(ix)}$$

which is the result obtained in Eq. [VII-27]. In obtaining the second term, show that $\psi_i^{(1)}$ affects the shape of the drop only when $v_1 = N_{Re}N_{Ca}$ and thus

$$\bar{N}_A^{(1)} - \bar{N}_B^{(1)} = -2\bar{H}^{(1)} \qquad \text{(x)}$$

To obtain the next term show that it is necessary to have $u = N_{Re}N_{Ca}$ and $v_2 = N_{Re}N_{Ca}^2$ and thus

$$\left(\bar{N}_A^{(2)} - \xi^{(1)} \frac{\partial \bar{N}_A^{(0)}}{\partial r} - \bar{N}_B^{(2)} + \xi^{(1)} \frac{\partial \bar{N}_B^{(0)}}{\partial r}\right)_{r=1} = -2\bar{H}^{(2)} \qquad \text{(xi)}$$

At this level the curvature $\bar{H}^{(2)}$ is affected both by $\psi_i^{(2)}$ and by the distortion $\xi^{(1)}$.

Note that it is not absolutely essential for u, v_1, and v_2 to have the above values. However, any other values would either lead to inconsistencies or would necessitate calculations of higher order terms in ψ_i or ξ. In the latter case some of the terms are repeated and become redundant.

Lastly, it is necessary to stress that in the asympototic expansions for ψ_i, ξ, and H, the order of a term like $u\psi_i^{(2)}$ is given by the small quantity u and the order of $\psi_i^{(2)}$ itself is of the order one. That is, the functions $\psi_i^{(j)}$, $\xi^{(j)}$, $\bar{H}^{(j)}$, $\bar{N}_A^{(j)}$, $\bar{N}_B^{(j)}$, etc., have been chosen to be of the order of one and hence are unaffected by the limits N_{Re}, u, $N_{Ca} \to 0$. One exception here is $\bar{H}^{(0)}$, which from (ix), should have been $N_{Ca}\bar{H}^{(0)}$ instead.

(c) Rework parts (a) and (b) as well as the other boundary conditions using the series

$$\psi_i \sim \psi_i^{(0)} + N_{Re}\psi_i^{(1)} + N_{Re}N_{Ca}\psi_i^{(2)} + \ldots$$

$$\xi \sim N_{Re}N_{Ca}\xi^{(1)} + N_{Re}N_{Ca}^2\xi^{(2)} + \ldots$$

and show that no inconsistencies arise. Show as well that it is necessary to make the assumption $N_{Re}/N_{Ca} \to 0$ as N_{Re}, $N_{Ca} \to 0$ and hence provide the orders of the next terms in the two series.

VII-2. A jet leaves a tube of radius a under creeping flow.

(a) What would be the profile of the jet as the first approximation?

(b) Assuming that the solution to this problem is known (see J.S. Vrentas and J.L. Duda, 1967, *AIChE J.* **13**, 97), how would a normal stress balance be used to obtain the first correction to the assumed profile? What is the necessary condition under which this correction is valid?

VII-3. Consider the flow of a thin film down a wall as shown in Figure VII-8. Write the relevant equations of motion under the lubrication theory approximation including the effect of surface tension. Obtain the thickness h_∞ far from the entrance. If h is the thickness at any distance from the entrance which can be approximated as

$$h \sim h_\infty + \varepsilon h' + \ldots$$

where $\varepsilon = (h_0 - h_\infty)/(h_\infty)$ is small, obtain a differential equation for the correction term h'. Here h_0 is the film thickness at the entrance. What are the boundary conditions for h'? Integrate the differential equation subject to the boundary conditions to obtain h' as completely as possible.

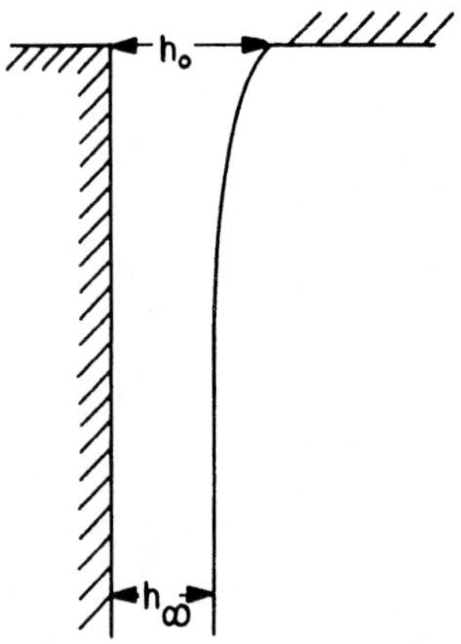

Figure VII-8. The shape of the falling film discussed in Problem VII-3.

VII-4. An air bubble is subjected to extensional flow characterized by $v_z = Gz$ and constant p, as shown in Figure VII-9. The bubble is long and slender, the surface of which is given by $r = R(z)$.

(a) From the continuity equation and the fact that the normal velocity $(v_r - \frac{dR}{dz} v_z)$ on the bubble surface is zero, obtain v_r.

(b) From the normal stress balance at the surface, show that

$$z \frac{dR}{dz} - R\left(\frac{p}{2\mu G} - 1\right) = -\frac{\gamma}{2\mu G} .$$

Integrate the equation to obtain R(z).

VII-5. In the problem discussed in Section 6 on the role of surface active agents on the terminal velocity of a drop, if the surface active agent is insoluble in both phases and resides only at the interface between the drop and the infinite medium, how will Eq. [VII-59] be modified?

The term e can be determined through the following detailed procedure. Starting with the solution

$$\psi_B = \sin^2\theta \left(\frac{Ar^4}{10} - \frac{1}{2} Br + Cr^2 + \frac{D}{r}\right)$$

$$\psi_A = \sin^2\left(\frac{Er^4}{10} - \frac{1}{2} Fr + Gr^2 + \frac{H}{r}\right)$$

satisfy the condition of boundedness at $r = 0$ on ψ_A and conditions (1) - (3) given after Eq. [VII-23]. Satisfy next Eq. [VII-19] where $\nabla_s \gamma \cdot \mathbf{t} = a^{-1}(\partial\gamma/\partial\Gamma)(\partial\Gamma/\partial\theta)$. Use now a theorem by L.E. Payne and W.H. Pell (*J. Fluid Mech.* 7, 529, 1960) to calculate the drag force as

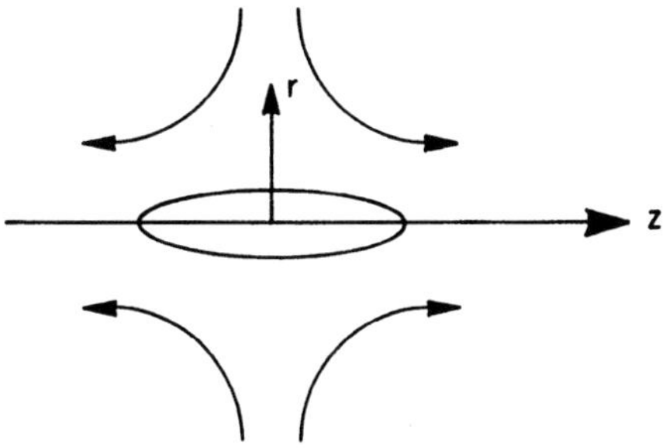

Figure VII-9. The shape of the air bubble discussed in Problem VII-4.

$$\lim_{r \to \infty} \frac{8\pi\mu_B\, r(\psi_B - \psi_B^\infty)}{\omega^2}$$

where ω is the radial direction perpendicular to the direction of U, i.e., $r \sin\theta$ in this case, and ψ_B^∞ is the value of the stream function at $r \to \infty$ (boundary condition (1)). Hence calculate U and e. Is the normal stress balance satisfied? (The pressures are $p_B = [p_{Bo} + (\mu_B B \cos\theta/r^2) - \rho_B rg \cos\theta]$ and $p_A = [p_{Ao} + \mu_A \cos\theta\,(-2E)r - \rho_A rg\cos\theta]$).

VII-6. In the same class of problems as above, the surface concentration of $\Gamma_0 + \Gamma'$ may be controlled by the diffusion from the infinite fluid to the interface. Under those conditions

$$j^* = \frac{D\Delta c}{\delta}$$

where δ is the thickness of the boundary layer, and $\Delta c = c_0 - c(a)$. Since the rates of adsorption-desorption are not controlling, the surface concentration is in equilibrium with c(a). Thus

$$\Delta c = \left(\frac{\partial c}{\partial \Gamma}\right) \Gamma'$$

where $\left(\frac{\partial c}{\partial \Gamma}\right)$ is the slope of the adsorption isotherm. Obtain the expression for e in this case. Evaluate e when the adsorption is given by the Langmuir isotherm. Is there a maximum or a minimum in e with respect to c_0? Assume an ideal solution.

VII-7. Can the effect of surface active agents as given in Section 6 describe the observations made in the deep channel viscometer experiments? Discuss.

VII-8. As discussed in Chapter III, measurement of the velocity of a charged colloidal particle in an applied electric field can be used to estimate the surface potential and hence the magnitude of the electrical interaction between nearby particles in a colloidal dispersion.

(a) Suppose that an electric field E_x is applied in a direction parallel to a flat charged solid surface. The tangential component of the equation of motion leads, in the absence of a tangential pressure gradient, to the following equation applicable in the double layer region

$$0 = \rho_e E_x + \mu \frac{d^2 v_x}{dz^2} \qquad (i)$$

where z is the coordinate in the direction perpendicular to the surface. Poisson's equation in this case is

$$\nabla^2\psi \simeq \frac{\partial^2\psi}{\partial z^2} = \frac{-4\pi\rho_e}{\varepsilon} \qquad (ii)$$

We note that $(\partial^2\psi/\partial x^2)$ has been neglected in this equation since the potential ψ varies rapidly in the z-direction in the double layer region. Substitute (i) into (ii) and integrate twice to show that the velocity v_x at the outer limit of the double layer is given by

$$v_x = \frac{-\varepsilon E_x \zeta}{4\pi \mu} \qquad (iii)$$

where ζ is the potential at the z-coordinate where $v_x = 0$, the "plane of shear".

(b) Use the result of (a) to estimate the steady velocity of a spherical non-conducting particle of radius a when $\kappa a >> 1$, i.e., in the limit of a very thin double layer. Use a moving coordinate system such that the fluid flows past the fixed sphere with a velocity U far away from sphere as shown in Figure VII-2. The general solution for the fluid flow around a solid sphere for this case can be obtained from Problem VII-5 with the boundary condition at $r \sim a$ given by (iii). That is, the local tangential velocity just outside the (thin) double layer is given by (iii) with $E_x = -\partial\psi/\partial x$ the local tangential field. Note that the potential distribution outside the double layer is found by solving $\nabla^2\psi = 0$ subject to the condition that the electric field is uniform far from the particle.

Calculate the drag force on the sphere using the method given in Problem VII-5. From physical arguments assign an appropriate value to this drag force and obtain U in terms of known parameters. Neglect the role of gravity. The result is the electrophoretic velocity in the von Smoluchowski limit referred to in Chapter III.

(c) Rearrange the physics of this problem as solved in the parts (a) and (b) above to show that this is a matched asymptotic solution with $(\kappa a)^{-1}$ as the small parameter.

VII-9. A capillary tube of radius a is placed vertically in a large pool of liquid (see Fig. I-10). Assuming perfect wetting and that, at any time t during the approach to equilibrium, the average velocity is given by the expression for steady state flow in a circular tube, derive a differential equation to predict how liquid height h depends on t from the time the tube is initially inserted until equilibrium is reached. Specify the boundary condition or conditions and show that the result is

$$1 - \frac{h}{h_e} = \exp\left[-\left(\frac{h+\theta}{h_e}\right)\right]$$

where $\theta = \frac{\rho g a^2 t}{\mu}$, a = tube radius, h_e = equilibrium height.

VII-10. A drop is spreading slowly on a solid surface. At large times of spreading, the drop appears to be thin and flat and the lubrication theory approximation can be used. Obtain the appropriate equations of motion (do not neglect gravity).

(a) Integrate the continuity equation to show that the integral form is

$$\frac{\partial h}{\partial t} = \frac{1}{r}\frac{\partial}{\partial r}\left(r\int_0^h v_r dz\right)$$

where r is the radial direction and z is the direction normal to the solid surface in a cylindrical coordinate system. z = h (r,t) describes the shape of the drop.

(b) From the above equations obtain a partial differential equation describing the profile h. Write the initial and the boundary conditions to solve for the case where surface tension effects can be neglected and the drop spreads under its own weight. Assume that the solution is of the form $r_o^{-2} z\ (r/r_o)$ where $r_o(t)$ describes the position of the contact line. Show that the pressure gradient $(\partial p/\partial r)$ at $r = r_o$ is infinite. Why?

(c) When the drops are small, the effects of gravity can be neglected and the drops are driven by surface tension. In that case show only that

$$\frac{dr_o}{dt} \sim t^{-(9/10)}$$

For a figure of revolution

$$2H = \left\{\frac{\frac{d^2h}{dr^2}}{\left[1+\left(\frac{dh}{dr}\right)^2\right]^{3/2}} + \frac{\frac{1}{r}\frac{dh}{dr}}{\left[1+\left(\frac{dh}{dr}\right)^2\right]^{1/2}}\right\}$$

VII-11. Derive Eq. [VII-82]. Consider Figure VII-7 where it can be assumed that $h \ll 2b$. Solve the equations of motion for the air flowing through the slit. Since the viscosity of air is much less than the viscosity of the liquid, ignore the thickness h and assume no slip at both the walls. Hence obtain the tangential stress at the liquid-air interface. With these boundary conditions, follow the procedure of Problem VII-10 to obtain the equations of motion under the lubrication theory approximation. Neglect gravity, derive the appropriate integral continuity equation as in Problem VII-10 and non-dimensionalize to obtain Eq. [VII-82].

VII-12. Consider the situation discussed in Problem VII-12 and Section 9, Figure VII-7. Suppose that there is no air current but that the liquid contains a volatile species which evaporates and gives rise to a variation in surface tension. As a first approximation, assume $(\partial\gamma/\partial x)$ to be a constant and formulate the problem under the lubrication theory approximation as in Problem VII-11. Nondimensionalize the resulting equation for the profile using $[(1/\gamma)(\partial\gamma/\partial x)]^{-1}$ as the macroscopic length scale. Solve the problem using the method of matched asymptotic expansions as in Section 9, both when $(\partial\gamma/\partial x)$ is positive and when it is negative.

VII-13. The concept of order is very important in asymptotic analysis since one is always involved in comparing the magnitudes of various terms. Consider functions $f(\varepsilon)$ and $g(\varepsilon)$, when as $\varepsilon \to 0$, $f \to 0$ and $g \to 0$. It is necessary to find which is "greater".

If in the limit $\varepsilon \to 0$, $f/g \to \infty$, then ord (f) > ord (g) or $g \sim o(f)$. If in the limit $\varepsilon \to 0$, $f/g \to$ finite value, then ord (f) = ord (g) or $f \sim O(g)$ and $g \sim O(f)$. If in the limit $\varepsilon \to 0$, $f/g \to 0$, than ord (f) < ord (g) or $f \sim o(g)$.

Show that $\exp[-1/\varepsilon] \sim o(\varepsilon^{\nu})$ and $\varepsilon^{\nu} \sim o([\ln|1/\varepsilon|]^{-1})$, for $\nu > 0$, however large or small be the values of ν. Consequently one has $\exp[-1/\varepsilon]$ "transcendentally small" and $[\ln|1/\varepsilon|]^{-1}$ "transcendentally large" compared to ε.

Index